Sustainability in the Development of Water Systems Management

Sustainability in the Development of Water Systems Management

Editor

José-Luis Molina

MDPI • Basel • Beijing • Wuhan • Barcelona • Belgrade • Manchester • Tokyo • Cluj • Tianjin

Editor
José-Luis Molina
University of Salamanca
Spain

Editorial Office
MDPI
St. Alban-Anlage 66
4052 Basel, Switzerland

This is a reprint of articles from the Special Issue published online in the open access journal *Sustainability* (ISSN 2071-1050) (available at: https://www.mdpi.com/journal/sustainability/special_issues/sustainability_development_water_systems_management).

For citation purposes, cite each article independently as indicated on the article page online and as indicated below:

LastName, A.A.; LastName, B.B.; LastName, C.C. Article Title. *Journal Name* **Year**, *Article Number*, Page Range.

ISBN 978-3-03943-202-8 (Hbk)
ISBN 978-3-03943-203-5 (PDF)

Contents

About the Editor

José-Luis Molina has a degree in Civil Engineering, obtained in 2015 (University of Salamanca), a degree in Environmental Sciences obtained in 2002 by University of Granada (Spain) and three Masters degrees related to the environment and water and hydraulic management. In addition, Dr. Molina obtained a Ph.D. in Water Management in 2009 from the Geological and Mining Institute of Spain (IGME) and the University of Granada for the thesis entitled "Integrated analysis and management strategies of aquifers in semi-arid zones: application to the case study of the Altiplano (Murcia, SE Spain)". He holds a position as an Associate Professor (Hydraulic Engineering Area) at the University of Salamanca. He also currently belongs to the Editorial Board of the *Journal of Hydrology* (ELSEVIER) as an Associate Editor. He is the author of more than 40 JCR research papers and several books.

sustainability

MDPI

Article

Intelligent System for the Predictive Analysis of an Industrial Wastewater Treatment Process

Luis Arismendy [1], Carlos Cárdenas [1], Diego Gómez [1], Aymer Maturana [2], Ricardo Mejía [2] and Christian G. Quintero M. [1,*]

[1] Department of Electrical and Electronics Engineering, Universidad del Norte, Barranquilla 081007, Colombia; arismendyl@uninorte.edu.co (L.A.); ccarlosa@uninorte.edu.co (C.C.); dgomez@uninorte.edu.co (D.G.)

[2] Department of Civil and Environmental Engineering, Universidad del Norte, Barranquilla 081007, Colombia; maturanaa@uninorte.edu.co (A.M.); marchenar@uninorte.edu.co (R.M.)

* Correspondence: christianq@uninorte.edu.co

Received: 4 July 2020; Accepted: 27 July 2020; Published: 7 August 2020

Abstract: Considering the exponential growth of today's industry and the wastewater results of its processes, it needs to have an optimal treatment system for such effluent waters to mitigate the environmental impact generated by its discharges and comply with the environmental regulatory standards that are progressively increasing their demand. This leads to the need to innovate in the control and management information systems of the systems responsible to treat these residual waters in search of improvement. This paper proposes the development of an intelligent system that uses the data from the process and makes a prediction of its behavior to provide support in decision making related to the operation of the wastewater treatment plant (WWTP). To carry out the development of this system, a multilayer perceptron neural network with 2 hidden layers and 22 neurons each is implemented, together with process variable analysis, time-series decomposition, correlation and autocorrelation techniques; it is possible to predict the chemical oxygen demand (COD) at the input of the bioreactor with a one-day window and a mean absolute percentage error (MAPE) of 10.8%, which places this work between the adequate ranges proposed in the literature.

Keywords: artificial neural network (ANN); chemical oxygen demand (COD); wastewater treatment plant (WWTP)

1. Introduction

Pursuing the ideas outlined in the sustainable development goals (SDGs), countries have been showing concern for terrestrial ecosystems even more for the reuse and conservation of water quality. On this topic, one of the concerns that exists and will be resolved day by day is related to the contamination of liquid effluents that arise from industrial uses. According to standards established by the laws of most countries, industry must respond to certain requirements that allow for the reuse of the water products in its activity. Globally, the most common problem regarding the quality of effluent water in industries is eutrophication, the result of large amounts of nutrients (mainly phosphorus and nitrogen), which leads to the purity of the water being reduced [1]. Additionally, pH levels and the suspended solids index contribute significantly to water quality [2]. Thus, industry daily faces the challenge of treating wastewater as a result of its processes. The monitoring of this treatment yields a large volume of revealing data that can increase the efficiency in the removal of the contaminant load in the water. Faced with this problem, it is worth asking: Is it possible to create an intelligent system that can monitor the determining variables in the treatment of industrial wastewater? Can this intelligent system predict the parameters of water quality with a prudent margin of error? How could it check the operation of this system? This paper focuses on answering the previous questions.

Taking into account the exponential growth of industry at present and the amount of wastewater that its processes generate, it is essential for it to have an optimal treatment system for such effluents to mitigate the environmental impact generated by its discharges and comply with the environmental regulatory standards that increase their demand. This leads to innovation both in the treatment systems and in control and information management systems thereof to achieve a more efficient process, whose advantages have been evidenced in different developed countries [3]. The proposed approach is an intelligent system that uses the data from the biological stage of the process and makes a prediction of the behavior of bioreactors in a way that provides support in the decision making related to the operation of the wastewater treatment plant that can improve its operational efficiency. Implementing a continuous prediction of out-of-range values leads to taking timely preventive measures. As a result, water of a higher quality than required and bottleneck reduction because of the adaptation of microorganisms are some of the advantages obtained, which represent savings in operational costs.

A wastewater treatment plant (WWTP) is composed of different stages depending on the properties of the effluents to treat, but it most commonly takes advantage of either physical, chemical or biological treatments to take away pollutants [4]. The present work refers to industrial wastewater, which is that from the discharges of manufacturing industries [5], and uses data from the activated sludge process in the biological stage for developing an intelligent system, making use of machine learning algorithms that allow for automatic extraction of information from previous examples and infer about new data [6], achieving the forecasting of the chemical oxygen demand (COD), which is an indicator of water pollution and is a key variable to evaluate the efficiency of the WWTP process [7].

2. Related Works

Over the last decade, the amount and complexity of data have increased significantly thanks to the improvement in generation and storage of data, related to the cost reduction of them and the presence of more computational power [8]. Therefore, all this data now available can produce valuable information leading to better phenomenon comprehension, modeling and reproduction capable of providing some advantages and improvements to industrial processes [9]. Referring to water treatment plants, they integrated programmable logic controllers, supervisory control and data acquisition systems at the beginning of the XXI century [3]. Residential, agricultural, commercial and industrial effluents can be treated by WWTPs, each with its characteristics [10]. In the present research, mostly industrial effluent source studies are presented as the main topic of interest.

The analysis of the process of a WWTP can be classified as a complex control problem, which behaves as a nonlinear dynamic process [11]. Taking into account the nature of the process, the implementation of real-time optimal control is a challenge. Thus, predicting the effluent quality of this operation would help to control some parameters to prevent disasters and make the challenge less complex. Understanding the WWTP's complex nature depends on microbial, chemical and physical features, which are important to improve the effectiveness of the process [12]. These factors vary with time and physical attributes, such as weather, season, influent water, pH and bacteria amount, among others. However, using the problem background, statistical analysis and computational techniques reduces the complexity that a human being must understand in the WWTP process. The concept of "machine learning" has revolutionized analytics techniques to solve elaborate problems; as a result, experts in this area have taken advantage of the progress in these techniques to implement algorithms that describe the WWTP process to make the analysis more intelligible.

2.1. Related Works Description

In [11], a q-learning (QL) algorithm with an activated sludge model (ASM2d-guided) reward setting was proposed. The integrated ASM2d-QL algorithms equipped with a self-learning mechanism were derived for optimizing the control strategies (hydraulic retention time (HRT) and internal recycling ratio (IRR)) of the WWTP system. In reference [12], a Bayesian network-based approach was proposed for real-time prediction of a wastewater treatment system based on Modified Sequencing Batch Reactor (MSBR). Based on the framework of the modified sequencing batch reactor prediction analysis, a Bayesian network model was constructed to analyze an MSBR using training data and information provided by domain experts.

Work [13] is a synthesis of a new neuro-fuzzy controller with an online learning procedure and a simple algebraic formulation, making it easy to interpret by a human being to control a bioreactor without requiring any analytical representation. The authors in [14] focused on the Tabriz wastewater treatment plant (TWWTP), proposing an ensemble of fuzzy logic (FL), committee fuzzy logic (CFL) and supervised CFL to predict water quality parameters. In [10], three nonlinear models (feedforward neural network, adaptive neuro-fuzzy interference system and support vector machines (SVMs)) and a classical multilinear regression (MLR) were applied to predict the performance of the Nicosia wastewater treatment plant in terms of biochemical oxygen demand (BOD), COD and total nitrogen (TN). For paper [15], a data-driven intelligent monitoring system was implemented (using the soft sensor technique and data distribution service). A fuzzy neural network (FNN) was applied for designing the soft sensor model.

The paper [16] established two machine learning models—artificial neural networks (ANNs) and SVMs—to predict one-day interval TN concentration of effluent from a wastewater treatment plant in Ulsan, Korea. Reference [17] showed how machine learning models obtained better prediction results concerning traditional methods when increasing the size of the time-to-failure datasets. Four diverse machine learning approaches were implemented: ANN, SVM, random forest (RF) and soft computing methods. The reference [18] presented a data-driven anomaly detection approach based on deep learning methods and clustering algorithms to monitor influent conditions of WWTP, which affect treatment unit states, ongoing process mechanisms and product qualities. These techniques were recurrent neural networks (RNNs) and the function to delineate complex distributions from restricted Boltzmann machines (RBM), with various classifiers.

In work [19], multilayer perceptron ANN–genetic algorithm (MLPANN–GA) and radial basis function ANN–genetic algorithm (RBFANN–GA) models were successfully implemented for sludge volume index (SVI) prediction, taking into account that when sludge bulking appears, it causes poor settleability of sludge that results in poor effluent quality, loss of active biomass and increased costs and poses several environmental hazards. BOD, COD, nitrate, ammonia, TN, total phosphorus (TP), total suspended solids (TSS), total dissolved solids (TDS), mixed liquor volatile suspended solids (MLVSS), mixed liquor suspended solids (MLSS), SVI, dissolved oxygen (DO), pH and T (Celsius) were measured and used for the estimation. The study [20] performed a simulation of plant behavior over a wide range of influent disturbances. An artificial neural network (ANN) was trained on the available WWTP, comparing ANN and a mechanistic WWTP model's performances.

The study [21] proposed the Kohonen self-organizing map (SOM), a useful tool for illustrating the prevailing states of a process and their evolution, monitoring the alteration of wastewater quality and alerting in case of unusual behavior, such as increasing concentrations of harmful discharge components. The method provided an advanced and efficient way of monitoring and visualizing many measurements conducted in wastewater treatment. Article [22] emphasized the high potential of some promising techniques, such as spectral analysis, and discussed issues that could appear soon concerning control of anaerobic digestion (AD) processes. The authors in work [23] provided a critical outlook of the evolution of industrial process monitoring (IPM) since its introduction almost 100 years ago. Several evolution trends that have been structuring IPM developments over this extended period were briefly referred to, with more focus on data-driven approaches.

Work [24] is a survey of the feasibility of utilizing soft computing models in predicting emission factors (gaseous H_2S) based on five input parameters, namely, the total dissolved sulfides, biochemical oxygen demand (BOD5), temperature, flow rate and pH. Multivariate nonlinear autoregressive exogenous (NARX) neural networks were developed and applied to predict weekly

H_2S in four WWTPs. The paper [25] described an optimized extreme learning machine (ELM) based on an improved cuckoo search (ICS) algorithm for the design of the soft BOD measurement model.

Reference [26] is a review of developments in artificial intelligence technologies for environmental pollution controls, including prediction of removal efficiency, evaluation of fuzzy logic to the control of the WWTP aerobic stage and AI-aided soft sensors for estimation of hard-to-measure variables.

The study [27] performed different machine learning techniques to model a soft sensor to predict weather conditions such as SVMs, k-nearest neighbors (KNN), decision trees (DT), RFs and Gaussian naive Bayes (GND). With accurate weather prediction, an advanced control system can fit the parameters for better performance.

2.2. Variable Prediction

One of the early approximations to intelligent monitoring and the predicting system was presented in [28] and [13], where Bayesian networks and neuro-fuzzy logic were implemented to fulfill limitations of rule-based systems. Further works started to focus their attention on variable prediction using a variety of methods and a combination of them, taking the major advantages offered by each one. Reference [29] used iterative predictor weighting–partial least squares (IPW–PLS) boosted by weighted predictions of a collection of regression models used as an ensemble prediction to estimate some water quality parameters. It was tested in the field, and its results showed a high correlation of the prediction.

Several recent studies used fuzzy logic or neuro-fuzzy systems, such as [10,14,15], and some deep learning approaches, as in [16–18], which have provided high performance in prediction tasks. Studies like [19] used a hybrid artificial neural networks–genetic algorithm approach to optimize the ANN estimation of the sludge bulking present in the sedimentation stage, which directly affects the effluent discharge water quality. Reference [30] made a performance comparison between the autoregressive integrated moving average (ARIMA) and time-delay neural network (TDNN) with such times-series variables as BOD and TSS and achieved more accurate predictions for real-world wastewater data with TDNN.

2.3. Fault Detection

There is a research branch whose aim is the opportune fault detection in very stringent processes, especially when it is part of the operational critical path where any unexpected event that occurs leads to a stagnation. Depending on the type of fault detection, the prediction of the problem can be focused on:

- The system's ability to operate under some given circumstances.
- The time range in which equipment needs no maintenance and logistic support [17].

Regarding system operability, faults and potential causes can be found before they occur by analyzing some patterns in WWTP data. The data visualization is capable of showing patterns that are products of a possible anomaly, known as abnormal patterns. These are classified as isolated, sustained, transient and drift [3]. Each one provides a hint about a future fault. Thus, it is possible to get fault information by looking at data behavior. Reference [18] implemented data-driven unsupervised anomaly detection approaches based on deep learning methods and clustering algorithms. The aim was to monitor and detect anomaly conditions in WWTP operations. The results showed its ability to detect the vast majority of abnormal events reported by the operator [18].

On the other hand, basic reliability analysis focuses on the prediction of the period in which equipment needs no support. This technique allows for finding a probability function $R(t)$ to forecast the performance time of a component without failing until a given period t [17]. The work of [31] used an ANN to find the best cumulative failure distribution of mechanical components, which had a performance to fit a set of failure data and estimate its parameters, especially under poor data conditions. As a result, the networks with a momentum equal to 0.75 produced the best approximation 83.46% of the time [31].

2.4. Big Data Tools

Nowadays, since the world creates new data every single second, it has had to look for technologies to treat this data properly. In the market, some of them are Apache Hadoop and SciDB (open source) and others owned by supercompanies like Google, IBM, Amazon and Microsoft (frameworks) [32]. Each framework is specialized to do a particular task. A review [33] synthesized these frameworks as shown in Table 1 (adapted from [33]). Besides, the main languages for analytics, data mining and data science are R, SAS and Python. Each language has weaknesses and strengths. However, according to a Burtch Works poll (2019), computer scientists and engineers preferred using Python, as shown in Figure 1.

Table 1. Big data tools.

Area	Amazon	Microsoft	Google
Big data storage	S3	Azure	Google Cloud services
Big data analytics	Elastic MapReduce (Hadoop)	Hadoop on Azure	BigQuery
Relational database	MySQL or Oracle	SQL Azure	Cloud SQL
NoSQL database	DynamoDB	Table storage	App Engine Datastore
MapReduce	Elastic MapReduce (Hadoop)	Hadoop on Azure	App Engine
Streaming processing	Nothing prepackaged	StreamInsight	Search API
Machine learning	Hadoop + Mahout	Hadoop + Mahout	Prediction API
Data sources	Public datasets	Windows Azure marketplace	A few sample datasets
Availability	Public production	Some services in private beta	Some services in private beta

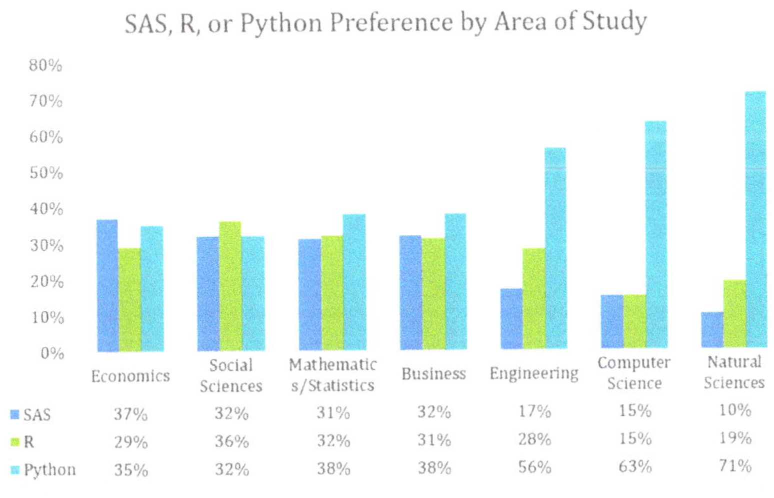

Figure 1. SAS, R or Python preferences.

2.5. Computational Techniques

According to related works, machine learning techniques have been implemented in several WWTP problems (Table 2). Around 64.71% of related work used an algorithm of ANN groups to develop forecasting models or a modified ANN to improve the analysis performance. Besides, support vector machines (SVM), fuzzy logic (FL), partial least squares (PLS) and principal component analysis (PCA) models were implemented by some authors. To clarify, percentages must not add up to 100% since some references used more than one algorithm. As shown in Table 3, last year, the ANN algorithm had significant participation in WWTP forecasting development in comparison with others.

Table 2. Related works.

Ref	Year	Method	Prediction	Error
[10]	2018	FFNN, ANFIS, SVM, MLR	BOD, COD, TN	DC, RMSE
[11]	2019	Q-learning	-	-
[12]	2012	Bayesian network	COD, TP, TN	-
[13]	2005	NFC	Dilution rate	-
[14]	2018	FL, SCFL, ANN	BOD, COD, TSS	MAPE
[15]	2018	FNN, PCA	BOD, COD, TSS, TP, NH_4-N	-
[16]	2015	ANN, SVM	TP, TSS, COD	R2, NSE, drel
[19]	2015	MLPANN–GA, RBFANN–GA	SVI	-
[20]	2006	ANN	BOD, COD, TSS, TN	R2
[21]	2013	SOM	-	-
[24]	2019	NARX	H_2S emission	MAPE, RMSE, GRI
[25]	2019	ICS–ELM, BP	BOD	-
[29]	2012	PLS, IPW–PLS, Boosting-IPW–PLS	COD, TSS, NTU	MinE, RMSEP, MaxE, R
[34]	2012	-	BOD, TSS, HRT, F/M	-

Table 3. Computational techniques used in wastewater treatment plant (WWTP) analysis from related works.

Algorithm	%	Algorithm	%
ANN	64.71	KNN	5.88
SVM	23.53	PCA	5.88
Fuzzy	17.65	PLS	5.88
BN	11.76	QL	5.88
RF	11.76	GND	5.88
DT	5.88	ICS	5.88

3. Materials and Methods

3.1. Model Design

COD is one of the most important variables in the process of a biological treatment since experts can make decisions based on the measurements of this variable. The objective of biological wastewater treatment is to perform a system to remove the pollutants present in water. Thus, this treatment is used overall because it is compelling and more efficient than numerous mechanical or compound procedures. In the bioreactor at this stage, a variety of microorganisms are used to break down organic matter in the water. However, the microorganisms are susceptible to change, depending on all the conditions in the tank.

For this reason, the present work proposes to use predictive analysis on COD to make decisions, knowing how contaminated the water will be in the tank. For studying how COD dynamics in the process are, a dataset was received from a WWTP from the Nantong, China plant with a daily data frequency for a total of 847 samples at different stages of the process, where a total of 22 variables were collected from 01/12/2017 to 24/05/2020. The COD dynamic can be observed in Figure 2.

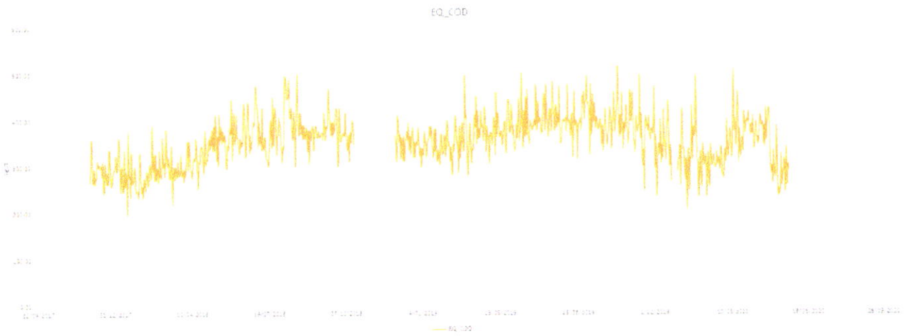

Figure 2. Chemical oxygen demand behavior.

Figure 3 shows the biological stages of the process in which the organic load of water is removed. Some important variables for the project that describe the WWTP process are represented as circles in blue and green. The blue circle is the output variable COD for the forecasting analysis, while green circles are input variables to design the intelligent system.

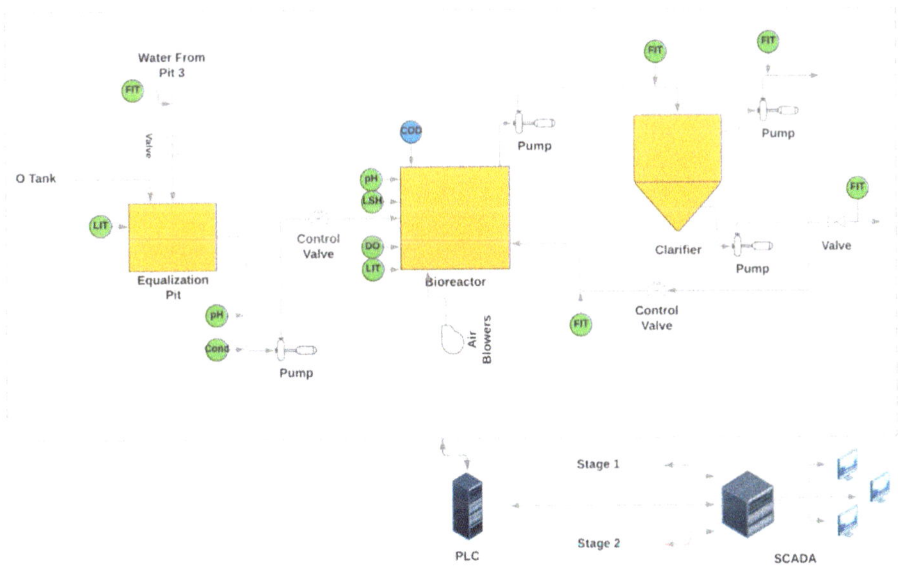

Figure 3. Biological WWTP process diagram.

For the development of the system, the selected technology was an ANN because of the state-of-the-art review supported by the complexity of the WWTP process. Figure 4 presents the flowchart that synthesizes the design process of the intelligent systems proposed, which started with the data collection and the use of different strategies for variable selection. Within the dataset, the main variables of the process were:

- Flow
- COD of influent water
- Suspended solids in influent water (SS)
- Mixed liquor suspended solids (MLSS)

- Mixed liquor volatile suspended solids (MLVSS)
- Nitrogen (N)
- pH
- Mixed liquor dissolved oxygen (DO)
- Food to microorganism (F/M)

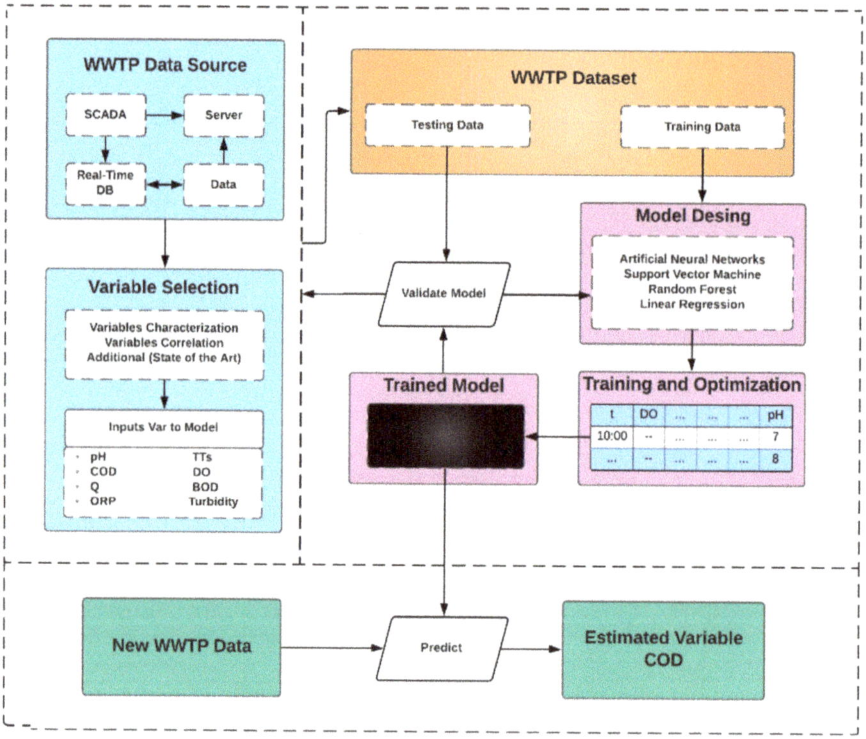

Figure 4. Model structure diagram.

Each characteristic can be repeated in one or more stages that are listed as below:

- EQ = Equalizer
- BIO = Bioreactor
- BT_N = Bioreactor Pit N
- BT_C = Bioreactor Pit C
- Clari = Clarifier
- OxT = Oxidation Tank
- D = Discharge Pit

After variable selection, the dataset is split into training, validation and test sets. However, in this case, the data was split into training and test sets since the number of samples was small in comparison with the amount of data used to train an ANN. It is important to note that a computational technique must be selected. As mentioned before in related works in Table 3, about 64.71% of the work of authors used an algorithm from the ANN group to develop forecast models. It has been verified that neural networks have suitable results in the area since the water treatment process is characterized by being nonlinear in behavior, so if they are used properly, they can represent the dynamics of this process very well. Once the model was selected, the model was trained and brought into operating condition to estimate COD. An error measure is necessary to support the performance of the model. Therefore, the MAPE), defined as shown in Equation (1), was chosen to quantify the ANN error. In this equation, x_i represents the actual point, which is intended to be predicted, $\hat{x}_i$ represents the predicted values of that observed point and N is the number of observed values that are intended to be predicted.

$$\text{MAPE} = \frac{100}{N}\sum_{i=1}^{N}\left|\frac{x_i - \hat{x}_i}{x_i}\right|, \tag{1}$$

Figure 5 shows in more detail how the model is conceived and how the COD forecasting is achieved. First, the objective variable taken from the dataset is studied using a time-series decomposition technique that transforms the variable into three additive components: trend, seasonality and residual. Leveraging an autocorrelation study over the components, the first two are estimated using their past values. On the other hand, the residual component is estimated using an ANN, which received exogenous variables selected from a correlation study and a past value of the same component. Finally, the addition of the three components provides the COD prediction. All data analysis and the intelligent system training were carried out by using Python, mainly taking advantage of Pandas, NumPy, Matplotlib, Statsmodels and TensorFlow libraries.

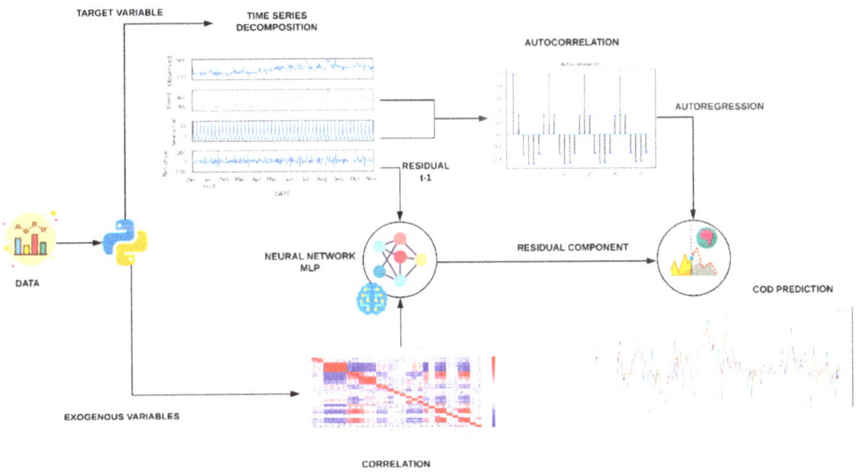

Figure 5. Model block diagram.

3.2. Platform Design

A web platform was designed to visualize all the variables of the WWTP dynamically, monitor the COD prediction provided by the forecast model and consult the historical measurements of the variables. Thus, the main sections of the platform were built as the real-time and historical data view. For this purpose, a model–view–controller schema was used to construct the platform using the technologies as Figure 6 shows. The technology that performed the view in the platform was ReactJS, responsible for rendering the visual content to interact with the user and make requests (frontend). ReactJS related to the master and brain of the platform, NodeJS, which controlled the logic responsible for managing all functions and methods that made the platform work (backend). Parallelly with NodeJS, TensorFlow.JS deployed the trained forecast model, which was developed to predict the COD at the beginning of the bioreactor. Besides, all the data and the information important to be the cog in this system were stored in a database schema settled in PostgreSQL. The interaction between those technologies allowed for reaching the objectives mentioned.

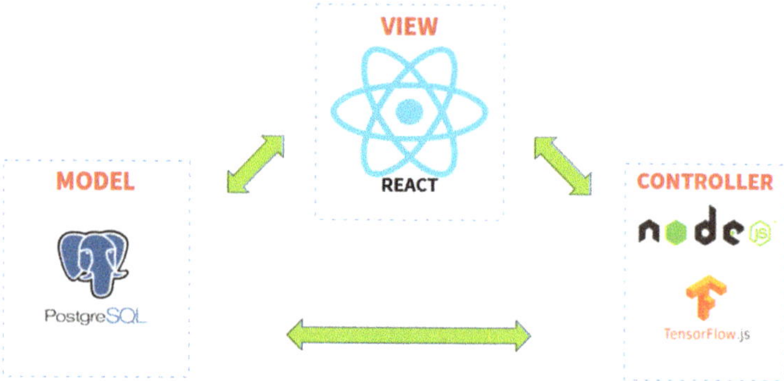

Figure 6. Platform schema.

4. Results

The experiments carried out were time-series decomposition, autocorrelation study and correlation study. Each one was to get the best performance of the model described below.

4.1. Time-Series Decomposition

For the time-series analysis of the target, the variable was made a component decomposition where the time series could be represented as a combination of trend, seasonality and residual components [35]. From this point, it was intended to forecast each component of the time series to obtain the objective series using the additive model stated by Pearson and presented in Equation (2) [36], where Tt refers to tendency or trend, St to seasonal movements, Rt to residuals or irregulars and Xt to the series observed.

$$Xt = Tt + St + Rt, \tag{2}$$

Figure 7 shows an example of how the equalizer's COD decomposition looks for the year 2019, where (a) shows the original COD variable, (b) the trend component, (c) the seasonal component and (d) the residual component.

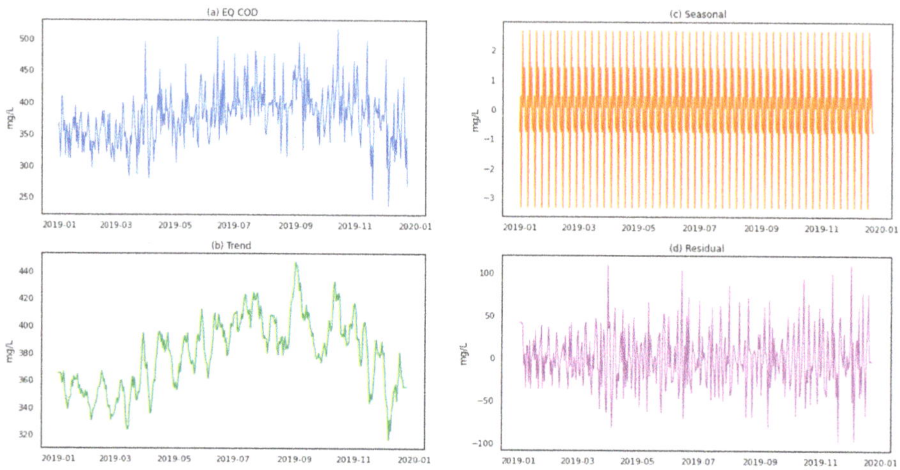

Figure 7. Equalizer chemical oxygen demand (COD) decomposition.

4.2. Autocorrelation Study

Analyzing the time-series decomposition, both autocorrelation and partial autocorrelation studies were made on residual, seasonal and trend COD to extract the important characteristics. From this analysis, it was possible to conduct an autoregressive estimation of the trend and seasonal component of the series. Figures 8–10 show the total and partial autocorrelation, respectively.

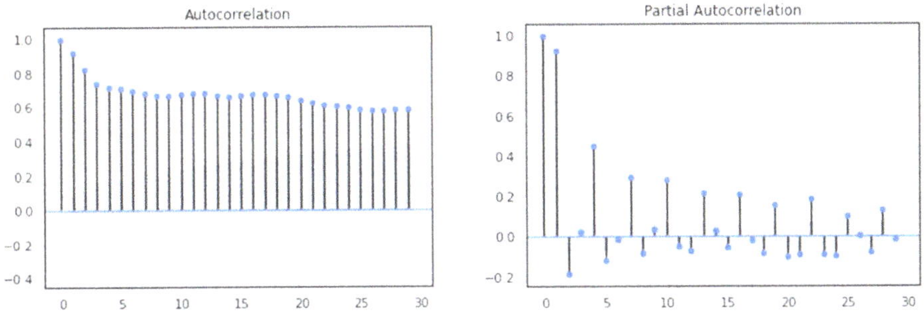

Figure 8. COD trend analysis correlation.

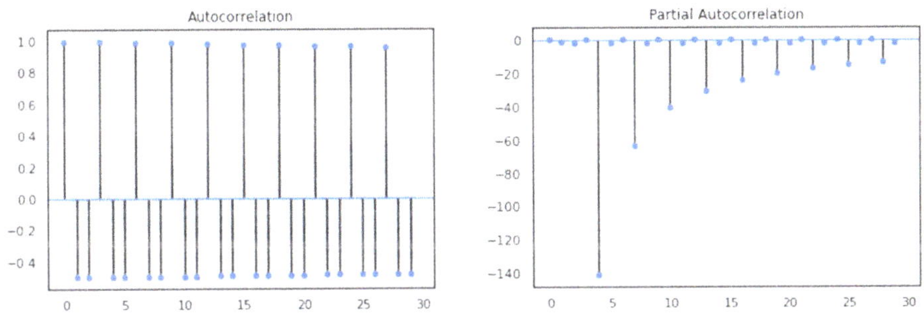

Figure 9. COD seasonal analysis correlation.

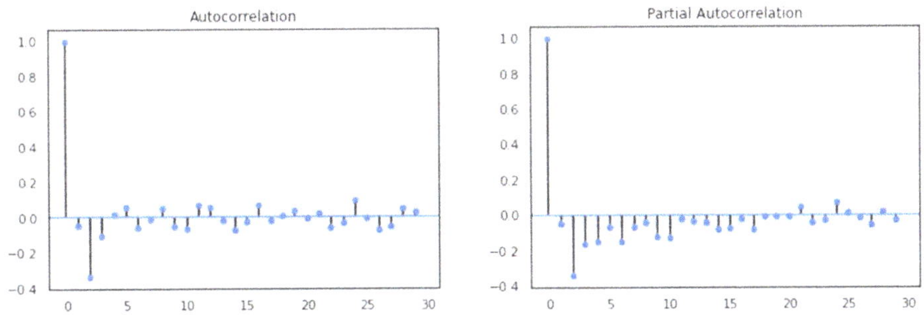

Figure 10. COD residual analysis correlation.

From Figure 8, it is clear how the past values were strongly correlated with the current COD trend value. Thus, the trend record provided significant information to the model on the dynamics of the COD. Additionally, Figure 9 shows the important effect of the seven past seasonal values. On the other hand, for the COD residual autocorrelation, the analysis was not very revealing, but it can be highlighted that for data with a validity of two days, there was a correlation of almost −0.35 with the current COD value.

4.3. Correlation Study

For determining which variables had a significant effect on the COD dynamic, a correlation study was used to decant characteristics and reduce the dimensionality of the model. Thus, the model could learn without the noise caused by raw characteristics. Besides, the variables with a high correlation improved system performance. The correlation selected for the analysis was the Pearson correlation since when exploring other types of correlations, the results were similar. The correlation results were carried out using the variable EQ_COD a day ahead of the target, considering that this was the purpose of this job. Figure 11 shows the correlation matrix, and focusing on the target, the suggested exogenous variables are below:

- BT_C_MLVSS
- D_SS
- BT_C_N
- EQ_N
- Clari_DO
- F/M

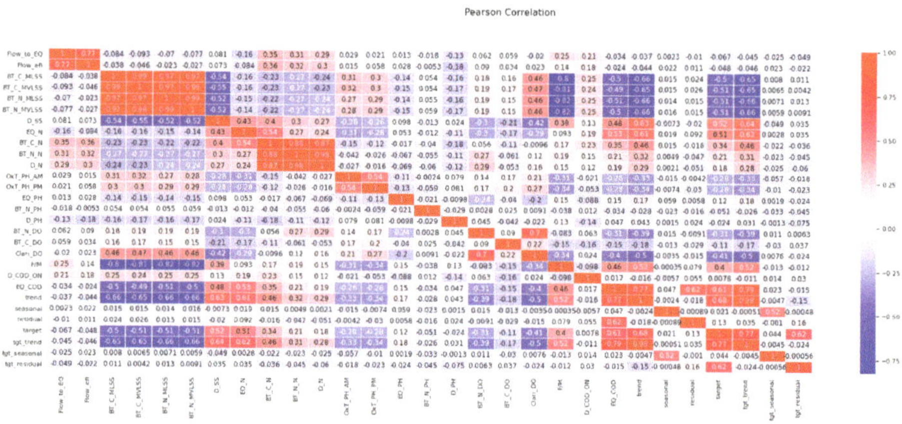

Figure 11. Correlation matrix.

Table 4 shows the correlation analysis summary focused on the target variable. To be noted, the selection threshold for the correlation was adjusted to 0.4, thus obtaining most of the variables suggested by the experts in the study area. However, BT_C_MLSS, BT_C_MLVSS, BT_N_MLSS and BT_N_MLVSS were highly related; therefore, the set could be represented by a single variable. In this case, BT_C_MLSS was selected, but any of the rest could be chosen. It is worth highlighting that EQ_COD on the correlation table refers to the current value of the variable.

Table 4. Correlation analysis summary.

Variable	Value
Flow_to_EQ	0.067
Flow_efl	0.048
BT_C_MLSS	0.50
BT_C_MLVSS	0.51
BT_N_MLSS	0.51
BT_N_MLVSS	0.51
D_SS	0.52
EQ_N	0.51
BT_C_N	0.34
BT_N_N	0.21
D_N	0.18
OxT_pH Morning	0.28
OxT_pH Afternoon	0.28
EQ_pH	0.12
BT_N_pH	0.051
D_pH	0.024
BT_N_DO	0.31
BT_C_DO	0.11
Clari_DO	0.41
F/M	0.40
D_COD_ON	0.0078
EQ_COD (t)	0.61

4.4. Artificial Neural Network

Utilizing selected variables from the correlation study, an artificial neural network was implemented to forecast the time-series residual. The architecture implemented was a multilayer perceptron (MLP) fully connected with 7 neurons in the input layer and 2 hidden layers, with 22 neurons each, and 1 neuron in the output layer to predict the residual component. The neural network was trained with approximately 80% of the samples, and 147 corresponding samples from the year 2020 were used for the test. During the 150 training periods, the training used the backpropagation algorithm to update the weights in the neurons, with the mean square error (MSE) as the loss function and Adam optimizer. Figure 12 shows the preliminary results, where the blue series is the real one and the orange is the predicted value.

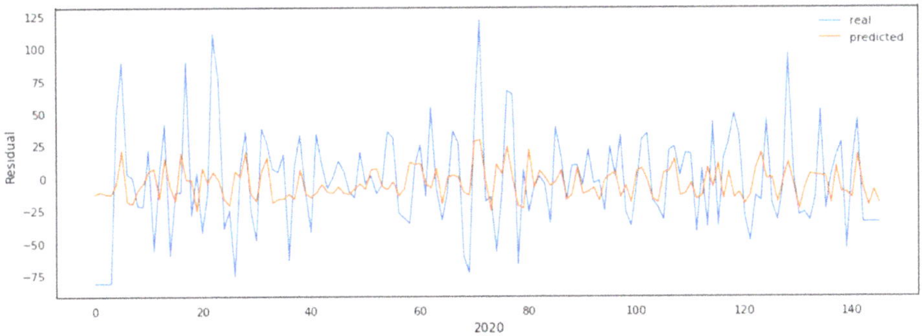

Figure 12. Residual prediction.

The number of neurons in each hidden layer of the neural network was obtained through a grid search, as shown in Figure 13, using training data.

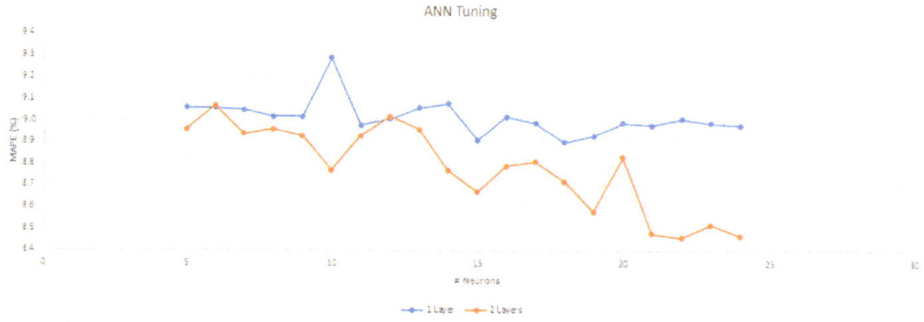

Figure 13. Artificial neural network (ANN) tuning.

Using the autoregressive estimation conducted on the trend, seasonal and the residual component obtained by the ANN, it was possible to forecast the equalizer COD (adding together the three components) as shown in Figure 14, obtaining a MAPE of 10.8%, which is appropriate with the values found in the literature, where similar works reported MAPEs between 4% and 11% as good forecasting performance.

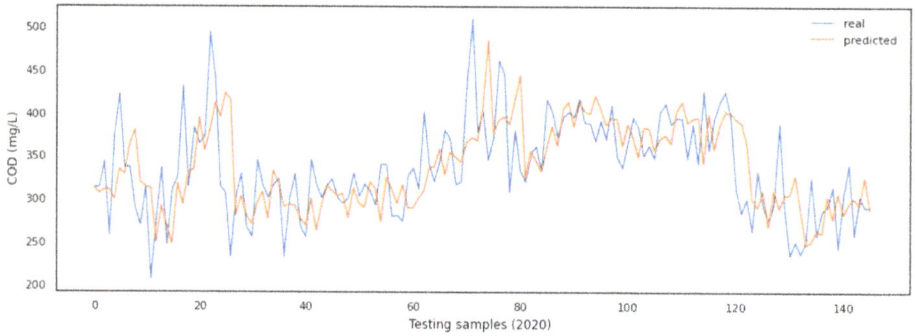

Figure 14. COD prediction.

The prediction achieved and presented above was made day by day, as was the error obtained. Pikes on the COD dynamic were not reached by the model. However, it was considered to increase the number of samples to improve the performance of the model in future work.

4.5. Web Platform.

The final result of the platform was designed so that a user could visualize all the variables of the WWTP dynamically, monitor the COD prediction and check the historical measurements of the variables (see Figure 15).

Figure 15. Real-time view.

This section hides a powerful backend behind its interface. The box where the current COD is displayed responds to the measurement that is currently being read from the COD variable at that moment. The box titled as Predicted COD is directly connected to the model that gives a prediction in response to the current COD input and the selected exogenous variables. To compare the behavior between the real and predicted COD, a window is available, as Figure 16 shows (this figure captured only behavior with training data). The prescription box is thought of and built for future work. On the other hand, there is a visualization of all the process variables and a condensed summary in a table of the measurement of each variable.

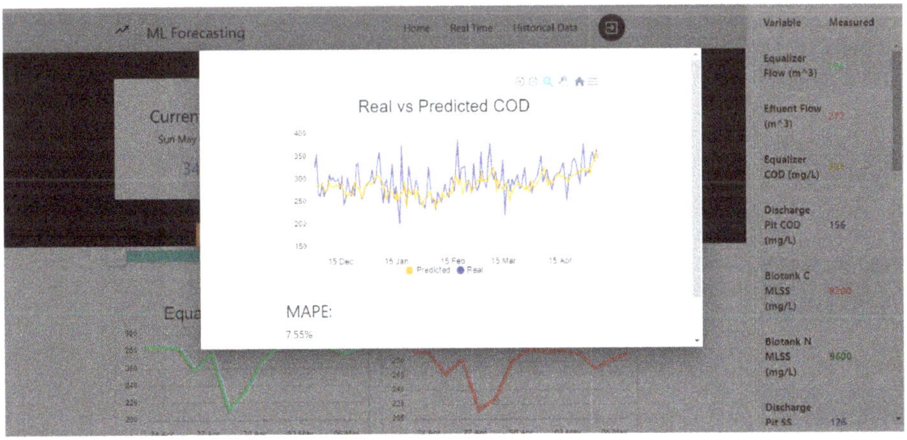

Figure 16. COD monitoring.

To have a visualization of the historical data, a section was developed with the corresponding graphs and a summary table to be able to choose a historical data point from the graphs and detail it in the right table. Figure 17 shows this result.

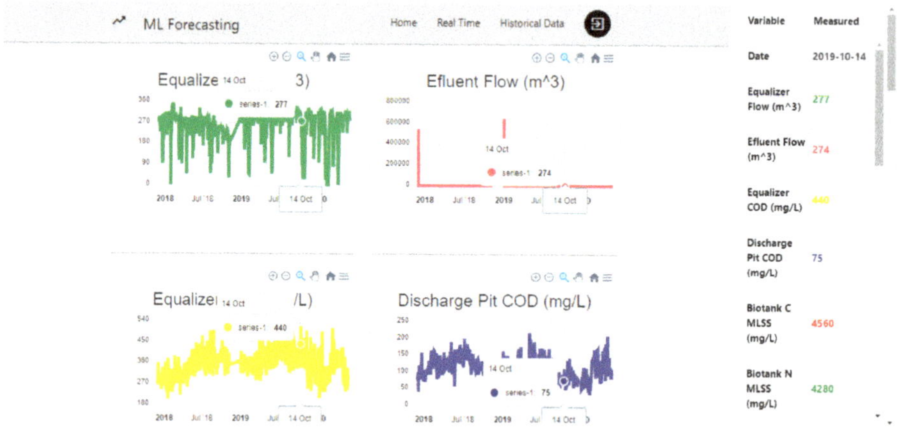

Figure 17. Historical data view.

5. Discussion

The selection and characterization of the most significant variables of the wastewater treatment process have been carried out satisfactorily using correlation analysis, autocorrelations and decomposition of the time series. With these variables, an intelligent system based on artificial neural networks was developed to be capable of giving an adequate prediction of chemical oxygen demand, one of the most suitable variables to measure the level of pollutant load in the water and make decisions. The results show that the model presented a MAPE of 10.8%, which supports its good performance according to historical data mentioned in [14], where the testing step ranged between 10% and 13%, predicting BOD, COD or TSS. Additionally, it is worth mentioning that this work presents as a novelty the use of time-series decomposition techniques to address the COD prediction and using an ANN, in comparison with the works presented in Section 2, whose summary can be seen in Table 2. This methodology can be useful to improve the prediction of some complex variables in which the ANNs do not have the desired performance. Finally, a platform was possible to design mainly to visualize available WWTP variables, monitor COD forecasting and consult the historical measurements.

In search of constant improvement of the industrial wastewater treatment process, it is considered for future works to scale the prediction of the system to other key variables of the process, obtain a larger amount of data considering newly available measurements in the process and increase the scope of the prediction.

Author Contributions: Conceptualization, A.M. and R.M.; methodology, C.G.Q.M.; software, L.A. and C.C.; validation, L.A., C.C. and C.G.Q.M.; formal analysis, L.A., C.C. and C.G.Q.M.; investigation, L.A., C.C. and C.G.Q.M.; resources, D.G.; data curation, L.A. and C.C.; writing—original draft preparation, L.A. and C.C.; writing—review and editing, L.A., C.C., D.G., A.M., R.M. and C.G.Q.M.; visualization, L.A. and C.C.; supervision, C.G.Q.M.; project administration, D.G.; funding acquisition, D.G. All authors have read and agreed to the published version of the manuscript.

Funding: This research was supported by the Colombian Ministry of Science and Technology, MINCIENCIAS, Investment Tax Benefits, Call No. 786.

Acknowledgments: This work was supported by the Universidad del Norte, Barranquilla, Colombia.

Conflicts of Interest: The authors declare that there are no conflict of interest regarding the publication of this paper.

Abbreviations

Abbreviation	Definition
ANFIS	Adaptive neuro-fuzzy inference system
ANN	Artificial neural network
BN	Bayesian network
BP	Backpropagation network
COD	Chemical oxygen demand
DC	Determination coefficient
DT	Decision tree
drel	Relative efficiency criteria
ELM	Extreme learning machine
F/M	Food to microorganism
FFNN	Feedforward neural network
FL	Fuzzy logic
FNN	Fuzzy neural network
GA	Genetic algorithm
GND	Gaussian naive Bayes
GRI	Global Reporting Initiative
HRT	Hydraulic retention time
ICS	Improved cuckoo search
IPW	Iterative predictor weighting
KNN	K-nearest neighbors
MAPE	Mean absolute percentage error
MLPANN	Multilayer perceptron ANN
MLR	Multilinear regression
MSE	Mean square error
MLSS	Mixed liquor suspended solids
MLVSS	Mixed liquor volatile suspended solids
NARX	Multivariate nonlinear autoregressive exogenous
NFC	Neuro-fuzzy controller
NH_4-N	Ammonium
NSE	Nash–Sutcliffe efficiency
O&G	Oil and grease
PCA	Principal component analysis
PCC	Pearson correlation coefficient
PLS	Partial least squares
QL	Q-learning
R	Correlation coefficient
R2	Coefficient of determination
RBFANN	Radial basis function ANN
RF	Random forest
RMSE	Root mean square error
RMSEP	Root mean squared error of prediction
SCFL	Supervised committee FL
SOM	Self-organizing maps
SRM	Structural risk minimization
SVI	Sludge volume index
SVM	Support vector machine
TN	Total nitrogen
TP	Total phosphorus
TSS	Total suspended solids
UVE	Uninformative variable elimination
WWTP	Wastewater treatment plant

References

1. UNWWA Programme. *The United Nations World Water Development Report 3: Water in a Changing World*; UNESCO: Paris, France, 2008.
2. Sener, E.S.S.; Devraz, A. Evaluation of water quality using water quality index (WQI) method and GIS in Aksu River (SW-Turkey). *Sci. Total Environ.* **2017**, *584–585*, 131–144. [CrossRef] [PubMed]
3. Newhart, K.B.; Holloway, R.W.; Hering, A.S.; Cath, T.Y. Data-driven performance analyses of wastewater treatment plants: A review. *Water Res.* **2019**, *157*, 498–513. [PubMed]
4. Anjun, M.; Al-Makishah, N.H.; Barakat, M.A. Wastewater sludge stabilization using pre-treatment methods. *Proc. Saf. Environ. Prot.* **2016**, *102*, 615–632. [CrossRef]
5. Tchobanoglous, G.; Schroeder, E.E. *Water Quality: Characteristics, Modeling, Modification*; Addison-Wesley Publishing Company: Boston, MA, USA, 1985.
6. Lake, B.M.; Ullman, T.D.; Tenebaum, J.B.; Gershman, S.J. Building machines that learn and think like people. *Behav. Brain Sci.* **2017**, *40*, e253. [CrossRef]
7. V´ıtez, J.S.T.; Oppeltov´a, P. Evaluation of the efficiency of selected wastewater treatment plant. *Acta Univ. Agric. Silvic. Mendel. Brun.* **2012**, *60*, 173–180. [CrossRef]
8. Romero, J.M.P.; Hallet, S.H.; Jude, S. Leveraging big data tools and technologies: Addressing the challenges of the water quality sector. *Sustainability* **2017**, *9*, 12.
9. Sbroiavacca, A.; Sbroiavacca, F. Industry 4.0: The Exploitation of Big Data and Forthcoming Perspectives, Economic and Social Development. In *Book of Proceedings, Proceedings of the 35thInternational Scientific Conference on Economic and Social Development–Sustainability from an Economic and Social Perspective, Lisbon, Portugal, 15–16 November 2018*; ESD Publishing: Varazdin, Croatia, 2018; pp. 742–745.
10. Nourani, V.; Elkiran, G.; Abba, S.I. Wastewater treatment plant performance analysis using artificial intelligence—An ensemble approach. *Water Sci. Technol.* **2018**, *78*, 2064–2076. [CrossRef]
11. Pang, J.; Yang, S.; He, L.; Chen, Y.; Ren, N. Intelligent control/operational strategies in WWTPs through an integrated Q-learning algorithm with ASM2d-guided reward. *Water* **2019**, *11*, 927. [CrossRef]
12. Li, D.; Yang, H.Z.; Liang, X.F. Prediction analysis of a wastewater treatment system using a Bayesian network. *Environ. Model.Softw.* **2013**, *40*, 140–150. [CrossRef]
13. Haggege, J.; Benrejeb, M.; Borne, P. On the design of a neuro-fuzzy controller—Application to the control of a bioreactor. *J. Syst. Sci. Syst. Eng.* **2005**, *14*, 417–435. [CrossRef]
14. Nadiri, A.A.; Shokri, S.; Tsai, F.T.; Asghari Moghaddam, A. Prediction of effluent quality parameters of a wastewater treatment plant using a supervised committee fuzzy logic model. *J. Clean. Prod.* **2018**, *180*, 539–549. [CrossRef]
15. Han, H.; Zhu, S.; Qiao, J.; Guo, M. Data-driven intelligent monitoring system for key variables in wastewater treatment process. *Chin. J. Chem. Eng.* **2018**, *26*, 2093–2101. [CrossRef]
16. Guo, H.; Jeong, K.; Lim, J.; Jo, J.; Kim, Y.M.; Park, J.-P.; Kim, J.H.; Cho, K.H. Prediction of effluent concentration in a wastewater treatment plant using machine learning models. *J. Environ. Sci.* **2015**, *32*, 90–101. [CrossRef]
17. Alsina, E.F.; Chica, M.; Trawiński, K.; Regattieri, A. On the use of machine learning methods to predict component reliability from data-driven industrial case studies. *Int. J. Adv. Manuf. Technol.* **2018**, *5*, 2419–2433. [CrossRef]
18. Dairi, A.; Cheng, T.; Harrou, F.; Sun, Y.; Leiknes, T. Deep learning approach for sustainable WWTP operation: A case study on data-driven influent conditions monitoring. *Sustain. Cities Soc.* **2019**, *50*, 101670. [CrossRef]
19. Bagheri, M.; Mirbagheri, S.A.; Bagheri, Z.; Kamarkhani, A.M. Modeling and optimization of activated sludge bulking for a real wastewater treatment plant using hybrid artificial neural networks-genetic algorithm approach. *Proc. Saf. Environ. Prot.* **2015**, *95*, 12–25. [CrossRef]
20. Ráduly, B.; Gernaey, K.V.; Capodaglio, A.; Mikkelsen, P.S.; Henze, M. Artificial neural networks for rapid WWTP performance evaluation: Methodology and case study. *Environ. Model. Softw.* **2007**, *22*, 1208–1216. [CrossRef]
21. Liukkonen, M.; Laakso, I.; Hiltunen, Y. Advanced monitoring platform for industrial wastewater treatment: Multivariable approach using the self-organizing map. *Environ. Model. Softw.* **2013**, *48*, 193–201. [CrossRef]
22. Jimenez, J.; Latrille, E.; Harmand, J.; Robles, Á.; Ferrer, J.; Gaida, D.; Wolf, C.; Mairet, F.; Bernard, O.; Alcaraz-González, V.; et al. Instrumentation and control of anaerobic digestion processes: A review and some research challenges. *Rev. Environ. Sci. Biotechnol.* **2015**, *14*, 615–648. [CrossRef]

23. Reis, M.; Gins, G. Industrial Process Monitoring in the Big Data/Industry 4.0 Era: From Detection, to Diagnosis, to Prognosis. *Process* **2017**, *5*, 35. [CrossRef]
24. Zounemat-Kermani, M.; Stephan, D.; Hinkelmann, R. Multivariate NARX neural network in prediction gaseous emissions within the influent chamber of wastewater treatment plants. *Atmospheric Pollut. Res.* **2019**, *10*, 1812–1822. [CrossRef]
25. Yu, P.; Cao, J.; Jegatheesan, V.; Du, X. A Real-time BOD Estimation Method in Wastewater Treatment Process Based on an Optimized Extreme Learning Machine. *Appl. Sci.* **2019**, *9*, 523. [CrossRef]
26. Ye, Z.; Yang, J.; Zhong, N.; Tu, X.; Jia, J.; Wang, J. Tackling environmental challenges in pollution controls using artificial intelligence: A review. *Sci. Total Environ.* **2020**, *699*, 134279. [CrossRef]
27. Hernández-Del-Olmo, F.; Gaudioso, E.; Duro, N.; Dormido, R. Machine Learning Weather Soft-Sensor for Advanced Control of Wastewater Treatment Plants. *Sensors* **2019**, *19*, 3139. [CrossRef] [PubMed]
28. Sangüesa, R.; Burrell, P. Application of Bayesian Network Learning Methods to Waste Water Treatment Plants. *Appl. Intell.* **2000**, *13*, 19–40. [CrossRef]
29. Qin, X.; Gao, F.; Chen, G. Wastewater quality monitoring system using sensor fusion and machine learning techniques. *Water Res.* **2012**, *46*, 1133–1144. [CrossRef]
30. Dellana, S.; West, D. Predictive modeling for wastewater applications: Linear and nonlinear approaches. *Environ. Model. Softw.* **2009**, *24*, 96–106. [CrossRef]
31. Alsina, E.F.; Cabri, G.; Regattieri, A. A neural network approach to find the cumulative failure distribution: Modeling and experimental evidence. *Qual. Reliab. Eng. Int.* **2016**, *32*, 567–579. [CrossRef]
32. Siddiqui, T.; Al Kadri, M. Big data analytics on the cloud. *Int. J. Emerg. Technol. Comput. Appl. Sci. (IJETCAS)* **2015**, *24*, 61–66.
33. Siddiqui, T.; Al Kadri, M.; Khan, N.A. Review of programming languages and tools for big data analytics. *Int. J. Adv. Res. Comput. Sci.* **2017**, *8*, 1112–1118.
34. Valentín-Vargas, A.; Toro-Labrador, G.; Massol-Deyá, A.A. Bacterial community dynamics in full-scale activated sludge bioreactors: Operational and ecological factors driving community assembly and performance. *PLoS ONE* **2012**, *7*, e42524. [CrossRef] [PubMed]
35. Cryer, J.D.; Chan, K.-S. *Time Series Analysis*; Springer: New York, NY, USA, 2008.
36. Dagum, E. Time series modelling and decomposition. *Statistica* **2013**, *70*, 5.

sustainability

MDPI

Article

Towards a Sustainable and Adaptive Groundwater Management: Lessons from the Benalup Aquifer (Southern Spain)

Mercedes Vélez-Nicolás [1], Santiago García-López [1,*], Verónica Ruiz-Ortiz [2] and Ángel Sánchez-Bellón [1]

1 Department of Earth Sciences, Faculty of Marine and Environmental Sciences, University of Cádiz, 11519 Puerto Real, Spain; mercedes.velez@uca.es (M.V.-N.); angel.sanchez@uca.es (Á.S.-B.)
2 Department of Industrial and Civil Engineering, Higher Polytechnic School of Algeciras, University of Cádiz, 11202 Algeciras, Spain; veronica.ruiz@uca.es
* Correspondence: santiago.garcia@uca.es

Received: 18 May 2020; Accepted: 24 June 2020; Published: 26 June 2020

Abstract: Reversing the chemical and quantitative impacts derived from human activity on aquifers demands a multidisciplinary approach. This requires, firstly, to update the hydrogeological knowledge of the groundwater systems, which is pivotal for the sustainable use of this resource, and secondly, to integrate the social, economic and administrative reality of the region. The present work focuses on the Benalup aquifer, whose exploitation plays a major role in the economy of the area, based mainly on irrigated agriculture. This activity has had negative consequences for the aquifer in quantitative and chemical terms, leading to its declaration as in poor condition. The study presented here shows the results obtained from the application of hydrogeological techniques, remote sensing and citizen participation tools, which have allowed us to deepen and improve the current knowledge of the system's hydrogeological, geometric, administrative and social aspects. Additionally, the lessons learned from this case study are analyzed. The deficiencies detected are discussed, and alternatives aimed at the sustainable use of groundwater are proposed, such as the possibility of a joint use of surface and groundwater resources, the creation of a Water User Association responsible for the management of groundwater and the need for greater efforts aimed at educating and raising awareness of water conservation among citizens.

Keywords: aquifer management; water governance; irrigation; unauthorized use; barbate river basin; biocalcarenites; remote sensing; citizen surveys; groundwater

1. Introduction

The concept of sustainability was brought to the forefront by the World Commission on Environment and Development (WCED) held in 1987, and, since then, it has been commonly applied to the whole variety of resources exploited by human society, including groundwater. In fact, the European Water Framework Directive, enacted in December 2000, establishes that it is necessary to promote a sustainable water use. While there is a general consensus on this basic principle, its practical application in natural resource management is daunting [1]. Although the concept of sustainability has traditionally been based on 3 basic pillars (social, economic and environmental), its definition depends on the context, and its dimensions are still under discussion. This point of view is endorsed by numerous authors: Sophocleous (2000) [2] describes sustainability as a dynamic concept that has yet to be redefined and translated into feasible policies; Shamir (2000) [3] claims that the concept of sustainability is integrated by up to 10 dimensions, that include hydrology, ecology, politics, economy and inter/intragenerational aspects, among others; Mays (2007) [4] asserts that "because water impacts so many aspects of our existence, there are many facets that must be considered in water resources sustainability". The present work is going to focus mainly on the hydrological aspects of groundwater sustainability, following the criteria established by the U.S. Geological Survey, which defined this concept as the "development and use of groundwater in a manner that can be maintained for an indefinite time without causing unacceptable environmental, economic, or social consequences" [5]. In this regard, it must be noted the intergenerational perspective inherent to this concept, which, in the words of Sophocleous (2000) [2], implies that "water resources must be used in ways that are compatible with maintaining them for future generations, thus constraining our management of water".

Groundwater constitutes the main and safest source of water, supplying numerous urban and rural communities and their associated economic sectors, especially in areas of arid and semi-arid climate where surface resources are available intermittently and seasonally [6]. Besides, groundwater is a valuable strategic reserve, considering the uncertainty posed by climate change regarding the availability of water resources in the medium and long term. From a hydrologic perspective, the sustainable exploitation of aquifers was commonly understood as the adjustment of withdrawals to recharge, which implied an intrinsic consideration of groundwater as a renewable resource. Nevertheless, this conception is an oversimplification of the hydrogeological systems. To achieve a thorough understanding of groundwater bodies (GWB) and their functioning, it is necessary to have a global vision of the hydrological cycle. This is because the depletion of the water table (WT) of an aquifer subject to pumping can lead to alterations that not only affect the discharge areas (springs, wetlands and groundwater-dependent ecosystems), but can also impact the quantity and quality of river flows, and even reverse the river-aquifer hydraulic interactions.

Great progress has been made in the field of groundwater management and protection since the last decade of the 20th century, but there are still important problems specific to each site and hydrogeological system that the water authorities and public institutions must address [7]. Despite the numerous competences legally attributed to the administrations, in practice, their efforts fall short of guaranteeing an effective control and monitoring of the status and exploitation of GWB. This situation is the result of several factors, such as the poor knowledge of hydrogeological systems, the lack of material and human resources, the existence of inherited regulatory figures that hinder a sound management of the resource or the lack of guidance for farmers and landowners regarding the efficient use and protection of groundwater. In addition, it must be considered that the adoption of excessively restrictive measures by the administrations may lead to social conflicts. In line with this, Galloway & Pentland (2003) [8] claim that the administrative implementation of groundwater management objectives must rely on the science of hydrology and that an improved scientific understanding of the groundwater systems can also reduce management uncertainties and enhance the beneficial use of the resource.

All these considerations evidence that dealing with the multiple problems concerning groundwater requires a multidisciplinary approach that combines geological, hydrogeological, historical and social aspects to build up a detailed picture of the system.

This multidisciplinary approach has been endorsed by numerous works that, despite being focused on the field of hydrology, display a marked integrative vocation. Custodio et al. (2016) [9] reviewed hydrogeological, environmental, legal, administrative, economic and social aspects of groundwater exploitation in Gran Canarias and Tenerife. Molina et al. (2009) [10] and Molina et al. (2010) [11] applied the concept of Integrated Water Resource Management (IWRM) to aquifers in the province of Murcia incorporating the aforementioned aspects in Bayesian network models. Melloul and Collin [12] proposed an integrated hierarchy of hydrological aspects, land uses and social needs to achieve a sustainable use of water. Gleeson et al. (2012) [13] focused on the most sociological aspects of water management and advocated the pursuit of multi-generational sustainability goals and the implementation of adaptive management and backcasting methodologies. Gondwe et al. [14] conducted a multidisciplinary research combining piezometric measurements with geochemistry, geophysics and remote sensing techniques to study the karstic systems of the Yucatan Peninsula and promote the sound management of its resources.

Following this multidisciplinary approach, this study provides a comprehensive vision of the Benalup groundwater body (GWB 062.014), located in the province of Cádiz (Andalusia, Spain). On the one hand, geological, hydrogeological and geophysical techniques and laboratory tests were used to obtain information on the properties, spatial configuration and dominant structures controlling the functioning of the hydrogeological system. On the other hand, hydrochemical determinations were carried out to evaluate its qualitative status, which is closely linked to management practices and the functioning of the system itself. Finally, in order to provide further knowledge of water usage, remote sensing and citizen participation tools were implemented in the study area. The first was aimed at monitoring the evolution and distribution of irrigated crops (useful to quantify the exploitation of aquifers) and the latter at assessing the citizens' knowledge and perception regarding the local water resources. A brief revision of the administrative and socioeconomic scenario is also provided. Merging all this information enabled us to identify weaknesses and set guidelines for the sustainable management of groundwater. In this case study, the chemical and quantitative problems of the aquifer are mainly linked to the agro-livestock activity developed in the area and indirectly to the current social and administrative circumstances. In this regard, it is noteworthy that these problems are shared by 9 of the 14 (64%) GWB defined in the province of Cádiz by the regional government of Andalusia [15], all of them presenting similar hydrogeological characteristics. Thus, this paper presents the lessons learned from the case of the Benalup aquifer, which are applicable to other systems, as well as alternatives and considerations aimed at the sustainable use of the resource.

2. Case Study

The Benalup aquifer is located in the mid-western sector of the hydrographic basin of the Barbate River, in the region known as "La Janda" (province of Cadiz), near the Strait of Gibraltar (Figure 1). In administrative terms, this region belongs to the Guadalete-Barbate Hydrographic Demarcation. The aquifer studied has an area of 33 km^2 and encompasses the whole groundwater body Benalup (GWB 062.014), which extends across the municipalities of Medina Sidonia, Benalup and Vejer de la Frontera. In geological terms, the Benalup aquifer consists of Neogene and Quaternary detrital materials of considerable interest from a hydrogeological point of view. Owing to its location and characteristics, the aquifer constitutes a natural reservoir of great local importance for human and agro-livestock supply.

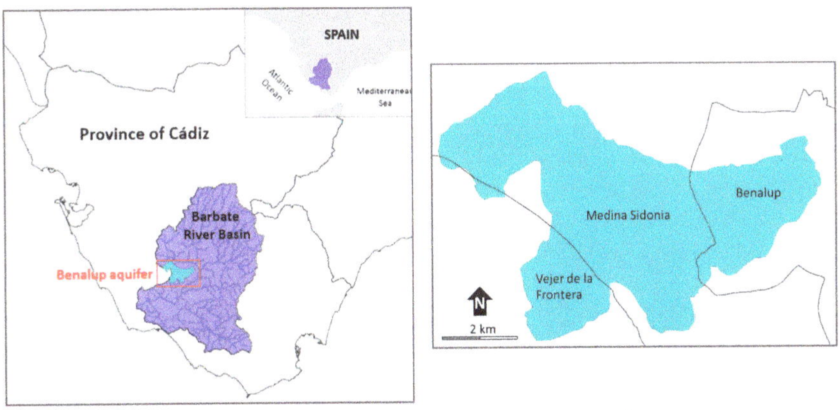

(**a**) (**b**)

Figure 1. (**a**) Location of the Barbate River Basin; (**b**) Detail of the Benalup aquifer and the municipalities that it encompasses.

The hydrographic basin that encompasses the Benalup aquifer has a Mediterranean climate with oceanic influence according to the Köppen-Geiger classification. The precipitation regime presents a marked seasonal character, with an average precipitation over the basin that exceeds 820 mm/year and concentrates mainly between November and March. The potential evapotranspiration (PET) reaches 930 mm/year, of which 77% takes place between May and October (UCA, 2019) [16], which conditions the irrigation water demand.

The productive model of La Janda region has traditionally been based on agriculture and livestock farming, while the industrial activity is almost inexistent. In particular, irrigated agriculture is of major importance and covers a surface of 15.000 ha, supplied with both surface water (84%) and groundwater (16%). Surface water comes from the Barbate, Celemín and Almodóvar reservoirs (joint capacity of 277 hm^3), which are exploited to supply mainly irrigation. The groundwater resources in La Janda region come from the Barbate GWB (GWB 062.013) and the aforementioned Benalup GWB (joint extension of 145 km^2); both aquifers display similar characteristics and are subject to intense exploitation. While the surface resources are managed in a planned and coordinated manner by a Water User Association (WUA), groundwater lacks an entity responsible for its control. This absence of a management entity responsible for the groundwater resources has led to an unplanned and uncontrolled exploitation of the aquifer.

Thus, the Benalup aquifer is exploited through several hundreds of boreholes, many of them of illegal character and without administrative authorization. Moreover, despite the efforts made by the administration to control the withdrawals, most of the large-flow boreholes are not equipped with water flow-meters or devices to control the operation time. The Benalup aquifer is exploited to serve two purposes: Firstly, to supply drinking water and secondly, for irrigation. This GWB supplies around 10.000 inhabitants distributed by several urban settings; Benalup, the most important (6.929 inhabitants, eastern sector of the aquifer), and other small settings of minor importance; Badalejos, Malcocinado (northern sector), Cantarranas (southern sector) and Naveros (outside the aquifer boundaries), whose annual consumption is about 0.85 hm^3/year (20% of the total extractions). To ensure this supply, there are seven boreholes operating in different sectors of the aquifer, which are managed by either the city council or specialized water companies. However, the transfer of surface water from the adjacent Guadalete river basin began in 2019, managed by the entity "Consorcio de Aguas de la Zona Gaditana". This water aimed at human consumption represents today $\frac{1}{4}$ of the urban demand previously covered with groundwater, but its proportion will gradually increase, so that exploitation for this purpose will become less important. Moreover, we should also mention the existence of important recreational facilities, with a golf course of about 50 ha and two equestrian facilities.

Agricultural pumping is carried out to irrigate the plots and farms located on the aquifer, being the only available source for this purpose. The authorized irrigated area within the aquifer is 785 ha, which represents 24% of the total surface of the system. The most commonly grown crops are potato and sweet potato, sunflower, carrot, leek, citrus and pasture for feeding cattle. The number of boreholes, the volume extracted and the actual irrigated area are difficult to determine. While many agricultural boreholes are no longer operating, owing to farm abandonment, other wells/boreholes with lower flow have been dug to supply second residences, which are used during the weekends or holidays and are usually linked to illegal land use. The area affected by these changes is more than 140 ha, with an average occupation between 20% and 30%. Indeed, in the framework of the present study, a total of 384 plots of this type and around 300–400 inhabitant-equivalent have been estimated. While urban areas have a sanitation network and purification and/or discharge systems to the river network outside the limits of the aquifer, many of the second residences lack wastewater collection systems, constituting potential sources of contamination. However, their effect is considered of minor importance compared to the agricultural diffuse pollution.

In summary, groundwater in the Benalup aquifer is extracted by individuals in an unplanned and uncoordinated manner, under poor administrative control. Moreover, the fact that some of the farms are covered by the 1879 Water Act and included in a "Catalogue of Private Waters" is another factor contributing to this chaotic situation. This uncontrolled exploitation has had detrimental chemical and quantitative effects on the aquifers. In fact, the Benalup aquifer was declared to have a poor quantitative and chemical status in the Hydrological Planning documents [15], which demand the adoption of measures to reverse this situation and achieve a sustainable use of the resource in the medium term. The poor quantitative status was declared because the rate of water extraction exceeded the limits set by the current legislation. The poor chemical status responds to the evidence of nitrate contamination linked to agricultural activities.

3. Administrative and Socioeconomic Framework

In southern Spain, as in other semi-arid and Mediterranean countries, the intensive exploitation of groundwater since the second half of the 20th century has given rise to what has been called a silent revolution, whose main beneficiary has been the sector of irrigated agriculture [17,18]. This phenomenon has contributed not only to achieve food security in the country, but also to a profound impact in micro and macroeconomic terms due to the important interrelationships of irrigated agriculture with other sectors of the economy. Likewise, groundwater exploitation has been fundamental for the development and consolidation of the population in rural areas. In Andalusia, the productivity of irrigated crops based on the exploitation of groundwater is remarkably higher than that obtained in crops that use surface water. Furthermore, farmlands based on groundwater exploitation create 3 times more employment per m^3 of water used [19]. This higher productivity can be explained, firstly, because water is extracted at the point of use, without transport over long distances, and secondly, because farmers can use it "on demand", maximizing its efficiency by reducing the energy costs derived from pumping [20].

However, this silent revolution has traditionally been developed without an effective control by the Hydraulic Administration, hindering the adequate management of groundwater [21] and resulting in impacts of different natures on aquifers. In line with what was stated above, the legal status of groundwater in Spain has undergone a significant evolution in recent decades, with numerous laws and doctrinal opinions related to the social, political and economic conflicts of the moment. Until 1985, the regulatory regime of groundwater was comprised in two norms: The 1879 Water Act and the Civil Code, which recognized the right to exploit groundwater as a consequence of land ownership [22], enabling the extraction of the resource without administrative requisites.

More than a century later, the enactment of the Water Act of 1985 was a turning point that changed the legal regime of water decisively. In the first place, it proclaimed the inclusion of groundwater in the public domain and suppressed the right to appropriate this resource through wells and galleries. Secondly, it conferred to the River Basin Agencies (Hydrographic Confederations), which belong to the State Administration, the competence for managing groundwater and granting authorizations and concessions for its exploitation. These Hydrographic Confederations are also responsible for the inspection and monitoring of the water resources, for the enforcement of the conditions set by the authorizations and for carrying out plans, programs and actions to promote water saving and an efficient use of the resource in economic and environmental terms. In the case of those hydrographic basins that belong entirely to an Autonomous Region, such as Andalusia, the competences attributed to the Hydrographic Confederations would be transferred to the Autonomous Government (Junta de Andalucía) through the corresponding organisms. They would also keep a Registry of Public Waters and would be responsible for declaring aquifers as over-exploited, having to formulate management plans aimed at their recovery.

The radical change in water ownership introduced by the Water Act of 1985 meant that the new regime had to be made compatible with the rights acquired by owners with the previous law. To achieve this, the law considered two possibilities: (i) that groundwater owners could benefit from the resource for 75 additional years, by virtue of administrative concessions or authorizations for the occupation or use of the public domain, or (ii) in a 3-year deadline, owners who did not have the corresponding administrative authorization had to regularize their situation by registering in a Catalogue of Private Waters [23]. Nowadays, the legal regime of groundwater is set up by the Act 10/2001 concerning the National Hydrological Plan, which emphasizes the role of users in water management, and by the European Directive 2000/60/EC (Water Framework Directive). However, after more than two decades, the groundwater situation remains chaotic; both the Water Registry of Public Waters and the Catalogue of Private Water are considerably incomplete, and hundreds of thousands of new wells and boreholes have been dug since the Water Law of 1985 without the corresponding authorization and outside the administration [24].

As stated above, in La Janda region the only regulatory entity, apart from the regional Water Authority, is a Water User Association that manages the surface water (rivers and reservoirs); there are no groundwater management entities in the area. The Water Law foresees the constitution of User Communities that incorporate all the beneficiaries of a concession. Said communities are public law corporations dependent on the corresponding River Basin Agency, having juridical personality and its own patrimony. The constitution of these User Communities is mandatory in the case of GWB in poor status; contravening this disposition would lead to the loss of the concessional rights. Water resources can be used by the community members with different purposes (supply, irrigation, industrial use). When most of the hydric resources are invested in irrigation, these entities are termed Irrigation Communities. Among their competences are the subsidiary execution of agreements breached by users, the expropriation and imposition of easements and the performance of works and facilities that the administration may require. In this sense, the current legislation advocates self-regulation.

Finally, some authors (Stephano et al., 2015) [25] identify four general stages in the groundwater exploitation development in Spain, which are closely related to the national legal and economic scenario: (1) In the first stage (silent revolution), the number of wells increased in a paced manner, and the administrative efforts were focused on surface water. (2) The second stage coincided with the approval of the 1985 Water Act and the intensification of drilling activity. These circumstances, coupled with the severe drought of 1991-1995, the EU irrigation subsidies and the inability of water authorities to curb unauthorized groundwater extractions resulted in a dramatic drop in WT. At the same time, the consolidation and expansion of irrigation were strongly supported by the authorities responsible for land-use policies. (3) In the third stage, users began to associate to protect their activities from sanctions and well closure, to achieve their legal recognition as water users and to lobby for new water resources to complement the shrinking or increasingly expensive groundwater resources. (4) The fourth stage, termed "technological fix", was marked by state-subsidized programmes for the modernization of irrigation systems and the search for additional water to complement groundwater resources and ensure the access to cheap, good-quality water. This situation in Spain has enabled the partial recovery of the water table in some aquifers. However, other factors, such as the occurrence of exceptionally rainy periods, that have favored the recharge of aquifers, or the fact that many farmers have hit the ceiling in their capacity to sell their products in a competitive way, choosing to consolidate their activity instead of increasing their pressure on the resource, must be considered.

Figure 2a describes the general trend of WT in Spain through the four stages considered by De Stephano et al. [25] and the most important features/events of each period. Figure 2b shows the general trend of the average piezometric level in the Benalup aquifer, according to historical data from the local monitoring network (11 piezometers with monthly records). Both trend lines display similarities, but the recovery of the piezometric levels during the last decades appears delayed in the Benalup aquifer, a circumstance that is more related to the agricultural land abandonment than to the technological fix stage, as will be discussed in subsequent sections.

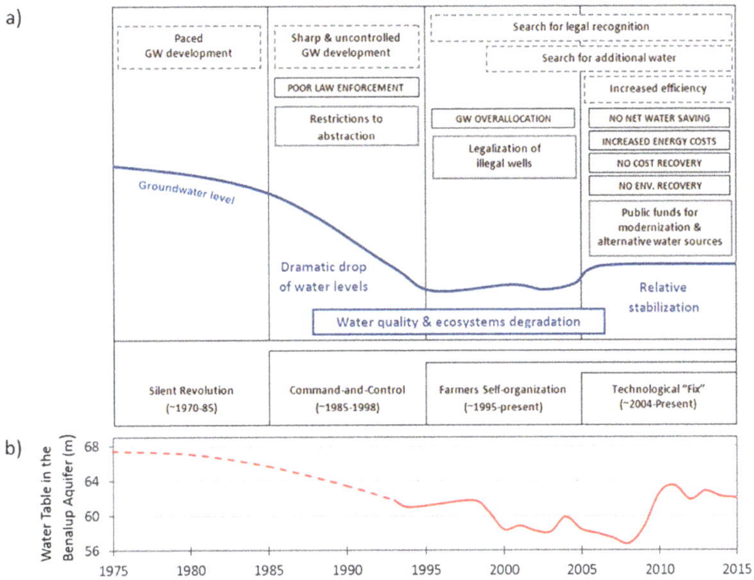

Figure 2. (**a**) General trend of the water table (WT) in Spain during the four stages considered by De Stephano et al. (2018), with the most important features/events of each period indicated (Source: [25]). (**b**) General evolution of the average WT of the Benalup aquifer. The dotted line indicates the period in which there are no quantitative records, but historical evidences. The solid line shows the evolution of the WT since historical records are available.

4. Materials and Methods

4.1. Data Collection and GIS Implementation

Cartographic databases (Digital Terrain Model (DTM) and digital cartography), geological and hydrogeological information from pre-existent monitoring networks and other relevant data from scientific and technical documents were collected. This information and that generated in the framework of this study were combined in a Geographic Information System (GIS) in order to be managed and analyzed, and to obtain graphical outputs.

4.2. Hydrogeological Methods

4.2.1. Piezometric and Hydrochemical Monitoring

An inventory of boreholes was made with information gathered from field visits and in situ observations. This information was subsequently used to design a piezometric monitoring network and a hydrochemical one.

Three piezometric control campaigns were carried out in successive years (2017, 2018 and 2019) at the beginning of the dry season. These campaigns allowed us to get to know the functioning of the aquifer regarding aspects such as the recharge and discharge areas, the main flow directions, the hydraulic relation between the aquifer and the river network, the influence of pumping and the compartmentalization in subsystems with a relatively independent functioning. In each campaign, the average density of control points was 1 point/km^2. These piezometric records were subsequently compared with the historical data from the pre-existing control networks that have been operating since 1993, to identify changes in water storage occurred in the last 25 years and their spatial distribution.

Furthermore, 7 automatic logging devices were installed in some of the wells distributed throughout the aquifer to obtain continuous records of water table and temperature fluctuations since June 2018. This information has not only enabled us to determine the storage status of the aquifer at each moment, but has also been useful to understand: (i) the effect of recharge due to rain infiltration; (ii) the behavior differences among the aquifer's sectors depending on their hydrogeological properties; and (iii) the exploitation regime of nearby boreholes and the effect of pumping in distant boreholes. In addition, the records from the logging devices were coupled with precipitation records from the nearby meteorological station of Vejer de la Frontera (station no. 6 of the Agroclimatic Network of Andalusia), located 6.5 km south of the centroid of the aquifer. This enabled us to monitor the response of the aquifer to recharge and to quantify the infiltration rate.

Two hydrochemical sampling campaigns were carried out in December 2018 (recharge period) and May 2019 (beginning of the dry period) to determine the hydrochemical facies and the quality of the groundwater resource. Samples from about 20 control points (which included springs and uninstalled/exploitation boreholes) were taken. Physical-chemical parameters were determined in situ (temperature, conductivity, pH, redox potential and dissolved oxygen). The hydrochemical determinations for each sample were carried out in the laboratories of CEHIUMA (University of Malaga) and included (i) major ions (HCO_3^-, CO_3^{2-}, Cl^-, SO_4^{2-}, Na^+, K^+, Mg^{2+} and Ca^{2+}), (ii) nitrogen compounds (NO_3^-, NO_2^- and NH_4^+), (iii) minor ions (PO_4^{3-}, F^-, Br^-, Sr^2+, Ba^{2+}), (iv) trace elements (Be, Al, P, V, Cr, Mn, Fe, Co, Ni, Cu, Zn, As, Se, Sr, Cd, Sn, Ba, Hg, Pb) and (v) stable isotopes ($\delta^{18}O$, δD, $\delta^{13}C$). These analyses are important not only to corroborate the chemical status of the aquifer, but because apart from being exploited for irrigation, its water is used for human consumption.

In addition, the Chloride-Mass Balance Method (CMB) was applied to estimate the effective infiltration into the aquifer, assuming that: (1) chloride in the groundwater originates only from direct precipitation on the aquifer, (2) chloride is conservative in the system, (3) the chloride-mass flux has not changed over time and (4) there is no recycling or concentration of chloride within the aquifer [26,27]. For the application of the CMB method, precipitation was systematically sampled between December 2018 and May 2019.

4.2.2. Geophysical Campaigns

A geophysical campaign was carried out to improve the current knowledge of the geometry of the aquifer. This campaign included the application of electrical and seismic techniques. The electrical works comprised 5 soundings with 585 m extension Schlumberger arrays. The seismic ones comprised 5 MASW (Multichannel Analysis of Surface Waves) seismic profiles with cable lengths between 110 and 120 meters and 12 4.5 Hz geophones. The information obtained from both geophysical surveys was combined with that from borehole columns and the Digital Terrain Model (DTM) to deduce the arrangement and thickness of the aquifer materials and, indirectly, to estimate the groundwater reserves.

4.2.3. Determination of Hydraulic Properties

Finally, to characterize the hydraulic properties of the aquifer, laboratory tests on the grain size, porosity and permeability were conducted. Additionally, pumping tests were carried out at some exploitation wells to determine the transmissivity (T) and storage coefficient (S) of the materials.

4.3. *Remote Sensing*

Remote sensing-derived information was used to identify the irrigated plots over the aquifer. Comparing this information with the official databases of irrigation authorizations allowed us to identify (i) the unauthorized areas where groundwater was being applied and (ii) the authorized irrigation areas that were not being cultivated.

To do so, satellite images from the ESA Copernicus program were used. These images were obtained from the MultiSpectral Instrument (MSI) aboard the Sentinel-2 mission, which comprises a constellation of two "twin" satellites (2A and 2B) placed in the same solar synchronous orbit. The processing level chosen for the images was an orthorectified Bottom-Of-Atmosphere (BOA) reflectance (Level-2A).

The Normalized Difference Moisture Index (NDMI) was used to spot the plots that were irrigated during dry periods. The NDMI was selected because it provides information about the level of water stress in crops [28]; in principle, high values of this index indicate soil moisture and, therefore, irrigation. The product that indicates the moisture level of crops was obtained from the combination of the NIR-Near InfraRed reflectance and the SWIR-Short Wavelength InfraRed reflectance:

$$NDMI = (NIR - SWIR)/(NIR + SWIR) \quad (1)$$

The remote sensing campaign of this pilot study comprised the hydrologic years 2016/17 and 2017/18. The analysis was focused on rainless periods, since during rainy months soil moisture and, therefore, the NDMI index are generally high, and the absence of water stress in crops makes irrigation unnecessary. Accordingly, a total of 20 images were selected from two temporal windows coinciding with dry periods. The irrigated areas obtained in this study are referred to average values. Finally, it should be mentioned the intrinsic difficulty in the identification of irrigation in fruit tree crops of permanent canopy (citrus), which has been estimated at 34 ha.

4.4. *Communication With Stakeholders*

This study included the implementation of several communication channels and participation tools:

(i) Personal communications and field visits: In general, these activities have contributed to a better knowledge of the physical and social environment and to contextualize the circumstances of the aquifer. These tasks consisted of conversations with landowners about their concerns, problems and difficulties in developing their activity. Moreover, the authors could have access to private properties and facilities. It should be noted that some landowners were reluctant to provide information or access to their farms, owing to (in many cases) the illegality of some aspects of their activity. In any case, these tasks were fundamental to take measurements in wells and water samples and to verify on site the remote sensing products.

(ii) Informative sessions and round tables: Two official informative sessions and several meetings were conducted, congregating a wide spectrum of participants and stakeholders of the Barbate River Basin area (water users and managers, directors of the adjacent natural parks, researchers of different profiles, administration technicians etc.). These acted as a forum where some of the preliminary results obtained by the authors were presented and where the participants shared their knowledge, expertise and concerns about different aspects of the study area and its resources.

(iii) Survey campaign: A survey campaign was carried out among the locals in 2019 to gather information about the water users' perception and knowledge about the local groundwater resources and the impacts to which they are subject. The questionnaire consisted of 9 multiple choice (closed-ended) questions (see Table A1) aimed at exploring citizens' perception regarding water resources, pressures and other adverse impacts related to climate change. These 9 questions were selected from a more extensive questionnaire. The method used was a simple random sampling and the population studied (N) comprised 10.000 inhabitants, approximately, in 2019. Surveys were classified according to different criteria: municipality, age intervals, gender and occupation. The total number of questionnaire surveys completed was 98, which implies a confidence level of 95%, a 10% margin of error and $p = 0.5$ [29]. All the interviews were conducted personally by the authors, and the respondents were allowed to suggest alternative answers.

5. Results

5.1. Geometry and Hydraulic Properties

The Benalup aquifer constitutes a tabular relief that rises several tens of meters above the surrounding river network. Most of the formation is flanked by a perimetral scarp, and, according to physiographic criteria, it can be divided into three compartments, whose hydrogeological behavior is independent and will be discussed later: (i) Mesa Alta, located to the west (6 km^2), (ii) Mesa Baja, that constitutes the major part of the aquifer (25 km^2) and (iii) Dehesa de Espartinas, located in the southern area of the central sector (2 km^2). The combined analysis of the DTM, the geological map of the area (Figure 3) and observations in situ allowed us to define the arrangement of the geological formations and revealed some significant tectonic accidents.

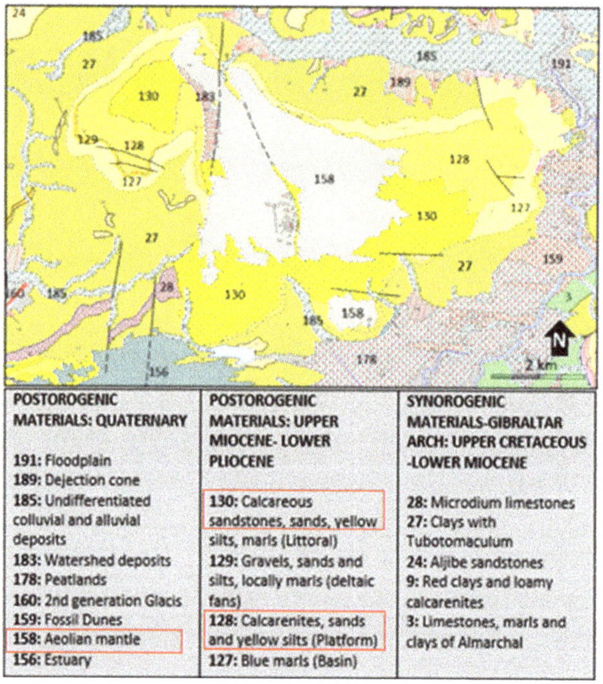

Figure 3. Geological map of the Benalup aquifer area. The main aquiferous materials that constitute the system are indicated in red. Source: Continuous Digital Geological Map of Spain (GEODE) [30].

As Figure 2 shows, the aquifer is made up of Neogene and Quaternary detrital materials, from bottom to top: (i) cross-bedded Upper Miocene biocalcarenites with levels containing fossil fragments, (ii) yellowish-white calcareous sands of the Pliocene age and (iii) Quaternary aeolian sands of special interest regarding system recharge.

The sedimentary formations that constitute the aquifer are scarcely deformed, adopting a subhorizontal position or forming a smooth syncline, except for the western sector, next to the great tectonic accident that separates Mesa Alta from Mesa Baja, where layers dip 45° E. Said steep dips are attributable to the drag effect of the fault.

The impervious wall of the aquifer comprises either blue marls from the Upper Miocene or turbiditic materials from the Gibraltar Arc. The latter are intensely deformed and consist of highly cemented sandstone, clays, marls and calcareous marlstones of very low permeability. The wall thickness ranges between 20 m at the southwestern sector and 130 m at various points in Mesa Baja

(Figure 4), where a significant unevenness was observed. This is attributable to the morphology of the paleorelief of the deposit basin. The average thickness has been estimated at 81.5 m and the volume of the aquifer at 2.650 hm^3.

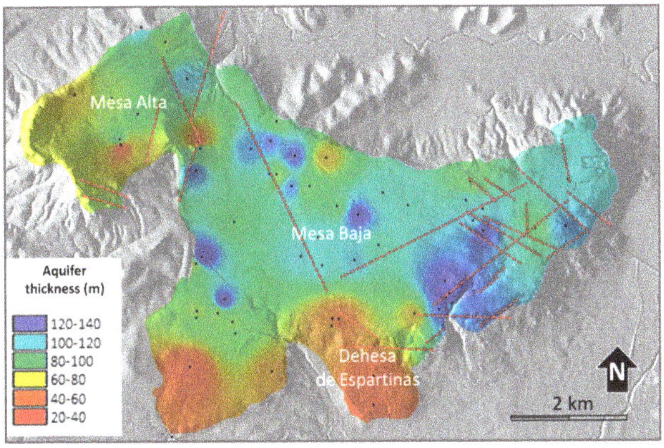

Figure 4. Thickness map of the aquifer. The red dotted lines indicate the presence of tectonic faults, and the black dots the locations where the depth of the impervious wall is known (through borehole columns and geophysical techniques).

Of the three geological formations that constitute the aquifer, the lower one (Upper Miocene calcarenites) is the most productive due to its hydrogeological characteristics. These calcarenites are a sedimentary formation of marine origin and coarse grain (coarse sand with conglomeratic levels), rich in calcareous fossils, that can reach a thickness greater than 100 m. Although the levels of coarse grain are abundant, lateral changes of facies and interspersed levels of finer grain (fine sand, silt and clay) have been detected. These materials initially present interstitial porosity, but the fossil-rich levels have undergone dissolution processes that produced secondary porosity. Based on the laboratory determinations, the total porosity (n), although highly variable, can present values of 0.28, with an effective porosity (n_e) of 0.06 and a hydraulic conductivity (K) of 11 m/day.

Over said calcarenites, there is an intermediate formation that mainly consists of yellowish sands of littoral origin, from the Pliocene age. Its base is practically indistinguishable from the aforementioned calcarenites, except by dating criteria. Upwards, the grain size decreases and brings about lateral changes of facies, with fine sands and interspersed silts that can reach a thickness of several meters. The thickness described for these materials is less than 20 m. Locally, they can act as aquitards and produce semi-confinement conditions. Laboratory tests carried out on samples of silty fine sand levels have given n values of 0.45, a n_e of 0.01 and a K of 0.04 m/day.

The most superficial and recent formation (Quaternary age) is constituted by loose aeolian sands that form a more or less continuous mantle over the underlying materials. These are well-graded medium to fine sands, with an interstitial porosity that can locally present lagoon-related levels of clayey nature. In general, the thickness of this formation is less than 5 m. Laboratory tests on this type of material showed n values of 0.45, n_e of 0.04 and K of 3 m/day.

An initial approximation of the volume of gravific water within the aquifer was made using the values above and considering an intermediate n_e (0.035) between that determined for fine Pliocene sands and that of the Miocene calcarenites: the volume obtained was 80 hm^3.

In addition to the aforementioned laboratory tests, the results of previous works and the methodologies carried out within the framework of this study have been compiled, displaying discrepant values (Table A2).

The K values varied by three orders of magnitude and were between 0.04 and 60 m/day, which undoubtedly responds to the grain size of the material and, in the case of the calcarenites, to the greater or lesser presence of karstified levels in the aquifer formation. During the pumping tests, the maximum *K* values (30 to 60 m/day) were obtained in those sectors where preferential flow paths and conduits linked to dissolution processes are important. The lowest *K* values (0.04 m/day) were obtained during laboratory tests on silty sand samples of the Pliocene age. In any case, an average *K* value between 1 and 10 m/day can be representative of the whole formation, although the presence of silty materials can significantly reduce the permeability and generate multilayer-aquifer conditions if the system is analyzed in detail.

The *S* values of the aquifer also varied by two orders of magnitude, between 6×10^{-2} and 5×10^{-4}. In this case, *S* values can be grouped into two classes, that will be considered as representative of different hydraulic conditions: (i) values around 4.5×10^{-2}, related to an unconfined regime within the aquifer, and (ii) values of 1×10^{-3} or less, which may reflect the influence of low permeability levels (linked to the presence of silts, clays, etc.) that locally produce semi-confinement conditions in the system.

5.2. *Aquifer Hydrodynamics Piezometry*

As a whole, the Benalup aquifer functions as an unconfined system, although semi-confinement conditions associated with the presence of low permeability materials can be found locally, as previously stated. The aquifer recharge takes place due to rain and runoff infiltration through the permeable outcrops. Thus, the permeable formation of loose aeolian sands that extends over a large part of the aquifer and enhances rain percolation plays a decisive role in this process. Various authors have estimated the rain-associated recharge at 20–30% (although there are no rigorous studies of this phenomenon), which would mean inputs of 5.3 and 7.9 hm^3/year into the system [15,31,32].

The topographic position of the aquifer ensures the absence of hydraulic connections with the surrounding rivers. The natural discharge of the system occurred exclusively through the springs located in the perimeter of the aquifer, in ravines and troughs produced by erosion. Evapotranspiration takes place in those locations where the WT is close to the surface. The exploitation of the aquifer began in the 1960s–1970s and led to an adaptation of the system. At first, the piezometric level decreased progressively until its stabilization, which resulted in the desiccation of some springs. Today, only two springs that evacuate more than 0.1 hm^3/year remain; the first is located in the northern area, associated with the Mesa Alta compartment, and the second one is located in the south, associated with the Mesa Baja compartment. It should be noted that the return flows from some irrigated plots could mean an additional water input into the system, which, although not being quantified, is estimated as 10% of the irrigation quotas.

The results obtained in the 3 piezometric campaigns were similar, with differences of less than 0.5 m among them in the different areas of the aquifer, not presenting a clear trend. The highest WT levels were found at the western and eastern limits of the aquifer, with values greater than 100 m.a.s.l. (Figure 5). These sectors are placed at higher altitudes and coincide with the Miocene calcarenite outcrops. Groundwater flows toward the aforementioned discharge areas are located to the north (Rendón spring, about 50 m.a.s.l.) and south (Cucarrete spring, about 45 m.a.s.l.). The piezometry reveals that the three differentiated compartments of the aquifer (Mesa Alta, Mesa Baja and Dehesa de Espartinas) present leaps related to hydraulic discontinuities that isolate each subsystem. These discontinuities produce a significant increase in the hydraulic gradient and are related to tectonic accidents identified in situ and/or on the DTM, and may be the result of the thinning of the calcarenite formation, or the contact between materials with different permeability.

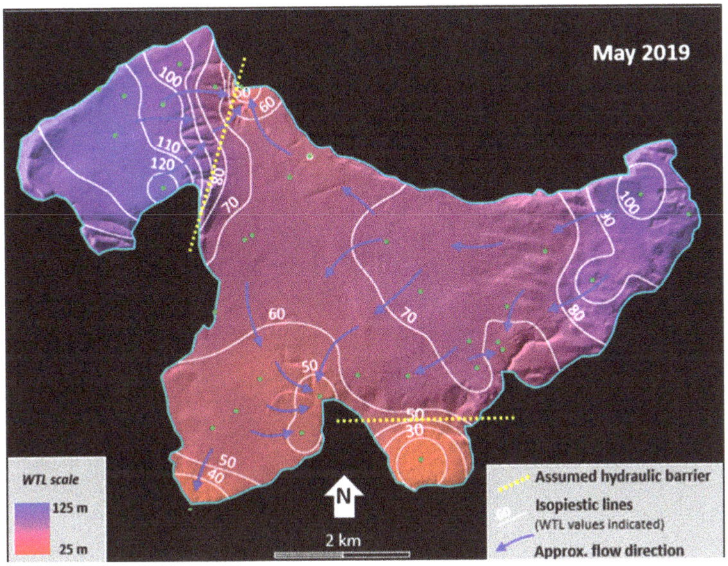

Figure 5. Piezometric map of the Benalup aquifer in May 2019, showing the water table (WT) level distribution. The blue arrows indicate the general flow direction in the porous media and the dotted yellow lines the assumed hydraulic barriers.

Figure 6 shows the changes on the WT levels over the last 25 years. While in the sectors closest to the natural discharge areas the WT was practically stabilized, with differences of a few decimeters, other sectors present significant drawdowns of 9 m (Mesa Alta compartment) and 3 m (N sector of Mesa Baja compartment).

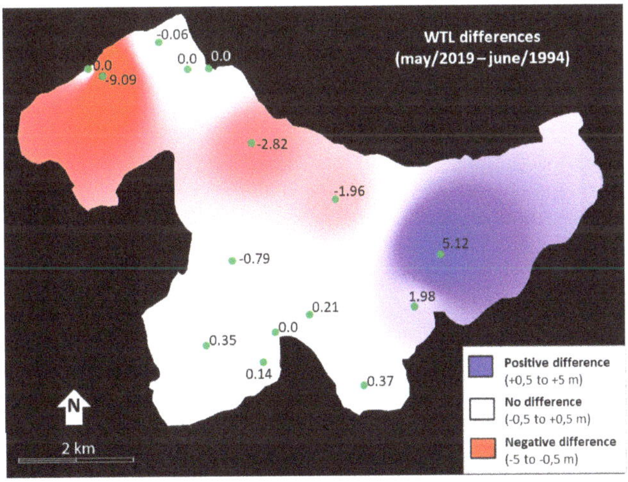

Figure 6. Differences in the water table level (WTL) recorded in May 2019 and June 1994 considering the coincident points of the monitoring network.

The largest drawdown is linked to the intensification of pumping with agricultural purposes. However, this is presumably a local effect, as its influence has not been detected in the nearby boreholes (see Figure 7).

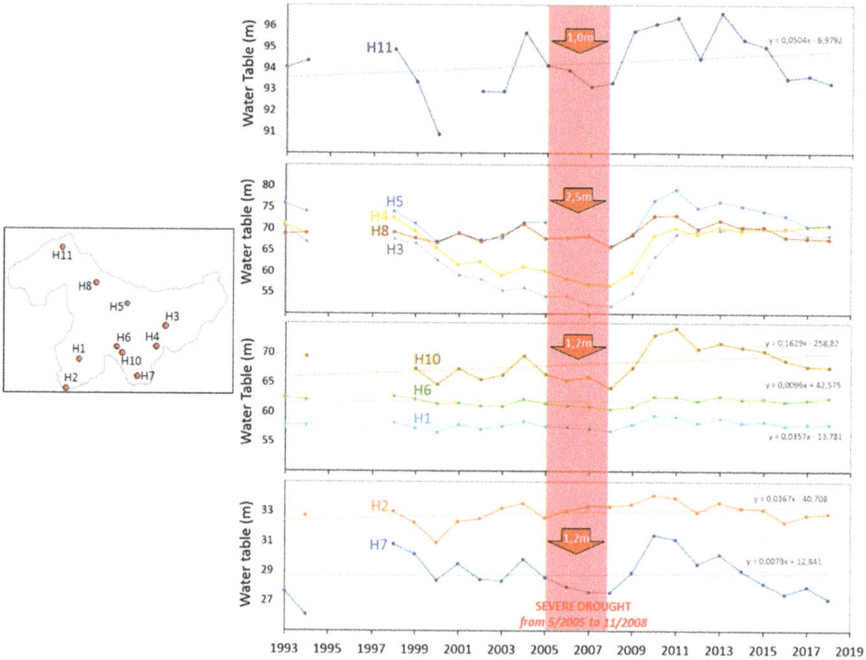

Figure 7. Evolution of the annual average WT level, obtained from monthly measurements at various points in the aquifer for the period 1993–2018 (25 years). The drought period 5/2005 to 11/2008 produced a general drawdown in the WT level of the aquifer.

Conversely, in the eastern sector of Mesa Baja compartment, a WT recovery of about 5 m was observed. This phenomenon is attributable to the different pumping rates and intensities associated to the boreholes from different farms and those exploited for urban supply. In any case, it should be reminded that the historical records start years after the beginning of the exploitation of the aquifer, which means that only some vestiges and qualitative references of the natural regime of the aquifer (dried-out springs and traditional wells, personal communications with older inhabitants, etc.) remain today.

Regarding the evolution of the WT during the last 25 years (Figure 7), a slight general trend of recovery has been detected in practically all the monitoring wells of the network for which there are records. This trend, however, is masked by the occurrence of drier periods and the dynamics of the aquifer itself, which is influenced by pumping at the local level. Thus, in most of the control points, it has been detected a piezometric increase between 0.2 and 1.25 m in the last 25 years. It is noteworthy that in one of these wells the level rise is noticeably larger, of almost 4 m for that period. Nonetheless, in the last 3 or 4 years, a slight drawdown trend has been detected, which is the result of several years with below-average precipitation. The 2005–2008 drought resulted in drawdowns between 1 and 2.5 m depending on the sector, more attributable to a reduced recharge than to an intensification of the exploitation, which gives an idea of the aquifer's natural regulation capacity.

The analysis of the piezometric level records obtained from the automatic loggers reveal that the aquifer's response to recharge is remarkably different among sectors. Figure 8 shows some of the records expressed as WT increments with respect to that existing at an initial instant, coupled with the daily precipitation registered.

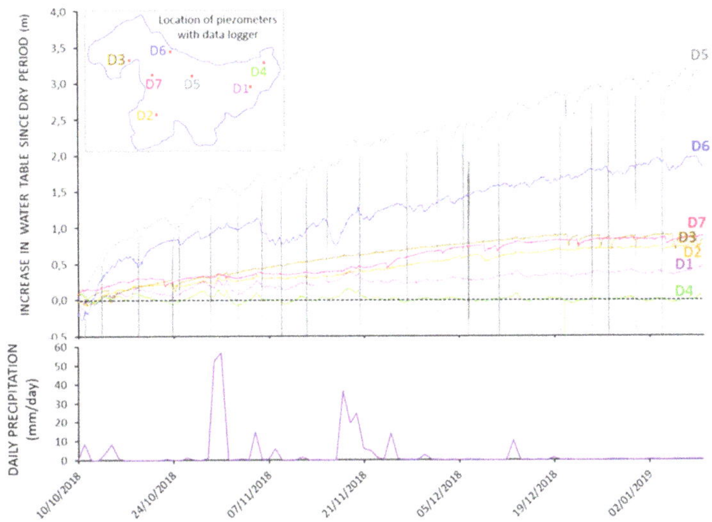

Figure 8. Evolution of the WT increases in the seven monitoring boreholes equipped with dataloggers during the recharge period (October 2018 to January 2019).

While in D4 the piezometric level rise in that period was practically nil, in D5 and D6 it was very sharp (3.22 and 1.94 m, respectively). The wells D2, D3 and D7 showed intermediate increases in their WT, around 0.8–0.9 m. Finally, D1 registered an intermediate piezometric evolution between the latter group and D4. Figure 9 shows the spatial distribution of the WT rises (attributable to the autumn rains of 2018) obtained from the interpolation of the values.

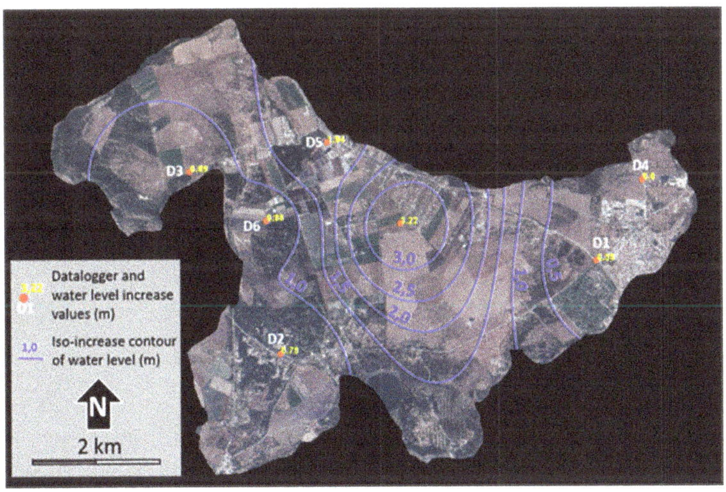

Figure 9. Spatial distribution of the rises in WTL during the recharge period (October 2018 to January 2019). Total precipitation: 284 mm.

Assuming that precipitation is homogeneously distributed in space and given that the topographic, edaphic and vegetation coverage conditions are very similar, the differences in the infiltration rate can be considered insignificant for the different monitored sectors of the aquifer. Thus, it can be inferred that the main factor conditioning the piezometric rise linked to recharge episodes is the n_e of the rock. The aquifer responds with high rises in sectors with low n_e, such as in the D5 sector, where the *S* value obtained during a pumping test was 0.001. In sectors with high n_e, such as at the eastern end of Mesa Baja, where there is evidence of dissolution conduits and rapid flow, the WT increases are reduced. Admitting an average n_e of 0.045 and an average rise of 0.85 m as a consequence of rainfall between October and January, an infiltration rate of 13.5% is obtained. Such infiltration rate is considered as a default value owing to the probable influence of some pumped boreholes.

5.3. Hydrochemistry

The analysis of the samples reveal that groundwater mineralization was medium. In the first sampling campaign (December 2018), the conductivity values ranged from 243 to 1299 µS/cm, with an average value of 706 µS. In the second campaign (May 2019), the conductivity values were slightly higher, ranging between 310 and 1383 µS/cm, with an average value of 792 µS/cm. In both cases, the highest conductivity values were found in the vicinity of Cantarranas (southwestern sector of the aquifer), while the lowest were found in a shallow piezometer on sands, with a vadose zone only 5 m thick. The pH was generally slightly alkaline, with an average value of 7.2 in the first sampling campaign and 7.4 in the second.

As can be seen in Figure 10, the predominant water type in the aquifer system is calcium bicarbonate, which is typical of porous media with abundant calcareous material (in this case fossil clasts). In the first sampling, calcium bicarbonate facies represented 73% of the total samples, and in the second 85%. In the sector of Cantarranas/Malabrigo, facies evidenced the dissolution of evaporites (gypsum, halite) or, perhaps, the influence of marine congenital waters, as they were of the sodium chloride or calcium-chloride-sulfate type.

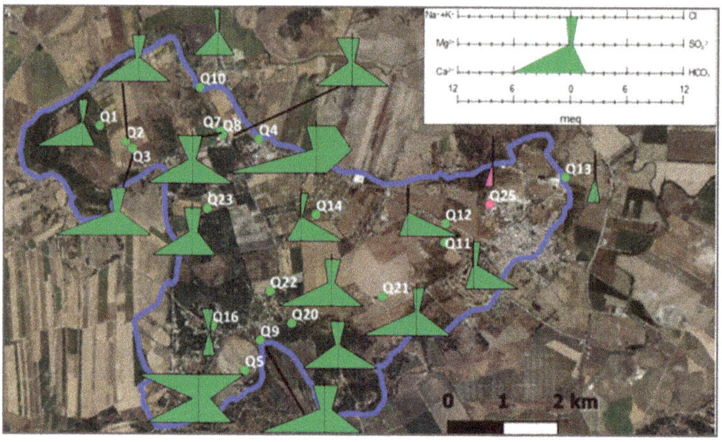

Figure 10. Stiff diagram displaying the groundwater facies in the sampling points.

The CMB method gave as result an effective infiltration of 18% of the rain for the control period. This value was obtained considering: (1) an average chloride concentration in rainwater of 5.6 mg/L (this value was weighted by the amount of precipitation) and (2) a mean value of Cl^- concentration in the aquifer of 30.9 mg/L. It should be noted that, to calculate the average value of chloride in groundwater, some well samples had to be discarded either due to their location in those sectors where the lithology contributed to the salt inputs (Cantarranas-Malabrigo sector) or due to their proximity to irrigated agricultural areas, since these did not meet the initial premises.

Nowadays, it is widely known that the agricultural activity represents the main source of nitrate inputs into the aquifers [33–35], an important problem that might lead to the declaration of these systems as being in poor condition. This is the case of the Benalup aquifer, declared in poor chemical condition for exceeding the permitted threshold values of this ion. For this reason, nitrate analyses were especially relevant in this study. Of the total of 35 samples collected during the two sampling campaigns, 17 (7 in December and 10 in May) exceeded the legal limit set at 50 mg/L, which represents about 50% of the samples. Figure 11 shows the spatial distribution of nitrate concentrations in the sampling campaign of December. The average NO_3^- concentration found was 44 mg/L, with values ranging between 0.8 and 250 mg/L. The latter value, which is extremely high, is attributed to point-source contamination from a poultry farm in Mesa Alta. Values of this magnitude were not detected during the second sampling campaign. In any case, during the second sampling (May), the nitrate average value was higher, 55.9 mg/L, with concentrations ranging from 0.6 mg/L (on a shallow borehole in a non-agricultural area) to 168.7 mg/L.

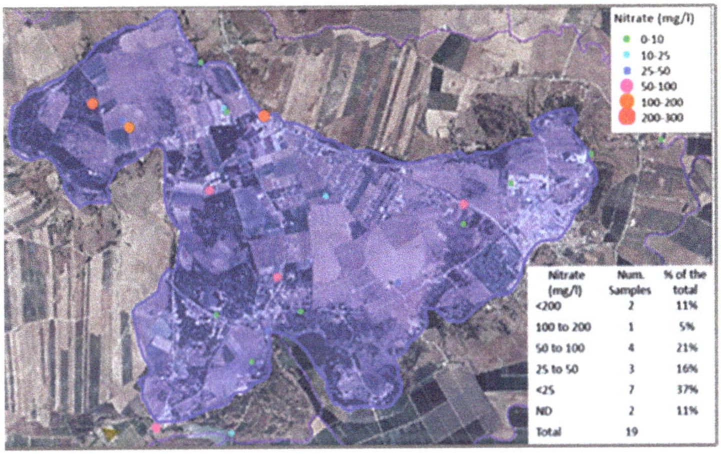

Figure 11. Nitrate concentrations for each monitoring point corresponding to the first sampling campaign (December 2018).

Figure 12 shows the concentration of metals and trace elements obtained during the first hydrochemical campaign. The threshold values set by the Spanish legislation for drinking waters were taken as reference for the elements Al, Cr, Mn, Fe, As, Se, Cd, Sb, Hg and Pb, as they are more restrictive. In the first sampling campaign, only three points exceeded the permitted concentrations of Mn (50 ppb) and two of them also exceeded the Fe limit (200 ppb). Another third point exceeded the maximum legal concentration of Se (10 ppb). During the second sampling campaign, only one of the previously mentioned points kept displaying high concentrations of Fe and Mn, which may be a result of the construction characteristics of the borehole itself, as it is a cased piezometer with low water renewal.

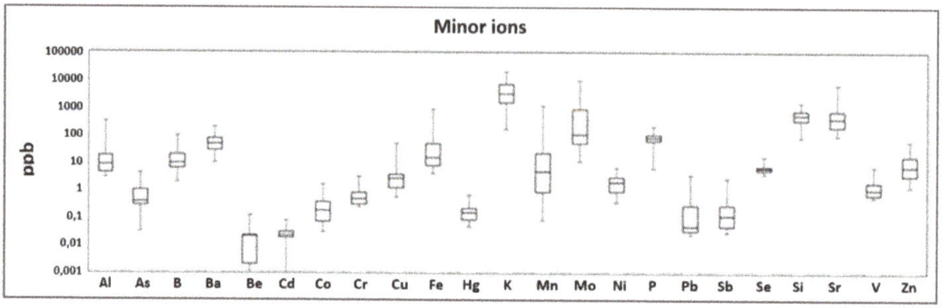

Figure 12. Boxplot showing the values for the median, maximum, minimum and upper and lower quartiles for minor elements corresponding to the first sampling campaign (December 2018).

5.4. Remote Sensing for the Identification of Irrigated Plots

Figure 13 displays as an illustrative example a selection of the results obtained in this study. Specifically, it shows the time-space evolution of the NDMI values obtained between June and September of 2018.

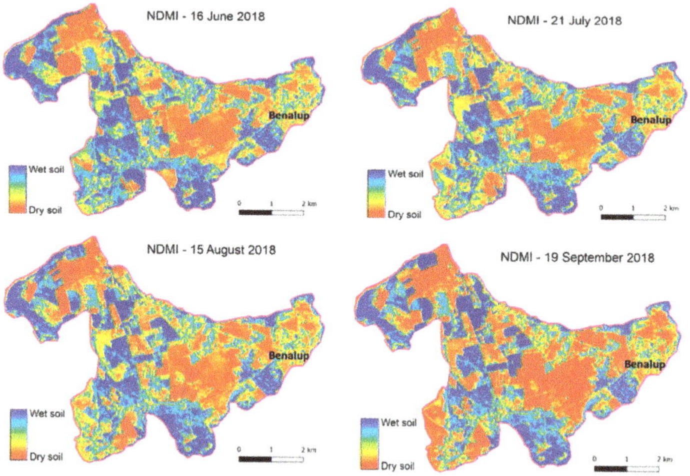

Figure 13. Time-space distribution of the NDMI values in the Benalup aquifer obtained from processed images from the Sentinel-2 mission.

On the one hand, the orange tones indicate low NDMI values, and therefore soils with a low moisture content and stressed vegetation. On the other hand, the blue tones point out high NDMI values that correspond to soils with high moisture content and/or non-stressed vegetation. Since there was no rainfall in the 30 days prior to the obtention of the processed satellite images, the high soil moisture levels are attributable to irrigation. It should be noted that some of the plots with permanently high NDMI values correspond to areas with natural ever-green vegetation that are not subject to water stress, due to being adapted to the local climatic conditions. Some areas covered by this type of vegetation can be observed in the western end of Mesa Alta compartment. In order to avoid the influence of these naturally vegetated plots in the analysis, all the areas that were not classified as agricultural land according to the official land use databases and presented constant and high NDMI values were excluded from the "irrigation plot" category. Taking all these aspects into consideration, and according to the NDMI index, the irrigated plots were identified and compared with the distribution of the irrigation licenses. The results obtained evidence a mismatch between the actual irrigated plots and those regularized by the administration. The total irrigated area in 2017 and 2018 was similar, with a monthly average of 296 ha. Of this area, 213 ha (72%) held irrigation licenses, whereas 83.5 ha (28%) affected land with no declared license comprising rainfed land and other agricultural uses. The maximum value of irrigated area (479 ha) was observed in May 2017 and represented 60% of the total area classified as irrigated land (786 ha).

5.5. Communication With Stakeholders

5.5.1. Personal Communications and Field Visits

The information obtained in this section is qualitative, sometimes subjective and difficult to systematize. However, some results can be cited:

- Identification of springs and wells that have disappeared as a consequence of pumping.
- Identification of the WT position prior to the exploitation in some sectors of the aquifer.
- On-site corroboration of farmland abandonment and its underlying reasons; low profitability and commercialization problems, population aging and lack of generational renewal, administrative restrictions, etc.
- Identification of unauthorized changes in land use and illegal groundwater abstractions.
- Recognition of the drinking water exploitation regime and the changes occurring because of it.

5.5.2. Informative Sessions and Roundtables

These initiatives enabled to contrast the water users' posture and that of the administration, as well as to seek possible alternatives combining law enforcement, economic development and a sustainable exploitation of the aquifer. An important outcome of these events was the idea that the hydrological knowledge of the aquifer should be improved in aspects such as the quantification of recharge, assessment of extractions, nitrate concentrations in groundwater, etc.

5.5.3. Citizen Surveys

The results obtained in the survey campaign highlight that most respondents know the origin of the supply water they consume. More specifically, 37.2% of the participants, mostly farmers belonging to the Water User Association, were supplied with water from the local reservoirs, while 47.8% claimed to use groundwater in their farms and households. In addition, 8.9% chose the option "other alternative sources", which could be attributed to a lack of knowledge about the origin of the water supplied to their own dwellings. It is noteworthy that most respondents supplied by the public network attributed its origin to surface water, when it is actually groundwater.

Water resources were generally regarded as adequate both in terms of quality and quantity (average scores of 6.5 and 6.3 out of 10, respectively). In line with this, almost 3/4 of the respondents pointed out the absence of conflicts arising from water exploitation. However, those who knew about the existence of conflicts referred to the high water demand of rice crops (outside the aquifer boundaries), illegal pumping wells or the need for authorizations for their exploitation. It is striking that none of the participants were aware of the declaration of the Benalup aquifer as having a poor quantitative and qualitative status and the administrative course of action that this may entail in terms of property transference and ownership. Likewise, respondents did not know of the existence of threshold values for the chemical parameters of groundwater or the maximum extraction rates that led to this categorization.

The surveys also revealed that the origin of groundwater is a still relatively unknown aspect, or at least vaguely defined by most of the users. While 40.7% rightly considered that the origin of groundwater was rainfall infiltration, 36.2% of respondents mistakenly conceived it as "underground rivers/lakes". The third most frequently chosen option (14.1%) was "I don't know". Again, this highlights that a significant fraction of the population is unaware of the functioning and dynamics of hydrogeological systems, which may have potentially adverse implications for the conservation of the resource owing to individual behavior.

Despite the aforementioned, most respondents acknowledged the susceptibility of groundwater to contamination and pointed out as pollution sources the diffuse spills of agrochemicals on the soil surface (53.6%) and the septic tanks of irregular settlements (22.8%), which are common in the study area. It is striking, however, that some of the respondents considered that aquifers are isolated from the land surface and susceptible to pollution (10.6%), that groundwater only becomes contaminated at well installations (8.1%) or that some of them did not even know it (4.9%).

Citizens' perception of changes regarding the presence of water and other related aspects was strongly influenced by the age of the respondents. The general feeling among older adults was that nowadays, there is less water in the region (38.7%). Most of them recalled springs and small streams that had disappeared, a situation that is verifiable and caused by the adaptation of the aquifer's natural discharge to pumping for more than 40 years. Conversely, 16.9% of the respondents (generally younger) did not perceive any change, presumably due to the stabilization of groundwater exploitation levels during the last years. These statements coincide with the evolution of the piezometric levels observed in the Benalup aquifer (Figure 1). It should be noted, however, that these personal perceptions might be somewhat biased by the extremely wet episodes of the 1960s and the severe droughts from 1998–2000 and 2005–2008.

On the other hand, most participants were familiar with the concept of climate change and related it to a decrease in the amount of rainfall (37.5%), an increased frequency of droughts (16.4%), water quality impoverishment (10.2%) or the desiccation of wells (8.6%). On the contrary, 19.5% of the respondents thought that there would be no changes beyond the cyclic character of climate. It should be considered, however, that these different outlooks might be influenced at some extent by mass media.

Finally, despite the uneven and sometimes remarkably poor knowledge about groundwater resources and their dynamics, most citizens agreed on their importance and the need to conserve them. A stricter control of water use by the administration (20.3%), limitations on groundwater withdrawals (17.7%) and further investment in hydraulic infrastructure (17.7%) were the most frequently proposed measures, along with the need for water awareness campaigns and citizenship education.

6. Discussion

6.1. Scientific Aspects

The achievement of a rational and sustainable use of groundwater must be based, first of all, on scientific knowledge. In this regard, this study has considerably improved the current knowledge about various aspects of the Benalup aquifer.

The geometry, nature and functioning of the aquifer shows the absence of hydraulic connection between the aquifer and the river network. This fact mitigates the environmental impacts derived from intensive exploitation. It can be affirmed that the groundwater depletion does not reverse the hydraulic relationship between the aquifer and the river and, therefore, does not appreciably reduce the surface flows or worsen the chemical quality of the surface water bodies. Nevertheless, the impacts of intensive exploitation cause the reduction of the natural discharge of small springs and a worsening of the chemical quality of drinking water.

Another important aspect for a sound management is the identified tectonic compartmentation that restricts the effects of the exploitation and the transport of pollutants in each subsystem.

Another noteworthy aspect is the wide variability in the hydrogeological parameters (K, T, S, n_e) obtained by different methods, which evidences an important heterogeneity in the aquifer and notable differences in the hydraulic conditions among sectors. Due to lithological and tectonic causes, the complexity of the aquifer is greater than that considered in previous studies [15,36]. Although its hydraulic regime can generally be considered as unconfined, the existence of strata/levels of fine grain size within the stratigraphic series (especially in Pliocene materials) can locally create confinement conditions. This implies that the WT drawdowns/rises attributable to the system exploitation and recharge affect very different volumes of water depending on the sector and, therefore, erroneous conclusions can be drawn regarding the degree of exploitation if only the water table level is considered.

The recharge values obtained in this study are lower than the ones from previous works [15,31], generally based on expert judgments without applying specific scientific methods. Infiltration is a complex phenomenon that is conditioned by multiple factors [37]. The CMB method provided a rainfall infiltration rate of 18%. Through a second technique, based on the comparison of the records of WT (obtained from data loggers) to precipitation inputs, an infiltration rate of 14% was recorded. This last result was highly conditioned by the value of n_e considered and was lower than that obtained with the CMB method, which is attributable to a period with rainfall contributions below the average year. It is advisable to extend the analysis period in order to obtain statistically relevant results in both methods.

The analysis of the piezometric evolution in the last 25 years has enabled us to identify a recovery trend in groundwater levels, although it is masked by climatic fluctuations. In fact, in the last three or four years, this trend seems to be slowly reversing at least in some monitoring wells, which is attributable to several consecutive years somewhat drier than the average. In any case, the levels are relatively stabilized, and the quantitative status of the aquifer is not getting worse. In this situation, it is striking that the administration has declared the aquifer as being in poor quantitative condition, despite being based on data with a high extent of uncertainty. To the aforementioned regarding the uncertainty in the evaluation of recharge, we must add the inherent difficulty in evaluating the exploitation of the aquifer: there is a lack of volumetric control devices in most of the boreholes with largest flow, and there are no gauging stations or other flow control mechanisms in the springs. Only the precautionary principle linked to administrative acts can justify such declaration, as long as more reliable information is not available. Obviously, this circumstance can negatively affect users and, if adequate measures are not taken, hinder the optimization of water management.

In addition, the pilot study developed with remote sensing techniques reveals that there is a significant mismatch between the agricultural plots authorized for irrigation and those that are actually being irrigated. The real irrigation surface over the aquifer has turned out to be less than the officially reported due to the abandonment of the agricultural activity, which means that the evaluation of groundwater withdrawals made by the administration is probably an overestimation. This reduction in the actual irrigated areas would justify the WT recovery in the aquifer in recent years. Thus, this recovery would not be the result of an improved management that some authors have termed "technological fix" [25] (See Figure 1), which consists in the implementation of subsidized programs for the modernization of irrigation systems and seeks suitable and affordable water resources that can complement groundwater. On the contrary, it would evidence a structural problem related to the low profitability of crops and the economic, social and administrative difficulties inherent to the agricultural activity in the area. In line with this, the declaration of the poor condition of the aquifer would result in restrictions on ownership transfer in the case of farms with groundwater use rights, leading to the abandonment of many agricultural farms.

The results obtained in the hydrochemical campaigns show an important contamination process related to the application of agricultural fertilizers. These data corroborate the results obtained by the administration from its own monitoring network, which only has three control points in this aquifer. The Spanish regulation justifies the declaration of the poor condition with the existence of a single control point/sample in the monitoring network that exceeds the maximum concentrations established for any of the constituents analyzed. Although this criterion is excessively restrictive, in the case of the Benalup aquifer, the declaration of poor chemical condition is amply justified, which makes it necessary to adopt measures to restore the quality of the water. Finally, the apparent lack of spatial autocorrelation in the distribution of nitrates and the substantial discrepancies observed at very close control points are also noteworthy. Among adjacent wells and boreholes of similar characteristics, those subject to intense pumping presented much lower nitrate concentrations than those scarcely pumped. Significant differences in the nitrate content were also found in spring waters during the first and second sampling campaigns. These findings evidence a significant stratification of the pollutant within the aquifer, with the existence of a vertical gradient. Thus, the highest nitrate concentrations appear close to the water table due to leaching processes as water infiltrates into the aquifer.

6.2. Management Guidelines

The management and governance of groundwater must encompass collective action to ensure a socially sustainable utilization and effective protection of groundwater resources, for the benefit of people and groundwater-dependent ecosystems [38].

Several lessons can be learned from this study in this regard; gearing up for resource management requires measures such as (i) the promotion of local management entities with first-hand knowledge of the system, (ii) information gathering and resource planning by establishing appropriate systems for groundwater monitoring on a regular basis (iii) sharing groundwater information, knowledge and experiences, (iv) supporting local water harvesting measures and the creation of extension services to promote high value crops with lower water consumption or water use efficiency measures.

According to Spanish water laws, the management of groundwater, especially in the case of those GWB at risk of not reaching a good quantitative or chemical condition, should be carried out by Water User Associations (WUA). These entities, which arise from the common use of the resource, should be responsible for a planned exploitation of the aquifer (with the support of the administration), based on a solid scientific knowledge and under the principles of transparency, participation and agreement among users. This type of self-regulation entails several advantages, such as the detailed knowledge of the physical environment by its members and their ability to gather information, monitor the resources and enforce the regulations. In addition, the resulting norms derived by water users themselves can be better tailored to local circumstances, with the right system of incentives and penalties [39].

Given the circumstances of the Benalup aquifer, declared as having a poor quantitative and chemical status during two hydrological planning cycles (12 years) and suffering from a lack of control in its exploitation, the best solution appears to be the creation of a groundwater user association that can represent the common interests of users, set rules and objectives and correct any deviant attitudes of the members. In fact, in the study area, we have as precedent the outstanding experience and results achieved by the WUA, that manages the surface waters (Celemín and Barbate reservoirs) in a planned and adaptive manner depending on the weather conditions and the needs of each moment.

Another aspect to consider is that the participation of citizens and stakeholders is essential for groundwater resource management and should be encouraged as an important contribution to groundwater conservation, management and protection [40]. This participation is especially necessary in the case of rural landholders, who have a major role in groundwater management and whose land use practices can have major impacts on groundwater quantity and quality. In this regard, groundwater resources tend to be undervalued and poorly understood in most of its aspects by the local population, a situation which, among other factors, favors its uncontrolled exploitation. As shown in the results section, the citizens in the study area assume that water resources in the region are sufficient both in quantity and quality and are not aware of the actual problems to which the aquifer is subject. Besides, their extent of knowledge about the dynamics and vulnerability of hydrogeological systems is uneven and usually remarkably poor; in fact, a certain fraction of the population does not consider groundwater a renewable resource, susceptible to be degraded in terms of quantity and quality. The widespread lack of knowledge about the declaration of the aquifer as being in poor condition and the consequences that it may entail was evident. Public education and outreach are therefore necessary not only to promote the community's interest in groundwater issues, but also to ensure the success of management goals and programs established to achieve sustainability [41].

Regarding the implementation of the aforementioned management, in a context of climatic, ecological and resource uncertainties, flexibility is needed. A flexible and adaptive management allows for a better use of the generally scarce economic resources, reduces administrative instability, helps to avoid decisions that may harm groundwater property rights and reduces administrative appeals [9]. Despite having an enormous intervention capacity through jurisdictional mechanisms, the administration often lacks the means to exercise the necessary control, and sometimes chooses to avoid social and legal conflicts. However, as Fornés et al. [21] point out, the lack of legal clarity regarding groundwater in Spain has resulted in the lodging of thousands of administrative appeals against the administration's decisions, exceeding the management capacity of the River Basin Agencies.

It is indisputable that, to achieve the sustainable and sound management of any hydrogeological system, the control and quantification of groundwater withdrawals is fundamental. In this sense, any regulatory instrument should include individual groundwater abstraction/use rights or licenses, either at a specified rate or allocation share, subject to periodic review and adjustment in light of the aquifer's behavior, trying to avoid the concept of "rights in perpetuity" [40].

An adequate quantification of the extractions can be made through the installation of control devices in all the pumping wells. Likewise, it is necessary to monitor croplands in order to identify the space-time distribution of the irrigated areas, the assessment of consumption and the determination of the areas affected by diffuse contamination. Remote sensing techniques have proven extremely useful tools for these tasks and for the identification of illegal irrigation [42]. The fact that both tools are within the reach of any entity (such as WUAs), from a technical and economic point of view, is an advantage that should not be overlooked.

Finally, water planning should promote the exploration of different management alternatives, such as the possibility of the conjunctive use of surface and groundwater. In the case studied, and given the poor condition of the aquifer, it would be advisable to delve into the option of transferring the surface water surplus to the aquifer through artificial recharge techniques. These techniques, despite being simple, are usually difficult to implement and often generate social conflicts. Scientific and technical questions about the aquifer's capacity to receive and store water require further research. Nevertheless, this is usually not the main obstacle. Often, the main problems are of economic, legal and organizational nature [43]. In spite of this, the knowledge gained about this system seems to endorse the possibility of applying artificial recharge techniques. The eastern sector of the Mesa Baja compartment, owing to its lithological and hydrogeological characteristics and its proximity to the Barbate reservoir, seems to be a particularly suitable sector for recharge. An increase in water inputs would enhance the natural discharge of the aquifer through springs and wetlands in the perimetral area, with the consequent improvement of the water-dependent habitats and the quality of the resource through the dilution of pollutants. Besides, a redesign of the pumping network may favor an increase in water storage by limiting or regulating natural discharge. In any case, the costs of the necessary infrastructure and operation should be shared among the users of the aquifer.

7. Summary and Conclusions

The hydrogeological research of the Benalup aquifer has enabled the identification of a series of relevant aspects that must be taken into account for an adequate management of groundwater resources, such as: (1) the geometry and storage volume of the aquifer, (2) the recharge of the system, (3) the hydraulic disconnection of the aquifer from the main hydrographic network, (4) the compartmentalization of the aquifer into three subsystems of relatively independent functioning, (5) the predominant directions of the groundwater flow, (6) the hydrochemical characteristics of the resource according to its contents of majority ions, nitrogenous compounds and trace elements, (7) the most suitable areas for the relocation of wells for the regulation of the underground reservoir or to carry out artificial recharge.

Although most of the exploitation of groundwater is used in agriculture, a significant part is also destined for urban supply and recreational uses. Currently, a gradual substitution of the groundwater sources for human supply by surface resources from a contiguous basin is taking place, which de facto implies an indirect artificial recharge and reduces the exploitation-associated pressure on the aquifer. This effect should not be compensated by an increase in pumping for other purposes. Instead, it should be considered as an opportunity to improve the quantitative status of the aquifer.

The analysis of the historical piezometric series available since 1993 has revealed that the storage level of the aquifer has slightly recovered since then. This is more attributable to the reduction of agricultural pumping linked to the abandonment of irrigated farmland than to a more efficient water use due to the technification of irrigation. In this regard, there is a significant discrepancy between the officially authorized irrigation plots and those that are actually irrigated, which evidences the current disorder in the exploitation of the aquifer. Many plots are irrigated exploiting unauthorized wells, which also makes it difficult to access reliable information on pumping. In this sense remote sensing techniques constitute useful low-cost tools for monitoring irrigation in real time and, indirectly, pumping water from the aquifer.

Despite the fact that in the last two planning cycles (2009–2015 and 2015–2021) the GWB Benalup has been cataloged as being in poor quantitative and chemical condition by the competent administration, the majority of the population of the region in general and of groundwater users in particular are unaware of this situation. There is a general lack of knowledge of the legal repercussions that this declaration may entail for water users and for the activities developed in the aquifer. Despite the elapsed time, no measures have been taken to reverse the GWB's poor condition. It is necessary to undertake education and outreach programs/campaigns to make the role and value of groundwater known and to engage citizens in the management and protection of this resource.

Likewise, citizen awareness is necessary to minimize the conflicts that may arise from the restrictions and rationalization in the use of the resource, aimed at achieving its sustainable exploitation. In this sense, the administration's inaction in reversing the state of the aquifer in recent years can be understood in light of the need to balance environmental sustainability with the economic and social requirements in the territory. Nevertheless, the management of groundwater resources and their exploitation in the study area is an indisputable need. In this sense, it is also necessary to implement initiatives aimed at self-regulation by water users and to promote effective water-management institutions, with a good knowledge of the aquifer, efficient and trained staff, good monitoring facilities and adequate economic resources and regulations.

Author Contributions: Conceptualization, M.V.-N. and S.G.-L.; formal analysis, M.V.-N., S.G.-L., V.R.-O. and Á.S.-B.; methodology, M.V.-N., S.G.-L., V.R.-O. and Á.S.-B.; investigation, M.V.-N., S.G.-L., V.R.-O. and Á.S.-B.; writing and original draft preparation, M.V.-N. and S.G.-L.; writing, review and editing, V.R.-O. and Á.S.-B.; visualization, M.V.-N., S.G.-L., V.R.-O. and Á.S.-B.; project administration, S.G.-L.; funding acquisition, S.G.-L. All authors have read and agreed to the published version of the manuscript.

Funding: This research was funded by the Biodiversity Foundation of the Ministry for the Ecological Transition through the research group RNM373-Geoscience-UCA of Junta de Andalucía.

Acknowledgments: This work has been developed within the framework of the REMABAR project, supported by the Biodiversity Foundation of the Ministry for the Ecological Transition. The authors also want to express their gratitude to the citizens who have kindly participated in the surveys.

Conflicts of Interest: The authors declare no conflicts of interest.

Appendix A

Table A1. Question wording and response set.

Code	Question	Response Set
1	Do you know where the water used in your activity/house comes from?	a. Surface water b. Groundwater c. Other sources (rain, reused water ...) d. I don't know
2	Rate from 0 to 10 the availability and quality of the water resources in the region.	a. Quality: b. Quantity:
3	Do you know if there are any conflicts derived from water use in the region?	a. Yes b. No c. I don't know
4	Do you know which is the origin of groundwater?	a. Underground river/lake b. Local mountain ranges c. It comes from the sea, due to salt filtration d. It comes from the rocks, it has always been there e. Rainfall infiltration/percolation f. I don't know g. Other
5	Do you think that groundwater can undergo contamination processes?	a. Groundwater is not susceptible to contamination b. Yes, when pollutants are poured into a well c. Yes, when pollutants leach from the surface d. Yes, due to the presence of septic tanks e. I don't know
6	Have you noticed any changes in the quantity or quality of the local water resources or in other related aspects during the last 20–30 years?	a. There is more water nowadays b. There is less water nowadays c. The quality has improved in recent years d. The quality has worsened in recent years e. There is more pollution nowadays f. Changes in the local flora and fauna g. There have been no changes h. Other
7	Do you think there will be problems with water in the next 20–30 years?	a. I don't think so b. Climate change will result in scarcer rain c. The quality will become worse due to pollution d. There will be more droughts e. Wells will run out of water f. Other

Table A1. *Cont.*

Code	Question	Response Set
8	Do you think it is necessary to take measures in order to ensure the future availability of water and its uses?	a. I don't think it is necessary b. Improve Administration's control over water use c. Improve the cooperation and transparency among water users d. Increase the investments in water infrastructure e. Control and set limits to groundwater pumping f. Allow more groundwater pumping g. Reinforce the protection of the natural environment h. Replace the current crops with other types that consume less water i. Other
9	Have you noticed changes in the local rainfall pattern in the recent years?	a. There have not been any changes b. Rainfall is becoming more unpredictable c. Rainfall is becoming more torrential d. I don't know e. Other

Table A2. Hydraulic parameters obtained by different authors and methods.

Source	Geological Material	Estimation Method	T (m^2/day)	K (m/day)	S (Dimensionless)
IGME, 1978 [44]	Upper Miocene calcarenites	Pumping test	210 to 2200	2 to 20	0.01
INTECSA-IGME, 1982 [31]	Upper Miocene calcarenites	-	20 to 100	0.3 to 2	
INTECSA-IGME, 1982 [31]	Upper Miocene calcarenites	Water table evolution	-	-	0.05
INTECSA-IGME, 1982 [31]	Upper Miocene calcarenites	Pumping test	200	2	
CGS-ITGE [45]	Upper Miocene calcarenites	Pumping test	4.000 to 6.000	40 to 60	$0.5{\cdot}10^{-3}$ to $1{\cdot}10^{-3}$
ITGE, 1996 [46]	Upper Miocene calcarenites	Pumping test	22 to 75		
Authors (UCA) [16]	Pliocene Sands/ Upper Miocene calcarenites	Pumping test	20 to 30	0.2 to 0.3	$1{\cdot}10^{-3}$
Authors (UCA) [16]	Upper Miocene calcarenites	Pumping test	3000 to 6000	30 to 60	0.045
Authors (UCA) [16]	Upper Miocene calcarenites	Laboratory test	-	11	0.06
Authors (UCA) [16]	Aeolian littoral dunes	Laboratory test		3	0.04
Authors (UCA) [16]	Pliocene silty sands	Laboratory test		0.04	0.01

References

1. Llamas, M.R.; Garrido, A. Lessons from intensive groundwater use in Spain: Economic and social benefits and conflicts. In *The Agricultural Groundwater Revolution: Opportunities and Threads to Development*; Giordano, M., Villholth, K.G., Eds.; CAB International: Wallingford, UK, 2007; pp. 266–295.
2. Sophocleous, M. From safe yield to sustainable development of water resources. The Kansas experience. *J. Hydrol.* **2000**, *235*, 27–43. [CrossRef]
3. Shamir, U. Sustainable Management of Water Resources. Transitive Towards Sustainability. In Proceedings of the Interacademy Panel, Tokyo Conference, Tokyo, Japan, 15–18 May 2000; pp. 62–67.
4. Mays, L. *Water Resources Sustainability*; McGraw-Hill Professional: New York, NY, USA, 2006.
5. Alley, W.M.; Reilly, T.E.; Franke, O.L. *Sustainability of Groundwater Resources*; U.S. Geological Survey Circular 1186: Denver, CO, USA, 1999.
6. Mianabadi, A.; Derakhshan, H.; Davary, K.; Hasheminia, S.M.; Hrachowitz, M. A Novel Idea for Groundwater Resource Management during Megadrought Events. *Water Resour. Manag.* **2020**. [CrossRef]
7. Simpson, H.C.; de Loe, R.C. Challenges and opportunities from a paradigm shift in groundwater governance. *Hydrogeol. J.* **2020**, *28*, 467–476. [CrossRef]

8. Galloway, G.; Pentland, R. Securing the Future of Ground Water Resourcesin the Great Lakes Basin. *Groundwater* **2005**, *43*, 737–743. [CrossRef]
9. Custodio, E.; Cabrera, M.; Poncela, R.; Puga, L.; Skupien, E.; del Villar, A. Groundwater intensive exploitation and mining in Gran Canaria and Tenerife, Canary Islands, Spain: Hydrogeological, environmental, economic and social aspects. *Sci. Total Environ.* **2016**, *557–558*, 425–437. [CrossRef] [PubMed]
10. Molina, J.L.; García-Aróstegui, J.; Benavente, J.; Varela, C.; de la Hera, A.; López-Geta, J. Aquifers Overexploitation in SE Spain: A Proposal for the Integrated Analysis of Water Management. *Water Resour. Manag.* **2009**, *23*, 2737–2760. [CrossRef]
11. Molina, J.; Bromley, J.; García-Aróstegui, J.; Sullivan, C.; Benavente, J. Integrated water resources management of overexploited hydrogeological systems using Object-Oriented Bayesian Networks. *Environ. Model. Softw.* **2010**, *25*, 383–397. [CrossRef]
12. Melloul, A.; Collin, M. A hierarchy of groundwater management, land-use, and social needs integrated for sustainable resource development. *Environ. Dev. Sustain.* **2001**, *3*, 45–59. [CrossRef]
13. Gleeson, T.; Alley, W.; Allen, D.; Sophocleous, M.; Zhou, Y.; Taniguchi, M.; Vandersteen, J. Towards Sustainable Groundwater Use: Setting Long-Term Goals, Backcasting, and Managing Adaptively. *Ground Water* **2012**, *50*, 19–26. [CrossRef]
14. Gondwe, B.R.N.; Lerer, S.; Stisen, S.; Marín, L.; Rebolledo-Vieyra, M.; Merediz-Alonso, G.; Bauer-Gottwein, P. Hydrogeology of the south-eastern Yucatan Peninsula: New insights from water level measurements, geochemistry, geophysics and remote sensing. *J. Hydrol.* **2010**, *389*, 1–17. [CrossRef]
15. Junta de Andalucía. Consejería de Agricultura, Pesca y Medio Ambiente. Plan Hidrológico del Guadalete Barbate 2009–2015. Tomo I (Memoria) y Tomos II, II, IV, V y VI (Anejos). 2009. Available online: http://www.juntadeandalucia.es/medioambiente/site/portalweb/menuitem.7e1cf46ddf59bb227a9ebe205510e1ca/?vgnextoid=88bc3adc6c46a310VgnVCM2000000624e50aRCRD&vgnextchannel=59fc92a482155510VgnVCM1000001325e50aRCRD (accessed on 13 May 2020).
16. UCA-Universidad de Cádiz. *Informe Sobre el Funcionamiento del Acuífero de Benalup y Posibilidades de Recarga Artificial. Project REMABAR: Análisis de Estrategias para Reducir las Pérdidas por Evaporación y Mejorar el Estado de los Acuíferos bajo Contexto de Cambio Climático en la Cuenca del Río Barbate*; Unpublished report; 2019; 79p.
17. Llamas, M.R.; Martínez-Santos, P. Intensive Groundwater Use: Silent Revolution and Potential Source of Social Conflicts. *J. Water Resour. Plann. Manag.* **2005**, *131*, 337–341. [CrossRef]
18. Chaturvedi, S.K.; Sandhu, J.S. Silent revolution in pulses production—India marching towards self–sufficiency. *Indian J. Agric. Sci.* **2020**, *90*, 17–24.
19. Llamas, M.R.; Custodio, E. Intensive use of groundwater: A new situation that demands proactive action. In *Intensive Use of Groundwater: Challenges and Opportunities*; Llamas, M.R., Custodio, E., Eds.; Swets and Zeitlinger: Balkema, The Netherlands, 2003; pp. 13–31.
20. Deb Roy, A.; Shah, T. Socioecology of groundwater irrigation in India. In *Intensive Use of Groundwater: Challenges and Opportunities*; Llamas, M.R., Custodio, E., Eds.; Swets and Zeitlinger: Balkema, The Netherlands, 2003; pp. 307–335.
21. Fornés, J.M.; de la Hera, A.; Llamas, M.R. La propiedad de las aguas subterráneas en España: La situación del Registro y del Catálogo. *Ing. Agua* **2005**, *12*, 125–136. [CrossRef]
22. Alcaín, M.E. La explotación de las aguas subterráneas en España. *Rev. Derecho Adm. Econ.* **2002**, *1*, 21–40.
23. Cantero, M.J. El régimen transitorio de la Ley de Aguas y los aprovechamientos preexistentes en la jurisprudencia del Tribunal Supremo. *Rev. Adm. Pública* **2002**, *159*, 221–256.
24. Molinero, J.; Custodio, E.; Sahuquillo, A.; Llamas, M.R. Groundwater in Spain. Overview and Management Practices. 2008. Available online: http://www.rac.es/ficheros/doc/00587.pdf (accessed on 15 May 2019).
25. De Stefano, L.; Fornés, J.M.; López-Geta, J.A.; Villarroya, F. Groundwater use in Spain: An overview in light of the EU Water Framework Directive. *Int. J. Water Resour. Dev.* **2015**, *31*, 640–656. [CrossRef]
26. Wood, W.W. Use and misuse of the chloride-mass balance method in estimating ground water recharge. *Ground Water* **1999**, *37*, 2–3. [CrossRef]
27. Marei, A.; Khayat, S.; Weise, S.; Ghannam, S.; Sbaih, M.; Geyer, S. Estimating groundwater recharge using the chloride mass-balance method in the West Bank, Palestine. *Hydrol. Sci. J.* **2010**, *55*, 780–791. [CrossRef]
28. Wilson, E.; Sader, S.A. Detection of forest harvest type using multiple dates of Landsat TM imagery. *Remote Sens. Environ.* **2002**, *80*, 385–396. [CrossRef]

29. Israel, G.D. *Determining the Sample Size. Fact Sheet PEOD-6. Florida Cooperative Extension Service*; University of Florida: Gainesville, FL, USA, 1992.
30. GEODE—Continuous Digital Geological Map of Spain, Scale 1:50.000. Available online: http://info.igme.es/cartografiadigital/geologica/Geode.aspx?language=en (accessed on 15 February 2020).
31. INTECSA-IGME. *Proyecto de Investigación Hidrogeológica de Vejer-Barbate (Sistema 33)*; Unpublished report; 1982; 126p + annexes.
32. De Andalucía, J. *Elaboración de un Plan de Gestión Integrada en las Masas de Agua Subterránea en mal Estado Químico y/o Cuantitativo Identificadas en las Demarcaciones Hidrográficas Andaluzas de Carácter Intracomunitario, con Objeto de Alcanzar los Objetivos Medioambientales Fijados en la Legislación Vigente en Materia de Aguas. Tomo I y II*; Unpublished report; 2013; 215 + 127 pp.
33. Wick, K.; Heumesser, C.; Schmid, E. Groundwater nitrate contamination: Factors and indicators. *J. Environ. Manag.* **2012**, *111*, 178–186. [CrossRef] [PubMed]
34. Lawniczak, A.; Zbierska, L.; Nowak, B.; Achtenberg, K.; Grześkowiak, A.; Kanas, K. Impact of agriculture and land use on nitrate contamination in groundwater and running waters in central-west Poland. *Environ. Monit. Assess.* **2016**, *188*, 172. [CrossRef]
35. Serio, F.; Miglietta, P.; Lamastra, L.; Ficocelli, S.; Intini, F.; De Leo, F.; De Donno, A. Groundwater nitrate contamination and agricultural land use: A greywater footprint perspective in Southern Apulia Region (Italy). *Sci. Total Environ.* **2018**, *645*, 1425–1431. [CrossRef]
36. IGME y Diputación de Cádiz. *Atlas Hidrogeológico de la Provincia de Cádiz*; Instituto Geológico y Minero de España y Diputación de Cádiz: Madrid, Spain, 2005; 264p.
37. Seiler, K.P.; Gat, J.R. *Groundwater Recharge from Run-Off, Infiltration and Percolation*, 1st ed.; Springer: Dordrecht, The Netherlands, 2007; pp. 31–67. [CrossRef]
38. Foster, S.; Garduño, H.; Tuinhof, A.; Tovey, K. *Groundwater Governance: Conceptual Framework for Assessment of Provisions and Needs*; GW-MATE Strategic Overview Series, 1; World Bank: Washington, DC, USA, 2010.
39. Lopez-Gunn, E.; Cortina, L.M. Is self-regulation a myth? Case study on Spanish groundwater user associations and the role of higher-level authorities. *Hydrogeol. J.* **2006**, *14*, 361–375. [CrossRef]
40. Garduño, H.; Foster, S. *Sustainable Groundwater Irrigation: Approaches to Reconciling Demand with Resources*; GW-MATE Strategic Overview Series 4; World Bank: Washington, DC, USA, 2010.
41. Custodio, V.E.; Kretsinger, M.R. LlamasIntensive development of groundwater: Concepts, facts and suggestions. *Water Policy* **2005**, *7*, 151–162. [CrossRef]
42. Castaño, F.S. Aplicaciones de la teledetección y SIG al control y cuantificación de las extracciones de agua subterránea. In *Medida y Evaluación de las Extracciones de agua Subterránea*; Ballester, A., Fernández, J.A., López-Geta, J.A., Eds.; IGME: Madrid, Spain, 1999.
43. Llamas, R.; Hernández-Mora, N.; Martínez-Cortina, L. *El uso Sostenible de las aguas Subterráneas*; Papeles del Proyecto Aguas Subterráneas; Fundación Marcelino Botín: Santander, Spain, 2000; p. 54.
44. Instituto Geológico y Minero de España (IGME). *Informe Técnico del Sondeo de Abastecimiento a Barbate de Franco (Cádiz)*; Unpublished report; 1978; 8p + annexes.
45. CGS-ITGE. *Apoyo a la Gestión de los Recursos Hidráulicos en Vejer-Barbate*; Technical Report nº 85/1991; Unpublished report; 1991; 112p + annexes.
46. ITGE. *Informe Técnico del Sondeo para Abastecimiento a la Población de Barbate (Cádiz)*; Technical Report 29/1996; Instituto Tecnológico Geominero de España: Madrid, Spain, 1996.

sustainability

MDPI

Article

Recognizing Crucial Aquatic Factors Influencing Greenhouse Gas Emissions in the Eutrophication Zone of Taihu Lake, China

Yulin Wang [1], Liang Wang [2,*,+], Jilin Cheng [2], Chengda He [1,*,†] and Haomiao Cheng [1]

1 School of Environmental Science and Engineering, Yangzhou University, Yangzhou 225127, China; wangyulin01986@126.com (Y.W.); yzchhm@yzu.edu.cn (H.C.)
2 College of Hydraulic Science and Engineering, Yangzhou University, Yangzhou 225009, China; jlcheng@yzu.edu.cn
* Correspondence: right628@126.com (W.L.); hcd@yzu.edu.cn (H.C.); Tel.: +86-15861568737 (W.L.); +86-13905275314 (H.C.)
† These authors contributed equally to this work.

Received: 14 August 2019; Accepted: 17 September 2019; Published: 20 September 2019

Abstract: Greenhouse gas (GHG) emissions, which are closely related to climate change and serious ecological instability, have attracted global attention. The estimation of crucial aquatic factors for the flux of GHGs in lakes is a key step in controlling and reducing GHG emissions. The importance of 14 aquatic factors for GHG emissions was estimated in Meiliang Bay, which is an eutrophication shallow bay in Taihu Lake in eastern China. The random forest (RF) method, which is an improved version of the classified and regression tree (CART) model, was employed. No distribution assumption on variables was required in this method and it could include nonlinear actions and interactions among factors. The results show significant positive correlations among the fluxes of CO_2, CH_4, and N_2O. The most crucial factor influencing CO_2 emissions is the water temperature (WT) followed by sulfate (SO_4^{2-}), alkalinity (Alk), dissolved oxygen (DO), and nitrate (NO_3^-–N). The important factors for CH_4 emissions are WT, SO_4^{2-}, DO, Alk, and NO_2^-–N. The outcome for N_2O, in which the key factor is NO_2^-–N, was slightly different from those of CO_2 and CH_4. A comprehensive ranking index (CRI) for the fluxes of all three GHGs was also calculated and showed that WT, NO_2^-–N, SO_4^{2-}, DO, and Alk are the most crucial aquatic factors. These results indicate that increasing DO might be the most effective means of controlling GHG emissions in eutrophication lake bays. The role of SO_4^{2-} in GHG emissions, which has previously been ignored, is also worth paying attention to. This study provides a useful basis for controlling GHG emissions in eutrophication shallow lake bays.

Keywords: GHGs; aquatic factors; random forest; water temperature; nitrogen; sulfate

1. Introduction

The emission of greenhouse gases (GHGs) to the atmosphere is closely related to climate change [1], resulting in significant disruption in biological living conditions and ecosystem instability [2,3]. Natural lakes, though representing only about 2% of the land surface area, are important sources of GHGs such as carbon dioxide (CO_2), methane (CH_4), and nitrous oxide (N_2O) [4–6], and the emission of GHGs in lakes has therefore attracted the interest of many researchers.

According to previous assessments, lakes contribute about 71.6 TgC CH_4 and 1943 TgC CO_2 to the atmosphere per year [5,7–9]. However, these data remain largely uncertain due to the spatial heterogeneity of emissions in waterbodies [9,10]. Furthermore, it is even harder to estimate the contribution of N_2O from global lakes [11]. In large developing countries, such as China, the problem may be more serious as data are lacking [12]. The flux of GHGs in lakes is also drastically different according to the distinct nutrient level zones. The CH_4 flux in the East Plain Lakes zone is about two times more than in the Tibetan Plateau and Inner Mongolia–Xinjiang Lakes zone [13]. GHG emissions may be different in the same lake. Previous observations have shown a one order of magnitude larger CO_2 flux in the overeutrophication zone compared with the eutrophication and mesotrophic zones in Taihu Lake in China [14]. The flux of N_2O in the emergent macrophyte-type area was also about 1.5 and 30 times larger than in the algae-type and submerged macrophyte areas, respectively, during summer [15], while in winter, the flux of the algae-type area was the largest [16].

Many factors affect the flux of GHGs. Inorganic nitrogen compounds such as nitrate [15,17,18] and ammonia nitrogen [19] are the factors controlling N_2O production, while total phosphorus and chlorophyll A promote CO_2 [14,20] and CH_4 [20–22] production in waterbodies. The water temperature, wind speed, water velocity, and turbulence are common factors influencing the three kinds of GHGs [23–25]. Other factors, for instance, pH, dissolved oxygen, and chloride ions may also affect the release of GHGs [16,26]. However, the roles of aquatic factors in the control of GHG emissions remain controversial because of their complex effects [15,27,28]. Identifying the main controlling factors and their roles is critical for further understanding the mechanisms of GHG emissions. Recognizing the aquatic variables affecting GHG emissions, especially under the nonlinear action and interaction effects of aquatic factors, is still an urgent problem to be solved.

China has 2700 lakes with a total area of 81,414.6 km^2 [29]. The carbon emission of lakes in China is larger than the mean of the world lakes in the temperate zone [7]. An initial assessment showed that the lakes in China release 3.0 TgC CH_4 per year [13]. Controlling the GHG flux from lakes, especially from the eutrophication lake bay, will play a key role for China in meeting their United Nations Framework Convention on Climate Change (UNFCCC) commitments. It is also very important for the sustainability of lake water in a social and environmental dimension [30]. Hence, it is necessary to recognize the crucial aquatic factors influencing the GHG flux at different nutrient level zones in China, especially the eutrophication zone.

Here, we provide a combination approach to identify important variables for GHG flux in Meiliang Bay, which is an eutrophication zone of Taihu Lake in eastern China. The statistic and seasonal characteristics of 14 aquatic factors and GHG emissions in this lake bay are performed and Pearson coefficients among them are also shown. The random forest (RF) method, which can take into account the nonlinear effects and interaction effects of factors, is employed to identify the most important factors influencing the flux of the three types of GHGs. A comprehensive ranking of the GHGs is also given.

The results showed dissolved oxygen, water temperature, alkalinity and nitrite are very important for the flux of GHGs. Sulfate, which may have been ignored by previous studies, also play a crucial roles in GHG emissions. Although this assessment is based on a specific shallow lake bay, it is a useful method and its result could easily be popularized to clarify the vital factors and their roles in GHG flux in other large shallow fresh water lakes, such as Chaohu Lake [31,32] and Hongze Lake [33], in eastern of China.

2. Methods

2.1. Data Sources

Taihu Lake, which is located in the north subtropical monsoon climate region, is the third largest freshwater lake in China. It has an area of approximately 2445 km^2 with a mean depth of about only 1.9 m [15]. There are about 100 million people living around the lake, contributing over 5 trillion dollars to the GDP in the year 2018. Shanghai, which is the most developed city in China, is also close to the lake. Hence, any research conducted on Taihu Lake could have potentially significant implications for China.

In recent decades, Taihu Lake has suffered from a eutrophication problem. The water quality in the north and east of the lake has improved, while it has deteriorated in other regions, especially from the 1990s to the 2010s [34,35]. The west eastern corner of the lake is Meiliang Bay, which is eutrophicated, and algae blooms occur frequently in spring and summer. Figure 1 shows the location of Taihu Lake and Meiliang Bay.

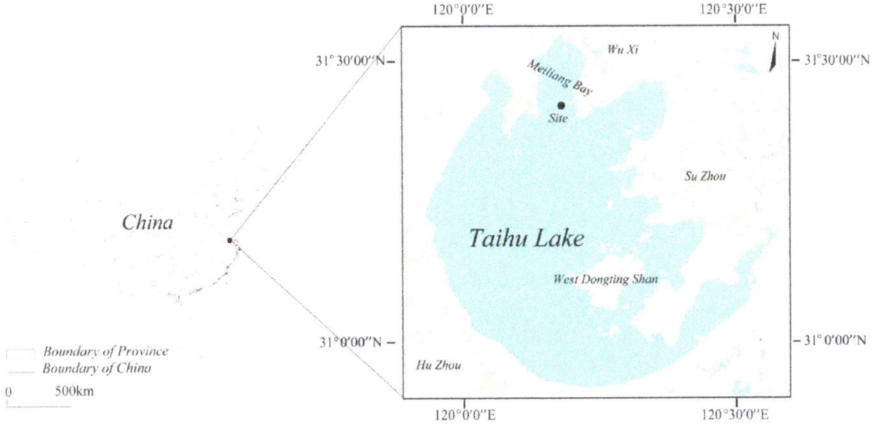

Figure 1. Location of Taihu Lake.

The research on GHG emissions carried out in this bay shows that there is considerable flux in GHGs that is significantly different from other regions in the lake [14–16]. However, only relationships between a few aquatic factors and the flux of GHGs were simultaneously considered in Meiliang Bay. The nonlinear relationships were also ignored because linear regressions and Pearson correlation analysis were applied. In this paper, the observation data of CO_2, CH_4, and N_2O in this bay (Figure 1 shows the observation site) are analyzed using 14 aquatic factors, including inorganic nitrogen (nitrate, NO_3^-–N; ammonia, NH_4^+–N; and nitrite, NO_2^-–N), phosphorus (phosphate, PO_4^{3-}; dissolved total phosphorus (dTP)), the response of nutrients (chlorophyll A ,Chl-a), physical indices of water (water temperature (WT); water depth (WD); and Secchi depth (SD)) and other chemical factors (dissolved oxygen (DO); sulfate, SO_4^{2-}; O_2 demand (COD_{Mn}); pH; and alkalinity, Alk). The data were sampled once per month from January 2004 to December 2004, and all data have been previously published [20,36].

2.2. Statistical Analyses

The RF method is an improved and robust version of the classified and regression tree (CART) method. It introduces the bootstrapping aggregation approach into CART [37–39] and calculates the predicted values by averaging the results of CART trees on bootstrap samples [40]. The variables used in the RF method should have an independent identical distribution [40]. However, different from the common bagging tree method, the RF method resamples prediction features at every split node to ensure independence among the selected features [41], and the scaled observations would reduce the difference in their distributions. Similar to many other statistical methods, the violation of the property of the identical distribution may not lead to serious consequences [42].

The biggest advantage of RF is that the nonlinear and interaction effects of independent variables can be included. RF has been demonstrated, through practice, to be a successful machine learning method for forecasting [43,44].

Although the RF model was proposed for prediction, it can be used for other purposes. The RF model can determine the quantitative importance of predictors using some indices. The most popular index, which is also used in this paper, is an increase in node purity. The node purity was measured by the Gini index [38,40]. All computations were completed using the R (3.6.1) language [45] with the randomForest package.

After calculating the importance of all aquatic factors for CO_2, CH_4, and N_2O, a comprehensive importance index was developed to further investigate, because the significant correlations between GHG emissions were shown. The comprehensive index for all three GHGs can be given by the following equation:

$$R_k = \sum_{i=1}^{3} \frac{1}{r_{i,k}} \qquad CRI = rank(R_k) \tag{1}$$

where R_k is the index whose rank is determined as the comprehensive ranking index (CRI) for the k-th factor and $r_{i,k}$ is the importance rank of the k-th factor for the i-th GHG. CRI is a simple, helpful, and widely used index to measure comprehensive importance [46].

3. Results and Discussion

As seen in Table 1, the flux of GHGs showed no significant differences to the observations in previous years for Meiliang Bay [14–16]. The minimum values of the three GHGs were lower than 0, indicating that, at some stages, the lake acts as a sink for these GHGs, which is also in agreement with previous studies [14,15,27]. This induced large fluctuations of GHG emissions compared with the aquatic factors. Based on the observations of CO_2, CH_4, and N_2O emissions, the coefficients of variation (CVs) were respectively 1.64, 1.50, and 1.50. These values imply that the fluxes were strongly influenced by the water environment.

Table 1. Statistics of Observations (n = 12).

Factors	Maximum Value	Mean Value	Minimum Value	Standard Deviation
CO_2 (mmol/m^{2}d)	200.67	39.62	−20.73	64.80
CH_4 (mmol/m^{2}d)	2.17	0.54	−0.18	0.81
N_2O ((mmol/m^{2}d)	0.27	0.06	−0.03	0.09
NO_3^--N (mg/L)	2.046	1.093	0.049	0.596
NO_2^--N (mg/L)	0.146	0.046	0.006	0.040
NH_4^+-N (mg/L)	1.456	0.507	0.024	0.577
PO_4^{3-} (mg/L)	0.020	0.006	0.001	0.007
dTP* (mg/L)	0.056	0.027	0.012	0.015
Chl-a*(mg/m^3)	39.26	16.19	2.46	11.67
WT*(°C)	29.9	16.7	4.2	9.0
WD*(m)	2.9	2.6	2.3	0.15
SD*(m)	0.80	0.44	0.30	0.14
DO* (mg/L)	12.43	8.98	6.42	1.99
SO_4^{2-} (mg/L)	103.60	75.99	51.60	15.26
Alk* (mmol/L)	2.60	2.16	1.76	0.26
COD_{Mn} *(mg/L)	5.94	4.88	4.10	0.60
pH	8.49	8.18	8.03	0.13

* dTP: dissolved total phosphorus, Chl-a: chlorophyll A, WT: water temperature, WD: water depth, SD: Secchi depth, DO: dissolved oxygen, Alk: alkalinity, COD_{Mn}: O_2 demand.

The values of aquatic factors showed that the water quality was not very poor, while the high mean concentration of Chl-a and the very low SD showed that eutrophication at the site was serious. The large CVs of NH_4^+–N and NO_2^-–N, which were 1.14 and 0.87, respectively, indicated that the release of GHGs might be affected by nitrogen. It was also observed that the concentration of SO_4^{2-} in the lake, which has previously been ignored, was very high. The minimum value of pH was larger than 8.0, meaning the water was weakly alkaline, which would also have affected the production of N_2O [47].

Figure 2 shows the scaled time series of the concentrations of the 14 aquatic factors and the flux in the three GHGs in the year 2004. All values were scaled by their mean and standard deviations.

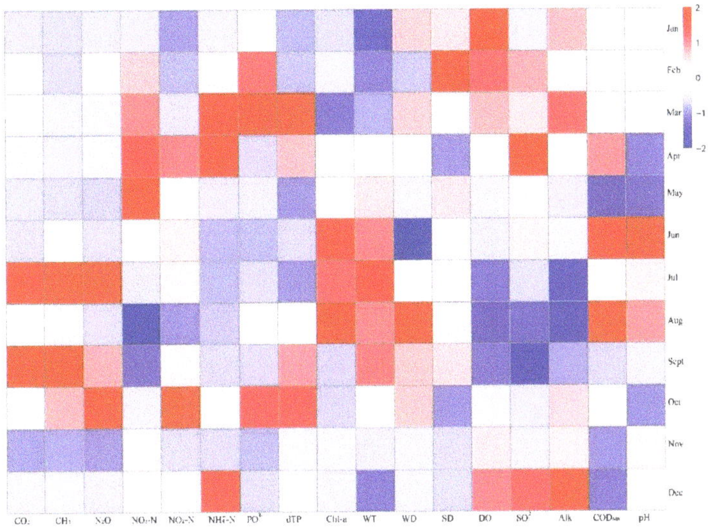

Figure 2. Scaled time series in the year 2004.

As shown in Figure 2, the GHG fluxes showed strong seasonal changes. Combined with the results of Table 1, they progressed from one extreme to the other from winter to summer. The maximum values of flux occurred in July, and the data for CH_4 and CO_2 in September also showed high values, while for N_2O, a second peak emerged in October. November was the only month that Meiliang Bay appeared to be a sink for all three GHGs. The observed values of N_2O showed slight differences to the data collected from Meiliang Bay in 2017 [15,16], but the characteristics of CH_4 were in agreement with those observed in Donghu Lake in China, which is similar to Taihu Lake [28]. This difference might be because N_2O is controlled by nitrogen, while CH_4 emissions are not. In addition, the relatively high concentration of NO_2^-–N in October might be the reason for the high concentration of N_2O observed in this month.

It was clearly observed that the water quality was better in winter, i.e., from November to January, and this can be explained by the low level of agricultural activity. The high concentrations of NH_4^+–N, NO_3^{2-}–N, and SO_4^{2-} in December were notable exceptions. Laboratory experiments showed that the low temperature would decrease the activity of nitrifiers and denitrifiers [48,49], and so both NH_4^+–N and NO_3^-–N accumulated. The reduction in SO_4^{2-} was also weakened as the sulfate-reducing bacteria (SRB) were also influenced by low WT. References indicate that SO_4^{2-} and SRB are closely linked to nitrogen cycling [50]; thus, the variation of NH_4^+–N and NO_3^{2-}–N in the time series showed the same pattern as that of SO_4^{2-}.

In addition to the time series, strong correlations were also observed between CO_2 and CH_4 and between CH_4 and N_2O, with Pearson coefficients being significant at $p < 0.01$ in Figure 3. Denitrification, acetate fermentation, and CO_2 reduction, which connect the production of CO_2, CH_4, and N_2O [51] could explain this outcome. The relationship between N_2O and CO_2 was a little less significant ($0.01 < p < 0.05$). This might be because the production pathways of N_2O and CO_2 were not involved with each other directly. SO_4^{2-} showed a negative correlation with the fluxes of CO_2 and CH_4 at $p < 0.1$. There have been few studies on the effects of SO_4^{2-} on GHG emissions, but those that have been done have suggested that SRB could take part in reactions with CH_4 [52,53]. SO_4^{2-} would also impact the nitrogen cycle [51], which may be another reason for the significant correlation among the three GHGs.

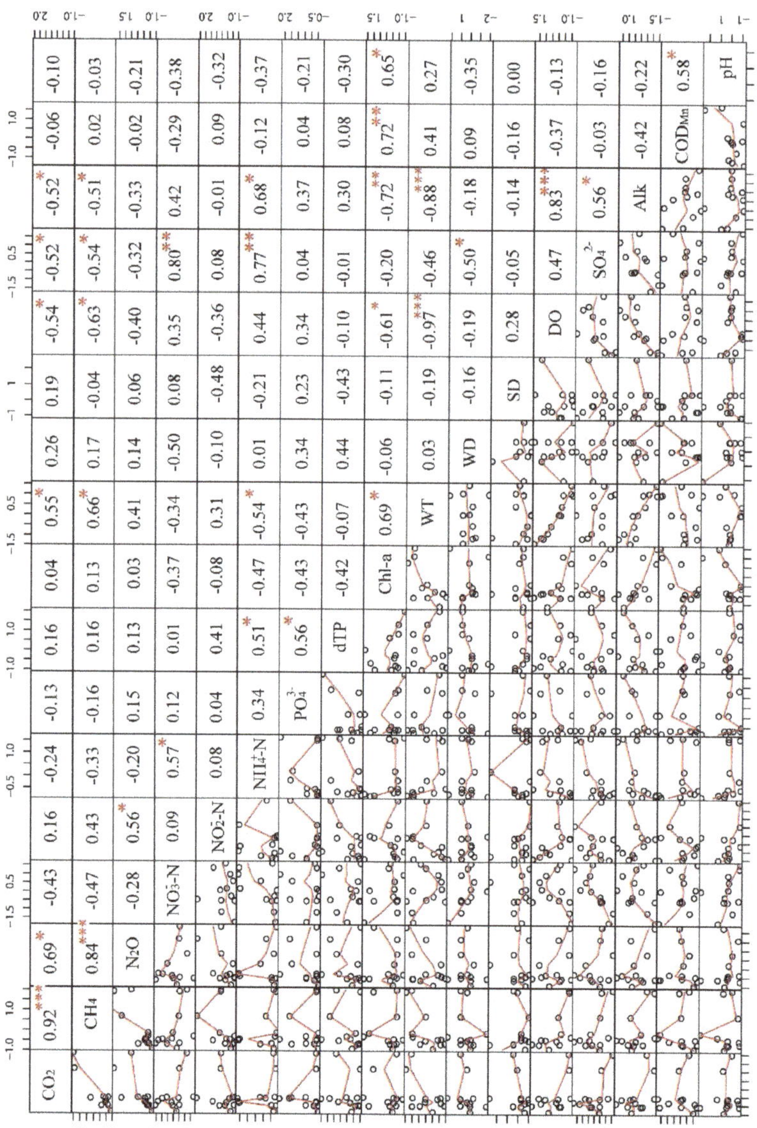

Figure 3. The correlations among factors. The upper matrix shows the Pearson coefficients, and results were significant at *** $p < 0.01$, ** $p < 0.05$, or * $p < 0.1$ as marked. The red solid lines in the lower matrix show a smooth regression between the two factors.

The crucial aquatic factors were similar for CH_4 and CO_2, with WT, DO, and Alk being recognized as key factors, which is in agreement with previous studies [20,23,26]. Chl-a was shown to have little effect on GHG emissions, which differs from previous research [14,20]. Having noticed that Chl-a also showed strong correlations with WT, DO, COD_{Mn}, pH, and Alk, the weak impact that was observed might be the result of complex interactions amongst the different factors.

Few factors show a correlation with the flux of N_2O, with the exception of NO_2^-–N. This should not be surprising considering the fact that NO_2^--N is the intermediate product of the denitrification reaction of nitrifying bacteria [54], which produces N_2O [55]. However, the effects of NO_3^- and NH_4^+ shown by some studies [15–17] may be masked by the effects of NO_2^--N under linear relationships. The influences of SO_4^{2-} on NO_3^--N and NH_4^+-N were very complex. The strong correlations could be explained by the effects of sulfur and sulfate on NO_3^--N reduction and NH_4^+-N oxidation [51]. We will describe and summarize these reaction details after the results of RF have been shown.

Figure 4 shows the importance of aquatic factors on the three GHGs, measured by an increase in the Gini index in the RF models. The explained variance of the three models was 80.4%, 86.2%, and 75.1%, respectively, for CO_2, CH_4, and N_2O, and these results imply that the RF models were adequate for exploring the crucial factors.

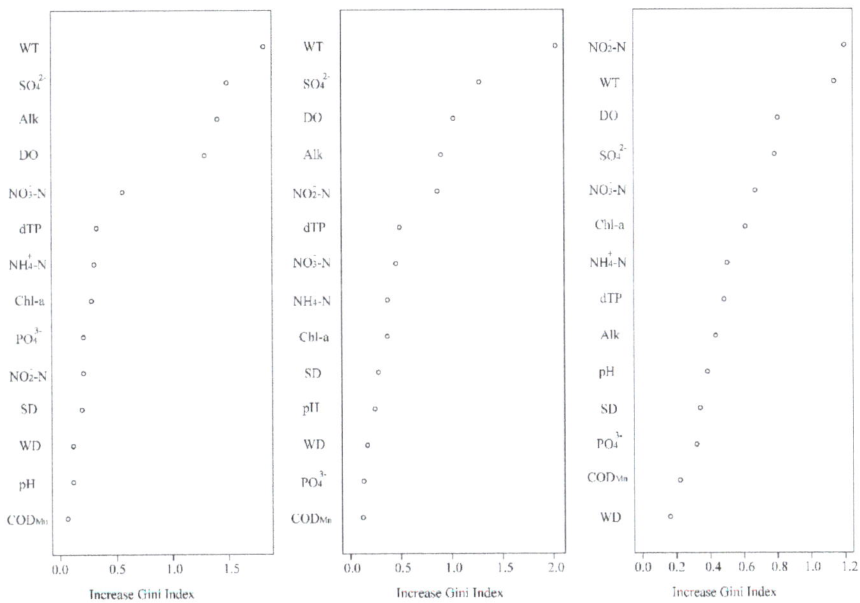

(a) CO_2 (b) CH_4 (c) N_2O

Figure 4. The importance order of aquatic factors for greenhouse gas (GHG) emissions using the random forest (RF) method.

As seen in Figure 4, some RF results were in agreement with the Pearson coefficients, while others were not. The results of RF showed that the five most important aquatic factors for the three GHGs were similar. WT was the first key variable implicated in the flux of CO_2 and CH_4, while it was also second most important for N_2O emissions. This is because methane bacteria choose different methanogen metabolic pathways [56,57] under different temperatures. The nitrifiers and denitrifiers are also sensitive to temperature [49]. This outcome was in agreement with the results of the Pearson coefficients (Figure 3) and were also analogous with the results of other field studies [23,24,58].

DO also played an important role in the emissions of the three GHGs. Methane bacteria are a diverse group of strict anaerobes [59] and are, therefore, greatly influenced by DO. The two main pathways for producing CH_4, acetate fermentation and CO_2 reduction, are both associated with CO_2 [51,57]. This may be part of the reason why DO also impacts CO_2 production. The results of the linear correlations and field observations also confirmed the effects of DO [20,56]. In addition, the observations showed that Alk would impact the carbon dioxide partial pressure [60] and anaerobic digestion [61], so Alk greatly influences the flux of CO_2 and CH_4. For N_2O emissions, although both nitrification and denitrification would produce N_2O, DO has a dominant influence in determining the pathway. This can explain the importance of DO for N_2O [15,16].

Nitrogen compounds, including NO_3^-–N, NO_2^-–N, and NH_4^+–N, were shown to be important for all three GHGs. It is easy to understand the effects of them on N_2O, while it should be noted that NO_2^-–N was shown to be more important. This may be because, in all four main pathways, the producing NO_2^-–N are closer to N_2O than NO_3^-–N [62]. Both denitrification and dissimilatory nitrogen reduction to ammonium (DNRA) oxidize organic matter and then produce CO_2 [51,59], so the importance of nitrogen for CO_2 can be rationalized. Methanotrophs outcompete nitrifiers for O_2 when CH_4 is sufficiently abundant, as more energy can be released from oxidizing methane than from oxidizing NH_4^+ [51]. This is a good explanation for the negative relationship between NH_4^+ and CH_4 shown by the Pearson coefficients, as well as the importance of nitrogen in the results of the RF method and field observations [21].

SO_4^{2-} was crucial in determining the flux of all three GHGs (Figure 4). This seemed a little strange as SO_4^{2-} has been often taken for granted when assessing GHG emissions from the lake during field studies. However, the result should not be surprising. On the one hand, NO_3^- may be used in the oxidation of reduced sulfur (S^0 or S^{2-}) and the production of SO_4^{2-}. These processes may occur in preference to DNRA and denitrification [51,59]. On the other hand, SO_4^{2-} reduction by SRB could also produce CO_2 [51]. Additionally, observations in freshwater wetlands indicated that SO_4^{2-} input would suppress CH_4 flux because of the higher energy alternative provided by SO_4^{2-} reduction [63,64]. In summary, SO_4^{2-} plays an important role in CO_2, CH_4, and N_2O production.

Generally speaking, the effects of nitrogen compounds, SO_4^{2-} and DO on GHG emissions is very complex and is summarized as concept models as follows.

As seen in Figure 5, the DO, WT, and ALK are conditions that affect the reactions. For example, DO determines if anaerobic or aerobic oxidation can take place, and it also chooses what type of reduction will happen. WT is also very important and it can affect the activities of SRB, nitrifiers, and denitrifiers. Different from these conditions, the NO_3^-–N, NO_2^-–N, NH_4^+–N, and SO_4^{2-} are involved with the production of GHGs directly. From the models, SO_4^{2-} participates in the production of all three GHGs simultaneously, which highlights its importance and its complex effects on GHG emissions.

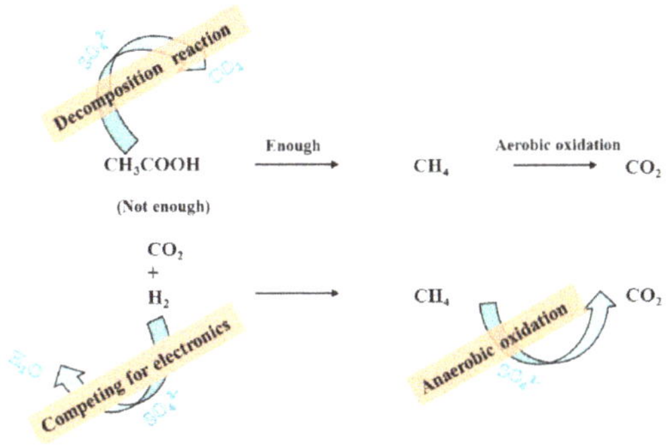

(a)

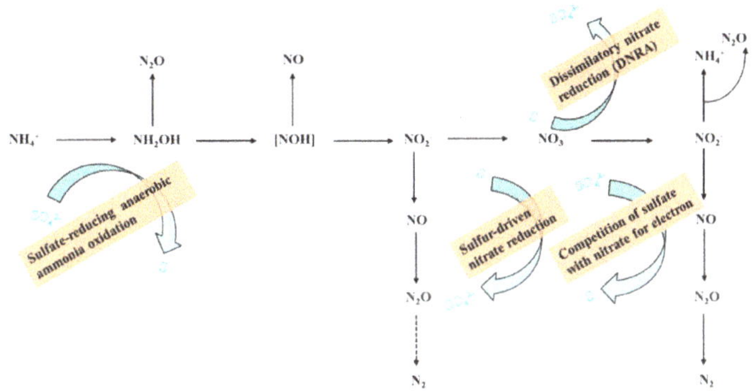

(b)

Figure 5. Concept models for GHG emissions. (**a**) Concept model for CH_4 and CO_2; (**b**) Concept model for N_2O.

The significant Pearson coefficients among CO_2, CH_4, and N_2O highlight the necessity for working out a comprehensive importance index for the flux of all three GHGs. The CRI values of the five most important aquatic factors calculated by Equation 1 are shown in Table 2.

Table 2. The comprehensive importance ranking index for GHG emissions.

Aquatic Factors	Index for CO_2	Index for CH_4	Index for N_2O	CRI*
WT	1	1	2	1
NO_2^--N	10	5	1	2
SO_4^{2-}	2	2	4	3
DO	4	3	3	4
Alk	3	4	9	5

* CRI: comprehensive ranking index defined in equation 1.

The results of CRI showed that WT, NO_2^-–N, SO_4^{2-}, DO, and Alk are the five most crucial aquatic factors influencing the flux of GHGs. The importance and positive relationships between WT and GHG emissions remind us that the largest flux should appear in summer. Perhaps the emissions will become larger and larger as global warming progresses [1]. NO_2^-–N is not very important for CO_2 and CH_4, while it is still the second crucial factors for GHG emissions because of its importance for N_2O production. Compared with WT and NO_2^-–N, SO_4^{2-} is the third key factors for the flux of GHGs because of its importance in all three. Maybe this is an indication that it is feasible to control GHG emissions by increasing the concentration of SO_4^{2-} in lake bays.

4. Conclusions

GHG emissions, which lead to serious ecological problems, have attracted widespread attention. The estimation of crucial aquatic factors in the flux of GHGs in lakes has played a key role in reducing GHG emissions. In this paper, RF methods, taking into account nonlinear effects and interaction effects of factors, were employed to identify the crucial factors among 14 aquatic variables in the flux of GHGs in a eutrophicated lake bay.

The results showed significant positive correlations between the fluxes of CO_2 and CH_4, which were shown to be affected by similar factors, while there was little difference for N_2O. WT, SO_4^{2-}, Alk, DO, and NO_3^-–N were identified as the five key factors in CO_2 emissions, while for CH_4, the key factors were WT, SO_4^{2-}, DO, Alk, and NO_2^-–N. The outcome that NO_2^-–N is the most crucial factor for N_2O emissions while NO_3^-–N is the fifth showed the importance of nitrogen in the flux of N_2O. Apart from these common factors, SO_4^{2-}, which has been previously ignored, was also shown to play an important role in GHG emissions. It is the second most influential factor for CO_2 and CH_4, and the fourth factor for N_2O. The concept models showed that SO_4^{2-} had very complex effects on the production of CO_2 and CH_4, as well as on the nitrogen cycle.

The outcomes of the comprehensive ranking index for the flux of all three GHGs have also been shown. WT, NO_2^-–N, SO_4^{2-}, DO, and Alk were found to be the five most crucial aquatic factors. Compared with WT and Alk, the remaining factors are easier to manage by engineered measures. A comprehensive analysis of the results show that increasing the DO might be the most effective way of controlling GHG emissions in eutrophication lakes. Apart from the direct benefits of increasing DO, such as reducing the fluxes of CO_2 and CH_4, N_2O emissions should also reduce, led by the decrease in the concentration of NO_2^-–N. It seems that a higher SO_4^{2-} concentration would also be good for decreasing GHG emissions, but this can be a dilemma for water quality managers because there is evidence that excess SO_4^{2-} can lead to black blooms in shallow lakes [65].

This study provides useful information for controlling GHG emissions in eutrophicated shallow lake bays. However, there is still work to be done. The quantitative mechanism model for water factors and GHG emissions in shallow lake bays is a very important topic for GHG emission reduction. This model will become more detailed as research continues. The smooth linear regression in Figure 3 suggests that there are threshold points for these relationships. The existence of threshold points indicates the necessity of investigating these crucial factors in GHG emissions using advanced methods. The role of SO_4^{2-} should also receive more attention in future studies.

Author Contributions: Yulin Wang conducted the research and completed the first manuscript of the paper. Liang Wang and Chengda He guided the research and improved the manuscript. Haomiao Cheng and Jilin Cheng collected the data and created the figures.

Funding: This work was financially supported by the National Key R & D Program of China (Grant No.2017YFC0403205), the National Natural Science Foundation of China (Grant Nos. 51909230 and 51809226), the China Postdoctoral Science Foundation Funded Project (Grant No. 2018M632390), the Jiangsu Planned Projects for Postdoctoral Research Funds (Grant No.2018K124C), and the Jiangsu Funded Recruitment of Postdoctoral Project (Grant No. 2018Z051).

Conflicts of Interest: The authors declare no conflict of interest.

References

1. Pachauri, R.K.; Allen, M.R.; Barros, V.R.; John, B. *Climate Change 2014: Synthesis Report. Contribution of Working Groups I, II and III to the Fifth Assessment Report of the Intergovernmental Panel on Climate Change*; IPCC: Geneva, Switzerland, 2014.
2. Panel, A.C.S. *What We Know: The Reality, Risks and Response to Climate Change*; American Association for the Advancement of Science: Washington, DC, USA, 2014.

3. Pimm, S.L. Climate Disruption and Biodiversity. *Cur. Biol.* **2009**, *19*, R595–R601. [CrossRef] [PubMed]
4. Abril, G.; Bouillon, S.; Darchambeau, F.; Teodoru, C.R.; Marwick, T.R.; Tamooh, F.; Ochieng Omengo, F.; Geeraert, N.; Deirmendjian, L.; Polsenaere, P. Technical Note: Large Overestimation of PCO_2 Calculated from pH and Alkalinity in Acidic, Organic-Rich Freshwaters. *Biogeoscience* **2015**, *12*, 67–78. [CrossRef]
5. Bastviken, D.; Cole, J.; Pace, M.; Tranvik, L. Methane Emissions from Lakes: Dependence of Lake Characteristics, Two Regional Assessments, and a Global Estimate. *Glo. Biol. Cyc.* **2004**, *18*, GB4009. [CrossRef]
6. Meybeck, M. Riverine Transport of Atmospheric Carbon: Sources, Global Typology and Budget. *Water Air Soil Pollut.* **1993**, *70*, 443–463. [CrossRef]
7. Bastviken, D.; Tranvik, L.J.; Downing, J.A.; Crill, P.M.; Enrich-Prast, A. Freshwater Methane Emissions Offset the Continental Carbon Sink. *Science* **2011**, *331*, 50. [CrossRef] [PubMed]
8. Aufdenkampe, A.K.; Mayorga, E.; Raymond, P.A.; Melack, J.M.; Doney, S.C.; Alin, S.R.; Aalto, R.E.; Yoo, K. Riverine Coupling of Biogeochemical Cycles between Land, Oceans, and Atmosphere. *Front. Ecol. Environ.* **2011**, *9*, 53–60. [CrossRef]
9. Tranvik, L.J.; Downing, J.A.; Cotner, J.B.; Loiselle, S.A.; Striegl, R.G.; Ballatore, T.J.; Dillon, P.; Finlay, K.; Fortino, K.; Knoll, L.B. Lakes and Reservoirs as Regulators of Carbon Cycling and Climate. *Limnol. Oceanogr.* **2009**, *54*, 2298–2314. [CrossRef]
10. Raymond, P.A.; Hartmann, J.; Lauerwald, R.; Sobek, S.; McDonald, C.; Hoover, M.; Butman, D.; Striegl, R.; Mayorga, E.; Humborg, C. Global Carbon Dioxide Emissions from Inland Waters. *Nature* **2013**, *503*, 355–359. [CrossRef] [PubMed]
11. Huttunen, J.T.; Alm, J.; Liikanen, A.; Juutinen, S.; Larmola, T.; Hammar, T.; Silvola, J.; Martikainen, P.J. Fluxes of Methane, Carbon Dioxide and Nitrous Oxide in Boreal Lakes and Potential Anthropogenic Effects on the Aquatic Greenhouse Gas Emissions. *Chemosphere* **2003**, *52*, 609–621. [CrossRef]
12. Yang, L.; Lu, F.; Zhou, X.P.; Wang, X.K.; Duan, X.N.; Sun, B.F. Progress in the Studies on the Greenhouse Gas Emissions from Reservoirs. *Acta Ecol. Sinic.* **2014**, *34*, 204–212. [CrossRef]
13. Yang, H.; Xie, P.; Ni, L.; Flower, R.J. Underestimation of CH_4 Emission from Freshwater Lakes in China. *Environ. Sci. Tech.* **2011**, *45*, 4203–4204. [CrossRef] [PubMed]
14. Fan, C.X.; Phillip, W.F.; Hu, W.P.; Qin, B.Q. Divergence of Carbon Dioxide Fluxes in Different Trophic Areas of Taihu Lake, China. *J. Environ. Sci.* **2003**, *15*, 433–442.
15. Zheng, X.L.; Wen, S.L.; Li, X.; Gong, W.Q.; Liu, D.H.; Zhong, J.C. Characteristics of N_2O Release and Influencing Factors in Grass-Type and Algae-Type Zones of Taihu Lake During Summer. *Environ. Sci.* **2018**, *39*, 2306–2315.
16. Zheng, X.L.; Liu, M.; Wen, S.L.; De Hong, L.; Zhong, J.C. Characteristics of N_2O Formation and Emission in Algae and Grasstype Zones in Lake Taihu in Winter. *Chin. Environ. Sci.* **2018**, *38*, 683–693.
17. Beaulieu, J.J.; Tank, J.L.; Hamilton, S.K.; Wollheim, W.M.; Hall, R.O.; Mulholland, P.J.; Peterson, B.J.; Ashkenas, L.R.; Cooper, L.W.; Dahm, C.N. Nitrous Oxide Emission from Denitrification in Stream and River Networks. *PNAS* **2011**, *108*, 214–219. [CrossRef] [PubMed]
18. Herrman, K.S.; Bouchard, V.; Moore, R.H. Factors Affecting Denitrification in Agricultural Headwater Streams in Northeast Ohio, USA. *Hydrobiologia* **2008**, *598*, 305–314. [CrossRef]
19. Zhao, Y.; Xia, Y.; Kana, T.M.; Wu, Y.; Li, X.; Yan, X. Seasonal Variation and Controlling Factors of Anaerobic Ammonium Oxidation in Freshwater River Sediments in the Taihu Lake Region of China. *Chemosphere* **2013**, *93*, 2124–2131. [CrossRef]
20. Lia, S.; Bush, R.T.; Santos, I.R.; Zhang, Q.; Song, K.; Mao, R.; Wen, Z.; Lu, X.X. Large Greenhouse Gases Emissions from China's Lakes and Reservoirs. *Water Res.* **2018**, *147*, 13–24. [CrossRef]
21. Wang, X.; He, Y.; Yuan, X.; Chen, H.; Peng, C.; Yue, J.; Zhang, Q.; Diao, Y.; Liu, S. Greenhouse Gases Concentrations and Fluxes from Subtropical Small Reservoirs in Relation with Watershed Urbanization. *Atmos. Environ.* **2017**, *154*, 225–235. [CrossRef]
22. Wang, X.; He, Y.; Chen, H.; Yuan, X.; Peng, C.; Yue, J.; Zhang, Q.; Zhou, L. Ch_4 Concentrations and Fluxes in a Subtropical Metropolitan River Network: Watershed Urbanization Impacts and Environmental Controls. *Sci. Total Environ.* **2018**, *622*, 622–623. [CrossRef]
23. Huang, W.; Bi, Y.; Hu, Z.; Zhu, K.; Zhao, W.; Yuan, X.; Huang, W.; Bi, Y.; Hu, Z.; Zhu, K. Spatio-Temporal Variations of GHG Emissions from Surface Water of Xiangxi River in Three Gorges Reservoir Region. *China Ecol. Eng.* **2015**, *83*, 28–32. [CrossRef]

24. Schrier-Uijl, A.P.; Veraart, A.J.; Leffelaar, P.A.; Berendse, F.; Veenendaal, E.M. Release of CO_2 and CH_4 from Lakes and Drainage Ditches in Temperate Wetlands. *Biogeochemistry* **2011**, *102*, 265–279. [CrossRef]
25. Zappa, C.J.; Raymond, P.A.; Terray, E.A.; McGillis, W.R. Variation in Surface Turbulence and the Gas Transfer Velocity over a Tidal Cycle in a Macro-Tidal Estuary. *Estuaries* **2003**, *26*, 1401–1415. [CrossRef]
26. Audrey, C.; Del Giorgio, P.A. Patterns in CH_4 and CO_2 Concentrations across Boreal Rivers: Major Drivers and Implications for Fluvial Greenhouse Emissions under Climate Change Scenarios. *Glob. Chang. Biol.* **2014**, *20*, 1075–1088.
27. Kling, G.W.; Kipphut, G.W.; Miller, M.C. The Flux of CO_2 and CH_4 from Lakes and Rivers in Arctic Alaska. *Hydrobiologia* **1992**, *240*, 23–36. [CrossRef]
28. Xing, Y.P.; Xie, P.; Yang, H.; Ni, L.Y.; Wang, Y.S.; Rong, K.W. Methane and Carbon Dioxide Fluxes from a Shallow Hypereutrophic Subtropical Lake in China. *Atmos. Environ.* **2005**, *39*, 5532–5540. [CrossRef]
29. Ma, R.H.; Yang, G.S.; Duan, H.T.; Jiang, J.H.; Wang, S.M.; Feng, X.Z.; Nong, L.A.; Kong, F.X.; Xue, B.; Wu, J.L. China's Lakes at Present: Number, Area and Spatial Distribution. *Sci. China Earth Sci.* **2011**, *54*, 283–289. [CrossRef]
30. Rui, C.M.; Cruz, N.F.D.; Pires, J. Measuring the Sustainability of Urban Water Services. *Environ. Sci. Policy* **2015**, *54*, 142–151.
31. Yang, L.; Lei, K.; Meng, W.; Fu, G.; Yan, W. Temporal and Spatial Changes in Nutrients and Chlorophyll-A in a Shallow Lake, Lake Chaohu, China: An 11-Year Investigation. *J. Environ. Sci.* **2013**, *25*, 1117–1123. [CrossRef]
32. Wang, Y.L.; Lin, H.Z.; Liang, W. Sensitivity Analysis of the Chaohu Lake Eutrophication Model with New Index Based on Morris Method. *Water Sci. Tech. Water Suppl.* **2018**, *18*, 1375–1387. [CrossRef]
33. Li, S.; Guo, W.; Mitchell, B. Evaluation of Water Quality and Management of Hongze Lake and Gaoyou Lake Along the Grand Canal in Eastern China. *Environ. Monitor Assess.* **2011**, *176*, 373–384. [CrossRef] [PubMed]
34. Wang, L.; Wang, Y.L.; Cheng, H.M.; Cheng, J.L. Estimation of the Nutrient and Chlorophyll a Reference Conditions in Taihu Lake Based on a New Method with Extreme-Markov Theory. *Int. J. Environ. Res.* **2018**, *15*, 2372. [CrossRef] [PubMed]
35. Qin, B.Q. *Lake Taihu, China-Dynamics and Environmental Change*; Springer: London, UK, 2008.
36. Qin, B.Q.; Hu, C.H.H. *Chinese Ecosystem Positioning Observation and Research Data Sets: Taihu Lake*; Agriculture Press: Beijing, China, 2010.
37. Speybroeck, N. Classification and Regression Trees. *Int. J. Pub. Health* **2012**, *57*, 243–246. [CrossRef] [PubMed]
38. Song, Q. *Environmental and Ecological Statistics with R*, 2nd ed.; CRC Press: Boca Raton, FL, USA, 2017.
39. Huo, S.; Ma, C.; Xi, B.; He, Z.; Jing, S.; Wu, F. Nonparametric Approaches for Estimating Regional Lake Nutrient Thresholds. *Ecol. Indic.* **2015**, *58*, 225–234. [CrossRef]
40. Breiman, L. Random Forests. *Mach. Learn.* **2001**, *45*, 5–32. [CrossRef]
41. Miguel Forte, R. *Mastering Predictive Analytics with R*; Packt Publishing: Birmingham, UK, 2015.
42. Hastie, T.; Tibshirani, R.; Friedman, J. *The Elements of Statistical Learning: Data Mining, Inference, and Prediction*, 2nd ed.; Springer: New York, NY, USA, 2009.
43. Arnab, B.; Surinder, S.K.; Debasish, S.; Harmandeep, S.; Anu, K.; Sandeep, S. Potential Indicators of Soil Health Degradation in Different Land Use-Based Ecosystems in the Shiwaliks of Northwestern India. *Sustainability* **2019**, *11*, 3908.
44. Svetnik, V.; Liaw, A.; Tong, C.; Culberson, J.C.; Sheridan, R.P.; Feuston, B.P. Random Forest: A Classification and Regression Tool for Compound Classification and QSAR Modeling. *J. Chem. Inf. Comput. Sci.* **2003**, *43*, 1947. [CrossRef]
45. R Development Core Team. *R: A Language and Environment for Statistical Computing*; R Foundation for Statistical Computing: Austria, Vienna, 2016.
46. Saltelli, A.; Ratto, M.; Andres, T.; Campolongo, F.; Cariboni, J.; Gatelli, D.; Saisana, M.; Tarantola, S. *Global Sensitivity Analysis*; The Primer: New York, NY, USA, 2008.
47. Chen, H.; Mothapo, N.V.; Shi, W. Soil Moisture and pH Control Relative Contributions of Fungi and Bacteria to N_2O Production. *Micro Ecol.* **2015**, *69*, 180–191. [CrossRef]
48. Wei, X.H.; Bi, X.J.; Yin, Z.X.; Zhou, X.L.; Xu, C.L.; Wen, J.G. Effects of Temperature and Dissolved Oxygen on Nitrification and Denitrification in MBBR System. *China Environ. Sci.* **2019**, *39*, 612–618.
49. Schulthess, R.V.; Kühni, M.; Gujer, W. Release of Nitric and Nitrous Oxides from Denitrifying Activated Sludge. *Water Res.* **1995**, *29*, 215–226. [CrossRef]

50. Zhu, J.; He, Y.; Zhu, Y.; Huang, M.; Zhang, Y. Biogeochemical Sulfur Cycling Coupling with Dissimilatory Nitrate Reduction Processes in Freshwater Sediments. *Environ. Rev.* **2018**, *26*, 121–132. [CrossRef]
51. Schlesinger, W.H.; Bernhardt, E.S. *Biogeochemsitry: An Analysis of Global Change*, 3rd ed.; Wiley: New York, NY, USA, 2013.
52. Raghoebarsing, A.A.; Arjan, P.; Pas-Schoonen, K.T.; van de Smolders, A.J.P.; Ettwig, K.F.; Rijpstra, W.I.C.; Stefan, S.; Damsté, J.S.S.; op den Camp, H.J.M.; Jetten, M.S.M. A Microbial Consortium Couples Anaerobic Methane Oxidation to Denitrification. *Nature* **2006**, *440*, 918–921. [CrossRef] [PubMed]
53. Blair, N.E.; Aller, R.C. Anaerobic Methane Oxidation on the Amazon Shelf. *Geochim. Cosmochim. Acta* **1995**, *59*, 3707–3715. [CrossRef]
54. Wrage, N.; Velthof, G.L.; Beusichem, M.L.V.; Oenema, O. Role of Nitrifier Denitrification in the Production of Nitrous Oxide. *Soil Biol. Biochem.* **2001**, *33*, 1723–1732. [CrossRef]
55. Li, C.S. Modeling Trace Gas Emissions from Agricultural Ecosystems. *Nutr. Cycl. Agroecosys.* **2000**, *58*, 259–276. [CrossRef]
56. Avery, G.B.; Shannon, R.D.; White, J.R.; Martens, C.S.; Alperin, M.J. Controls on Methane Production in a Tidal Freshwater Estuary and a Peatland: Methane Production Via Acetate Fermentation and Co_2 Reduction. *Biogeochemsitry* **2003**, *62*, 19–37. [CrossRef]
57. Meirong, W.; Rui, Z.; Jun, Z.; Xinxin, X.; Xiaoyu, Y.; Zhiying, Y.; Mingmin, G.; Tao, Z. Effect of Temperature on Methanogens Metabolic Pathway and Structures Predominant Bacteria. *J. Chem. Ind. Eng.* **2014**, *65*, 1602–1606.
58. Trumbore, S.E.; Chadwick, O.A.; Amundson, R. Rapid Exchange between Soil Carbon and Atmospheric Carbon Dioxide Driven by Temperature Change. *Science* **1996**, *272*, 393–396. [CrossRef]
59. Zehnder, A.J.; Stumm, W. *Biology of Anaerobic Microorganisms*; Wiley: New York, NY, USA, 1988.
60. Yue, X.P.; Fu, M.H.; Li, Y.X.; Cao, J.Z. Effect of Influent Alkalinty on Anaerobic Sequencing Batch Reactor. *J. Chem. Ind. Eng.* **2008**, *59*, 1257–1263.
61. González, J.S.; Rivera, A.; Borja, R.; Sánchez, E. Influence of Organic Volumetric Loading Rate, Nutrient Balance and Alkalinity: COD Ratio on the Anaerobic Sludge Granulation of an Uasb Reactor Treating Sugar Cane Molasses. *Inter. Biodeter. Biodegrad.* **1998**, *41*, 127–131. [CrossRef]
62. Stocker, T.F.; Qin, D.; Plattner, G.K. *Climate Change 2013: The Physical Science Basis. Contribution of Working Group I to the Fifth Assessment Report of the Intergovernmental Panel on Climate Change*; Cambridge University Press: Cambridge, UK, 2013.
63. Gauci, V.; Chapman, S.J. Simultaneous Inhibition of Ch 4 Efflux and Stimulation of Sulphate Reduction in Peat Subject to Simulated Acid Rain. *Soil Biol. Biochem.* **2006**, *38*, 3506–3510. [CrossRef]
64. Dise, N.B.; Verry, E.S. Suppression of Peatland Methane Emission by Cumulative Sulfate Deposition in Simulated Acid Rain. *Biogeochemsitry* **2001**, *53*, 143–160. [CrossRef]
65. Wang, L.; Wang, Y.; Cheng, H.M.; Cheng, J.L. Identifying the Driving Factors of Black Bloom in Lake Bay through Bayesian Lasso. *Int. J. Environ. Res. Pub. Health* **2019**, *16*, 2492. [CrossRef] [PubMed]

sustainability

MDPI

Article

A Multivariate Geomorphometric Approach to Prioritize Erosion-Prone Watersheds

Jesús A. Prieto-Amparán [1], Alfredo Pinedo-Alvarez [1], Griselda Vázquez-Quintero [2], María C. Valles-Aragón [2], Argelia E. Rascón-Ramos [1], Martin Martinez-Salvador [1] and Federico Villarreal-Guerrero [1,*]

[1] Facultad de Zootecnia y Ecología, Universidad Autónoma de Chihuahua, Chihuahua 31453, Mexico; jesus_prieto06@hotmail.com (J.A.P.-A.); apinedo@uach.mx (A.P.-A.); argelia.rascon@yahoo.com.mx (A.E.R.-R.); msalvador@uach.mx (M.M.-S.)

[2] Facultad de Ciencias Agrotecnológicas, Universidad Autónoma de Chihuahua, Chihuahua 31350, Mexico; gquintero@uach.mx (G.V.-Q.); valles.cecilia@gmail.com (M.C.V.-A.)

* Correspondence: fvillarreal@uach.mx; Tel.: +52-(614)-434-1448

Received: 15 August 2019; Accepted: 15 September 2019; Published: 19 September 2019

Abstract: Soil erosion is considered one of the main degradation processes in ecosystems located in developing countries. In northern Mexico, one of the most important hydrological regions is the Conchos River Basin (CRB) due to its utilization as a runoff source. However, the CRB is subjected to significant erosion processes due to natural and anthropogenic causes. Thus, classifying the CRB's watersheds based on their erosion susceptibility is of great importance. This study classified and then prioritized the 31 watersheds composing the CRB. For that, multivariate techniques such as principal component analysis (PCA), group analysis (GA), and the ranking methodology known as compound parameter (*Cp*) were used. After a correlation analysis, the values of 26 from 33 geomorphometric parameters estimated from each watershed served for the evaluation. The PCA defined linear-type parameters as the main source of variability among the watersheds. The GA and the *Cp* were effective for grouping the watersheds in five groups, and provided the information for the spatial analysis. The GA methodology best classified the watersheds based on the variance of their parameters. The group with the highest prioritization and erosion susceptibility included watersheds RH24Lf, RH24Lb, RH24Nc, and RH24Jb. These watersheds are potential candidates for the implementation of soil conservation practices.

Keywords: prioritization; geomorphometric parameters; compound parameter; geospatial distribution; GIS

1. Introduction

Soil erosion is considered one of the most important degradation processes in the world [1,2]. The soil resource is limited and its wide use is of utmost importance; it sustains biogeochemical processes and is the habitat for a great diversity of microorganisms [3]. Sustained soil development, conservation, and restoration is one of today's main challenges for humankind.

Hydric erosive processes affect the fertile soil layer, which is a key factor in the primary production of ecosystems [4]. The production of goods and satisfiers for the population such as wood, food, fiber, fodder, water, and recreational areas, among others, in addition to industrial expansion and the need for infrastructure facilities, have increased land-use/land-cover changes, increasing the pressure over the soil [5]. This has caused experts to pay more attention to the growing trend of soil erosion and the importance of water and soil conservation for achieving sustainable development.

Integrated watershed management is an alternative for soil management [6–8]. Watersheds are one of the spatial units that are used for the planning and management of soil resources [9]. Management implies the characterization of the ecosystems inside the watershed and the understanding of the relationships between uplands, lowlands, land use/land cover, geomorphic processes, slope, and soil [10, 11]. In watershed management, erosion control is one of the main components [3]. Thus, the hydrological planning and monitoring of a watershed is important for the development of environmental policies [12]. In this sense, the quantification of the watersheds' characteristics is fundamental to understanding their dynamics and degradation levels. This knowledge serves to define and implement strategies to prevent soil erosive processes and promote the conservation and restoration of watersheds [13].

Morphometry is used in the analysis of the watershed configuration [14]. Such methodology was developed by Horton [15] and then modified by Strahler [16], and provides information on the behavior of the basin [17]. It is an important tool for identifying and prioritizing eroded watersheds [18].

Nevertheless, monitoring soil erosion *in situ* is costly and time consuming in large watersheds. Thus, the analysis of geomorphometry is often carried out based on geographic information systems (GIS) [19–23]. On a spatial scale, geomorphometric parameters, e.g., the Gravelius compactness coefficient [24] and elongation ratio [25], among others, are important to know the hydrological configuration of watersheds. The relationships among these parameters are useful for developing hydrological models, which allow prioritizing watersheds based on their condition, such as erosion susceptibility. To determine the aforementioned relationships, statistical methods, such as multivariate techniques, have been widely used worldwide [26,27]. For instance, Gavit et al. [28] characterized 13 geomorphometric parameters in 11 watersheds located in the Godavari river in Maharashta, India. Youssef et al. [29] estimated the erosion risk by using remote sensing technology, GIS, and geomorphometric parameters in 11 watersheds located in Sinai, Egypt. Makwana and Tiwari [30] used seven geomorphometric parameters to characterize 19 watersheds in the region of Gujarat, India. Sharma et al. [31] applied the multivariate technique of principal component analysis (PCA) to 13 geomorphometric parameters from eight watersheds located in the Madhya district, India. Meshram and Sharma [32] and Farhan et al. [33] used PCA to analyze the geomorphometric parameters of a group of watersheds located in the Shakkar Catchment River in India and Jordan.

A large number of shape, relief, and hydrological parameters are associated with watershed geomorphometry [34,35]. The statistical techniques of PCA and group analysis (GA) [36–41], as well as multivariate analysis of variance (MANOVA) [41] and the ranking methodology known as compound parameter (*Cp*) [42] have been widely used in recent years for the analysis of environmental data from watersheds. These techniques assist with analyzing the spatial variability of watersheds, their structure, as well as the relationships existing among them.

The most important basin in the state of Chihuahua, as a runoff source [43], is the Conchos River Basin (CRB). Yet, this basin has experienced water stress conditions during the past years. According to Ordoñez [44], approximately 8000 km^2 (11.8%) of the basin high lands present strong erosion problems, which could impact waterflow and water quality. In these high lands, deforestation and land-use/land-cover changes had contributed to reduce the amount of infiltrated water, impacting on groundwater availability [43]. In the low lands of the basin, agriculture consumes 90% of the water harvested in the basin. Other consumers include the industrial and domestic sectors [43,45]. In addition, the international water trade between Mexico and the U.S. from 1944 commands Mexico to deliver annually from this basin a total of 432×10^6 m^3 of water to the U.S. [46]. Therefore, specific knowledge about water management and the status of the basin's soil erosion is required to implement strategies to solve water-related problems and to promote the sustainable development of the region.

The objective of this study was: (1) to describe the behavior of the 31 watersheds located along the CRB in the state of Chihuahua, Mexico, based on the values of their geomorphometric parameters; (2) to spatially classify the 31 watersheds into groups; and (3) to prioritize the groups of basins according to their erosion susceptibility. For the grouping, multivariate techniques and the compound parameter (*Cp*) were used and their efficacy compared.

2. Materials and Methods

2.1. Study Area

The CRB is located in the state of Chihuahua and Durango, Mexico, and is part of the 24th Rio Bravo-Conchos Hydrological Region [47] (Figure 1). The basin has an area of 67,800 km^2 [48], with an altitudinal distribution ranging from 841 m to 2348 m [49]. It presents a diversity of climates ranging from temperate in the upper, semi-arid in the middle, and arid in the lower parts of the basin [50]. The physiography of the upper basin belongs to the mountainous zone made up of temperate forests dominated by species of pines and oaks. The middle part of the Altiplano or central valleys is made up of transition zones where oaks and bushes are present. Regarding the lower part, it belongs to the arid region and is made up of shrublands and grasslands [51]. The basin has a precipitation regime from June to September, with July and August being the wettest months, and fluctuating from 200 mm to 700 mm [48].

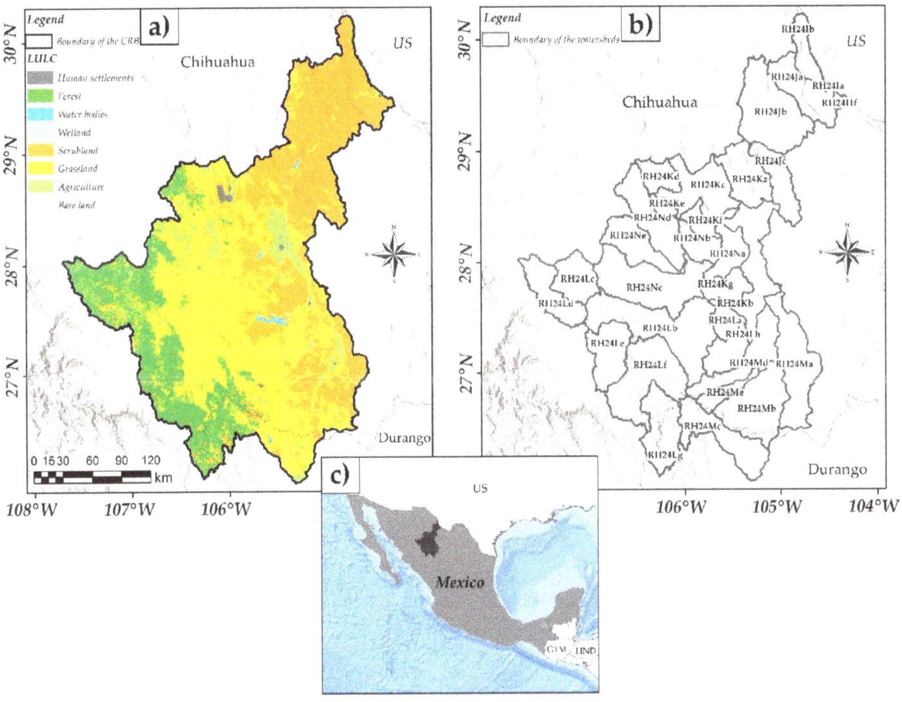

Figure 1. (**a**) Land-use/land-cover types of the Conchos River Basin (CRB), (**b**) Delimitation of the 31 watersheds of the CRB, (**c**) Location of the CRB in Mexico.

2.2. Data

Data of the CRB was obtained from the online GIS source of CONABIO [52]. Likewise, the data of the 31 watersheds composing the CRB (Figure 2) were obtained from the Watershed Water Flows Simulator [53]. The relief and hydrology type parameters were estimated by processing the necessary data from a Digital Terrain Model (DTM), with a resolution of 15 × 15 m, downloaded from INEGI [54]. The values of the basic parameters from the watersheds were obtained by using the Hydrology tool [55] of ArcMap© 10.3 (ESRI, Redlands, CA, USA; https://wwwesri.com/en-us/home). The values of the shape, relief, and linear type parameters were calculated from the equations listed in Table 1.

Table 1. Geomorphometric parameters.

Geomorphometric Parameter	Equation	References
	Basic Parameters [1]	
Area (A)	A = Watershed surface area [km^2]	Horton [15]
Perimeter (P)	P = Watershed perimeter [km]	Horton [15]
Length (Lb^2)	Lb^2 = Watershed length [km]	
Stream order (u)	u = Stream order [unitless]	Strahler [16]
Main Channel Length (Lc)	Lc = Main flow channel length [km]	
All Channel Lengths (Lu)	Lu = Length of all the flow channels in the watershed [km]	Horton [15]
Contour Length (Li)	Li = Contour lines' length [km]	
Number of Flow Channels (Nu)	Nu = Number of flow channels [unitless]	
Number of First-Order Flow Channels (No_1)	No_1 = Number of total first-order flow channels in the watershed [unitless]	
Maximum Height (H_{max})	H_{max} = Watershed maximum height [m]	
Minimum Height (H_{min})	H_{min} = Watershed minimum height [m]	
Medium Height (H_{med})	H_{med} = Watershed medium height [m]	
	Shape Parameters	
Gravelius Compactness Coefficient (Cc)	$Cc = P/2\sqrt{\pi A}$	Zavoianu [24]
Elongation Ratio (Re)	$Re = 1.1284\ (\sqrt{A}/Lc)$	Schumm [25]
Shape Factor (Rf)	$Rf = A/Lb^2$	Horton [15]
Elongation Index (Ia)	$Ia = Lb^2/W$ where: W = watershed width (Km)	Horton [15]
Unit Shape Factor (R_U)	$R_U = Lb^2/A^{0.5}$	Horton [15]
Circularity Ratio (Rc)	$Rc = 4\pi A/P^2$	Miller [56]
	Relief Parameters	
Mean Watershed Slope (J)	$J = (\Sigma Li\ E/A) \times 100$ where: E = equidistance among contour lines (km)	Horton [15]
Massivity Coefficient ($tg\alpha$)	$tg\alpha = H_{med}/A$	
Relief Relationship(Rh)	$Rh = H_{max}/Lb$	Schumm [25]
Relative Relief (Rr)	$Rr = H_{max}/P$	Schumm [25]
Orographic Coefficient (Co)	$Co = Hmed \times tg\alpha$	
	Linear parameter	
Drainage Density (Dd)	$Dd = \Sigma Lu/A$	Horton [15]
Mean Slope of the Main Channel (j)	$j = (H_{max} - H_{min})/L_c \times 100$	Horton [15]
Mean Distance (Am)	$Am = L_c/(\sqrt{A})$	Horton [15]
Sinuosity of the Main Flow Channel (Scp)	$Scp = L_c/Lb^2$	Mueller [57]
Kirpich Concentration Time (TcK)	$TcK = 0.066(Lb^2/j)^{0.77}$	Kirpich [58]
Average Peak Flow (Qp)	$Qp = 43A^{0.522}$	Sen [59]
Texture Ratio (T)	$T = N_u/P$	Horton [15]
Rivers Frequency (Fu)	$Fu = N_u/A$	Horton [15]
Resistance Number (Rn)	$Rn = H_{max} \times Dd$	Schumm [25]
General Flow Length (Lo)	$Lo = {}^1/_2 \times Dd$	Schumm [25]
Drainage Intensity (Di)	$Di = Fu/Dd$	Faniran [60]

[1] The basic parameters were calculated with the GIS software.

Basic parameters include the area (A), perimeter (P), watershed length (Lb^2), stream order (u), main channel length (Lc), all channel lengths (Lu), and contour length (Li). The area and perimeter are the most important parameters of the watersheds to understand their hydrological design and reflect the volume of water that can be discharged in a rainfall event [61].

Shape parameters include the Gravelius compactness coefficient (Cc), elongation ratio (Re), shape factor (Rf), elongation index (Ia), unit shape factor (Ru), and circularity ratio (Rc). The Gravelius compactness coefficient is the relationship between the perimeter of the watershed and the perimeter of a circle of area equal to that of the watershed [24]. Low values of the Gravelius compactness coefficient, shape factor, and elongation index indicate a more elongated watershed, while high values correspond to a less elongated watershed. Watersheds with less elongated shapes have a shorter maximum flow duration, while elongated watersheds correspond to watersheds with longer flow duration [24]. The elongation ratio is the diameter of a circle with an area equal to that of the watershed divided by the length of the watershed [32]. Those watersheds with values close to or greater than the unit correspond to flat watersheds, while values that move away from the unit are watersheds with pronounced relief [25]. The shape factor is the relationship between the area and the length of the watershed [15]. The elongation index is also a relationship between the length and width of the watershed [15]. The unit shape factor is the relationship between the length and the area of the watershed. Values less than 2 indicate that they have weak flood discharge periods, while values greater than 2 indicate that their flood discharge is strong [30]. The circularity ratio is defined as the relationship between the area and the perimeter of the watershed. If the circularity of the watershed is low, then discharge will be slow, and the susceptibility to erosion will be low [61].

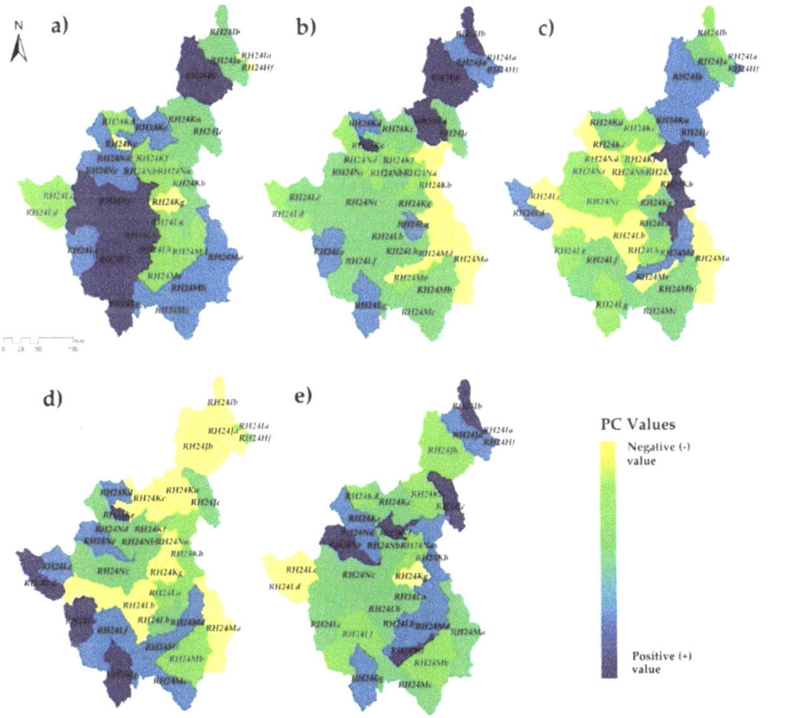

Figure 2. Classification of 31 watersheds based on the values of five principal components resulting from a principal component analysis. Conchos River Basin, Chihuahua, Mexico. (**a**) PC1, (**b**) PC2, (**c**) PC3, (**d**) PC4, (**e**) PC5. PC1 = Principal component 1, PC2 = Principal component 2, PC3 = Principal component 3, PC4 = Principal component 4, PC5 = Principal component 5.

Relief parameters include the mean watershed slope (*J*), massivity coefficient (*tgα*), relief ratio (*Rh*), relative relief (*Rr*), and orographic coefficient (*Co*). The mean watershed slope indicates the degree of the terrain roughness. As the slope increases, the watershed will be prone to erosion. The massivity coefficient represents the relationship between the mean watershed height and its surface area, which is expressed as a percentage. Small values of the massivity coefficient correspond to very mountainous watersheds, while large values correspond to watersheds from moderately mountainous to flat. The relief ratio is directly related to the slope of the currents and the watershed surface, being an indicator of hydrological processes and erosion. The relief ratio, similar to the relative relief, has a direct correlation with the watershed erosion processes [25].

Linear parameters include drainage density (*Dd*), mean slope of the main channel (*j*), mean distance (*Am*), sinuosity of the main channel (*Scp*), Kirpich concentration time (*Tc*), average peak flow (*Qp*), texture ratio (*T*), rivers frequency (*Fu*), resistance number (*Rn*), general flow length (*Lo*), and drainage intensity (*Di*). High drainage density values indicate a high current density, and therefore a rapid response to precipitation events [62]. The mean slope of the main channel indicates the slope of the longest channel in the watershed. The high values of this parameter indicate that sediment flow and entrainment will quickly exit the watershed [15]. The sinuosity of the main channel indicates the velocity of flow movement in the channels. The lowest values of sinuosity correspond to channels where the flow travels at greater speed, whereas the channels with the highest values of sinuosity indicate the slow movement of the flow [57]. The Kirpich concentration time is the time when a drop of water falls at the furthest point until it reaches the exit point [58]. Average peak flow is defined as the mean maximum amount of water passing through a specific section [59]. The texture ratio corresponds to the relationship between the total number of streams and the watershed perimeter. Rivers frequency represents the total number of streams of all orders per unit area [15]. The resistance number is used to measure the flood potential of rivers. It has a direct relationship with erosion, where increasing its value would represent an increment in erosion susceptibility [15].

2.3. *Watershed's Description and Classification*

Prior to the watersheds' classification, their geomorphometric parameters were correlated [6,63]. Correlation indicates when part of the information contained in a set of geomorphometric parameters is also contained in the remaining ones [32]. That served to reduce the number of parameters included in the subsequent analysis.

To describe the variability of the geomorphometric parameters, principal component analysis (PCA) was performed. The technique of PCA reduces the data dimensionality, simplifies the dataset, and makes it easier to explain through a set of new principal components (PCs) [64,65]. The first principal component is the linear combination of the original geomorphometric parameters that contributes the most to the total variance; the second principal component, not correlated to the first, contributes the most to the residual variance, and so on until the total variance is analyzed. The PCs are orthogonal variables that could be obtained by multiplying the original variables, which are correlated, with coefficients similar to the ones described in Equation (1):

$$z_{ij} = a_{i1}x_{1j} + a_{i2}x_{2j} + \ldots + a_{im}x_{mj} \quad (1)$$

where z represents the coefficient of the component, a represents the weight of the component, x represents the measured value of the variable, i corresponds to the component number, j represents the sample number, and m represents the total number of variables.

The PCA was applied to the values of geomorphometric parameters to calculate the correlation matrix and to obtain the PCs [66]. Both the correlation analysis and the PCA were performed in the SAS© 9.1.3 software (The SAS Institute Inc., Cary, NC, USA, https://www.sas.com/en_us/home.html). Then, the PCs were mapped for interpretation.

For the classification, the first method used for the watersheds was a hierarchical group analysis, which was based on the Ward criterion [67]. The Ward criterion was applied to the GA by using the square Euclidean distance to explore the clustering of the 31 watersheds. The definition of the watershed groups was performed based on the coefficient of determination (R^2) [68]. Finally, a multivariate analysis of variance (MANOVA) was used to analyze whether significant multivariate differences exist between the groups based on the values of their geomorphometric parameters [69].

The second classification method considered in this work was the compound parameter (*Cp*). Previous research has employed this approach for sustainable watershed planning and management [42]. Linear and shape parameters have been commonly used for this method, whereas the relief and basic parameters were additionally included in this study. Linear geomorphometric parameters have a direct relationship with erosion susceptibility, where high values are more likely to result where high erosion probabilities are present [61,70]. Thus, for watershed classification, the highest value of linear parameters was ranked as 1, the second highest was ranked as 2, and so on. Conversely, shape parameters have an inverse relationship with erosion [61,70], and low values are related to high susceptibility to erosion and vice versa. Then, the lowest value of the shape parameters was ranked as 1, the next lowest value was ranked as 2, and so on. Regarding the relief and basic parameters, the highest value was ranked as 1, the second highest value was ranked as 2, and so on. After this procedure was completed, the ranked values from each watershed were summed and then averaged to produce the *Cp* of each watershed. This average represents the collective impact of all the parameters, and is calculated according to Equation (2) [42]:

$$C_P = \frac{1}{n}\sum_{i=1}^{n} R \qquad (2)$$

where *Cp* is the compound parameter, *R* is the ranked value of each parameter, and *n* is the number of parameters.

Based on the *Cp*, the highest priority was assigned to the watersheds with the lowest *Cp* value, the second priority was assigned to the next higher *Cp* value, and so on. Then, the *Cp* was classified into five categories or groups of erosion susceptibility. The categories were defined as: Very High (Group 5), High (Group 4), Moderate (Group 3), Low (Group 2), and Very Low priority (Group 1), similar to classifications made in previous studies [71].

2.4. Comparison of the Classification Methods

To compare the grouping methods, the following procedure was followed.

First, an ANOVA was carried out for each geomorphometric parameter (independent variable), and separated for each grouping method. The source of variation or class was considered to be the group. Such analyses served to determine possible significant differences among the groups within each grouping method. After the ANOVA analyses were completed, the grouping method that achieved the highest number of significant *p*-values ($\alpha < 0.05$) was considered the most effective for grouping the watersheds of the CRB.

A list of abbreviations can be found in Appendix B (Table 3).

3. Results

3.1. Characterization of Conchos River Basin Watersheds

The values of the basic, shape, relief, and linear-type parameters from the 31 watersheds used in this study are shown in Table A1. Watershed RH24Ia has the smallest area and perimeter with 78.62 km^2 and 50.59 km, respectively. Meanwhile, the watershed with the largest area and perimeter is RH24Lb, with 5428.29 km^2 and 640.77 km, respectively. The watershed with the highest stream order is RH24Kb, while watershed RH24Lb has the longest main channel. First-order streams do not have tributaries, and their flow depends on the secondary surface contributions that converge to them [61]. The watershed with the highest number of channels of order one is RH24Ia, whereas the watershed with the lowest number of channels is RH24Jb. The watershed lengths vary from 11.17 km to 133 km. Watershed RH24Kb presents the largest elongation ratio value, indicating that it is the flattest, while watershed RH24Ib has the lowest value for this parameter, indicating that it has the steepest slope [25]. The values of sinuosity of the main channel vary from 0.05 km to 4.1 km. The watershed with the shortest Kirpich concentration time is RH24Kb, while watershed RH24Lb has the longest. The lowest average peak flow value corresponds to watershed RH24Ia, whereas watershed RH24Lb presents the highest. The texture ratio values are between 4.98–75.1, which are considered as moderate to high values; the low values correspond to watershed RH24Na, while the highest value belongs to watershed RH24Lg, so the former is not susceptible to erosion, while the latter is. The watershed with the lowest resistance value is RH24Na, while the watershed with the highest value is RH24Lg.

3.2. *Correlations and Principal Component Analyses*

The data matrix of the 31 watersheds and the 33 parameters, which included the basic shape, linear, and relief type parameters, were analyzed through a correlation analysis. A set of parameters showing high correlations were identified. From each pair of highly correlated parameters, only one parameter was chosen, and the rest were eliminated to reduce the data dimensionality. After the reduction, the final number of geomorphometric parameters was 26, as listed in Table 2.

The PCA was performed on the 26 parameters and showed that the first five principal components were the most important for explaining the data variance (Figure 2). The most important parameter was selected according to its contribution to the principal component, as it is shown by the values of the eigenvectors (in bold) of the correlation matrix (Table 2). The first five principal components accounted for 88.44% of the total variance of the dataset. The linear parameters (hydrology) are the ones mainly explaining the CRB behavior (Table 2).

Table 2. Principal components and geomorphometric parameters explaining the greatest portion of variance in the watersheds of the Conchos River Basin, Chihuahua, Mexico.

PC	Eigenvalue	Variance	Accumulated Variance	Geomorphometric Parameter (First)	Geomorphometric Parameter (Second)
1	9.0010	0.3462	0.3462	Average peak flow (*Qp*)	Channel lengths (*Lu*)
2	6.0848	0.2340	0.5802	Unit shape factor (*Ru*)	Drainage density (*Dd*)
3	3.4222	0.1316	0.7119	Mean distance (*Am*)	Elongation ratio (*Re*)
4	2.8620	0.1101	0.8219	Minimum height (H_{min})	Mean slope (*J*)
5	1.6246	0.0625	0.8844	Elongation index (*Ia*)	General flow length (*Lo*)

PC = Principal component, First = Primary importance, Second = Secondary importance.

3.3. *Watershed's Classification Based on Group Analysis*

Figure 3 shows the dendrogram, resulting from the group analysis (GA) of the 31 watersheds. Five groups were identified based on their basic, relief, shape, and hydrology-type parameters, and considering a value of $R^2 = 0.84$. Group 1, with seven watersheds, has the largest amount of low values for the shape, relief, and linear-type parameters, as can be verified in Table 3. Group 2, with eight watersheds, presents the lowest average values of drainage density, sinuosity of the main channel, and general flow length. Group 3, also with eight watersheds, showed the highest values of elongation ratio, drainage density, mean slope of the main channel, and general flow length. Group 4, with four watersheds, has the highest values in maximum and minimum height, mean slope of the watershed, mean distance, unit shape factor, resistance number, and drainage intensity. In this group, the lowest values correspond to the elongation ratio. Group 5, also with four watersheds, presents the largest amount of high values for the basic, shape, relief, and linear-type parameters. The multivariate analysis of variance (MANOVA) showed significant differences among the groups of watersheds, showing a value of Wilks' lambda equal to 0.0025, with a value of $p < 0.0001$.

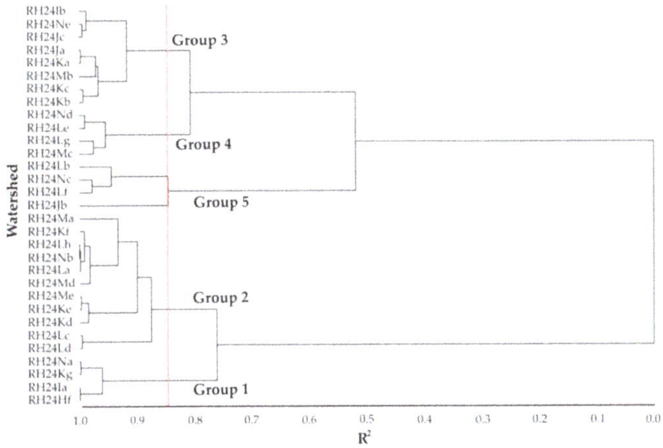

Figure 3. Dendrogram classifying 31 watersheds by group analysis. Conchos River Basin, Chihuahua, Mexico. The red line was drawn to define the groups.

Table 3. Average values of the geomorphometric parameters by group.

G_{id}	Lb^2	*Lc*	*Li*	*Lu*	Hmin	Hmax	*Cc*	*Re*	*Rf*	*Ia*	*J*	*tgα*	*Dd*
1	46.12	31.31	371.43	1038.50	1495	995	3.22	0.74	20.03	1.69	6.43	0.49	1.97
2	80.16	65.56	1820.63	2660.91	2489	1415	3.70	0.91	23.28	3.11	13.62	0.79	1.97
3	102.47	75.16	2765.79	6206.70	2294	1060	3.95	1.99	34.73	2.41	12.07	1.48	2.59
4	146.56	88.94	5556.08	6201.08	2970	1545	3.86	0.40	30.30	3.10	21.45	1.17	2.36
5	192.52	99.49	8176.13	12,155.08	2680	1230	4.07	0.44	51.66	2.20	17.31	2.45	2.58
G_{id}	*j*	*a*	*TcK*	*Scp*	*Qp*	*T*	*Ru*	*Fu*	*Rn*	*Rh*	*Rr*	*Lo*	*Di*
1	1.42	1.92	11.74	1.51	194.51	8.83	1.24	2.56	2673.67	67.01	13.48	0.99	1.30
2	2.44	2.17	18.48	1.37	313.28	16.48	1.69	3.03	4835.28	47.81	11.26	0.98	1.51
3	3.70	2.18	22.19	1.69	417.90	27.78	1.50	3.86	5915.17	34.49	6.97	1.30	1.51
4	0.98	2.87	31.18	1.68	433.61	37.17	1.73	4.87	7040.26	34.58	8.63	1.18	2.04
5	0.82	2.77	42.44	1.94	589.36	49.69	1.44	4.99	6806.71	28.86	5.50	1.29	1.95

G_{id} = Group identification, 1 = Group 1 (Very low erosion susceptibility), 2 = Group 2 (Low erosion susceptibility), 3 = Group 3 (Moderate erosion susceptibility), 4 = Group 4 (High erosion susceptibility), 5 = Group 5 (Very high erosion susceptibility), *Lc* = Length of main channel, Lb^2 = Length of watershed, *Li* = Length of contour lines, *Lu* = Length of channels, *Hmin* = Minimum height, *Hmax* = Maximum height, *Cc* = Gravelius compactness coefficient, *Re* = Elongation elation, *Rf* = Form factor, *Ia* = Elongation index, *J* = Mean slope of watershed, *tgα* = Mass coefficient, *Dd* = Drainage density, *j* = Mean slope of main channel, *a* = Medium distance, *TcK* = Kirpich concentration time, *Scp* = Sinuosity of main channel, *Qp* = Average peak flow, *T* = Texture ratio, *Ru* = Unitary shape factor, *Fu* = River frequency, *Rn* = Resistance number, *Rh* = Relief ratio, *Rr* = Relative relief, *Lo* = Length of general flow, *Di* = Drainage intensity.

The geospatial distribution of the groups is shown in Figure 4. Group 1 shows a homogeneous pattern in its distribution, which is concentrated in the central part of the study area. In contrast, Group 2 shows a dispersed distribution, mainly at the edges of the CRB. Group 3 is distributed in the northern, central, and southern parts, and is represented by small clusters of two or three watersheds. Group 4 corresponds to watersheds spatially dispersed over the basin. The watersheds of these groups are isolated from the other watersheds of the same group. Group 5 shows watersheds clustered in the southern part of the basin, except for the watershed RH24Jb, which is located in the northern region. The GA showed a clustered geospatial pattern for Groups 1, 3, and 5, who share characteristics in their parameters and space.

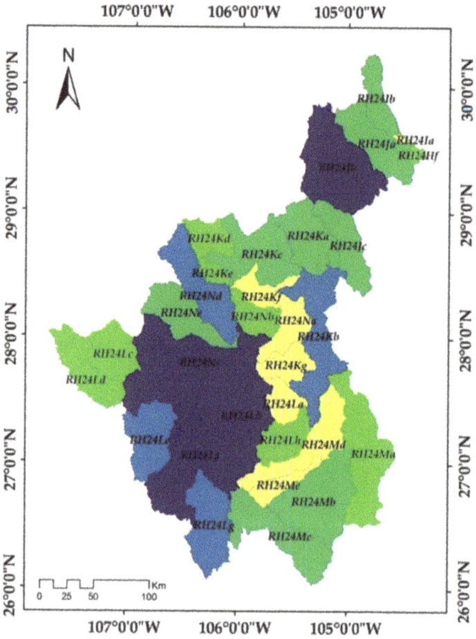

Figure 4. Geospatial distribution of watershed groups classified by group analysis. Very High (●), High (●), Moderate (●), Low (●), Very Low (●). Conchos River Basin, Chihuahua, Mexico.

3.4. Watershed's Classification Based on the Compound Parameter (Cp)

Considering the 26 geomorphometric parameters selected after the correlation analysis was performed, the value of the compound parameter (*Cp*) was calculated for the 31 watersheds of the CRB (Figure 5a). The watershed RH24Lg received the highest priority (1), followed by the watershed RH24Le (2). The watershed with the lowest priority (31) was watershed RH24Kg. A high priority is an indicator of a high degree of erosion susceptibility in the watershed.

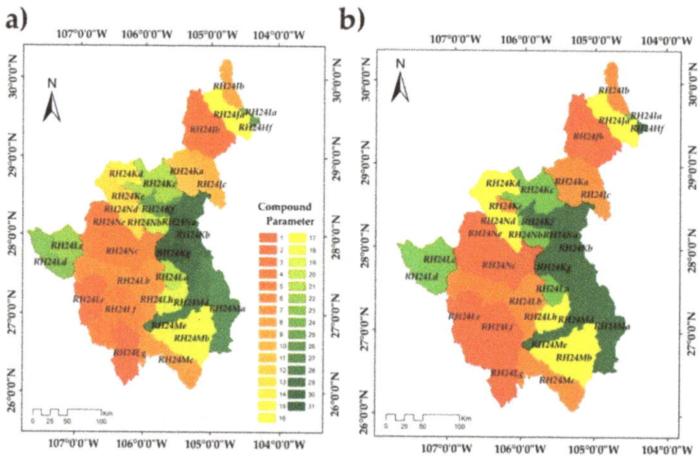

Figure 5. (**a**) Watersheds and their compound parameter (*Cp*). (**b**) Geospatial distribution of watershed groups by *Cp* reclassification. Very High 1 (●), High 2 (●), Moderate 3 (●), Low 4 (●), Very Low 5 (●). Conchos River Basin, Chihuahua, Mexico.

The resulting *Cp* map (Figure 5a) was reclassified into the following five categories: Very High, High, Moderate, Low, and Very Low (Figure 5b). The spatial distribution of the groups was reclassified by natural breaks [71]. The watersheds classified as Very High show a homogeneous pattern in their distribution in the southwestern part of the watershed. Meanwhile, the watershed RH24Jb is isolated in the northwestern part. The High, Moderate, and Low classes show a dispersed distribution, with at least two of their watersheds clustered in space. The Very Low class shows a homogeneous distribution in space in the southeastern part of the study area, with only one dispersed watershed (RH24Hf).

3.5. Comparison of the Classification Methods

Regarding the GA classification method, the results from the ANOVA analyses performed on 14 geomorphometric parameters, out of 26, detected significant differences among the groups defined by the method. In the case of the *Cp* classification method, the ANOVA analysis of only two parameters detected significant differences among the groups defined by this classification method (Table 4).

Table 4. *p*-values resulting from the ANOVA analyses performed to detect differences among the groups defined by each classification method.

	Geomorphometric Parameters												
	Lc	*Lb*²	*Li*	*Lu*	*Hmax*	*Hmin*	*Cc*	*Re*	*Rf*	*Ia*	*J*	*tgα*	*Dd*
GA	0.0091	0.0865	<0.0001	<0.0001	0.1954	0.3315	0.4011	0.2675	0.0005	0.4273	0.0750	<0.0001	0.0224
Cp	0.3711	0.3576	0.3207	0.1462	0.7602	0.1029	0.4368	0.4983	0.3214	0.8336	0.5236	0.1225	0.1563
	j	*a*	*TcK*	*Scp*	*Qp*	*T*	*Ru*	*Fu*	*Rn*	*Rh*	*Rr*	*Lo*	*Di*
GA	0.3316	0.7712	0.0356	0.5116	<.0001	<.0001	0.5327	0.0069	0.0004	0.1205	0.0086	0.0224	0.0265
Cp	0.5902	0.6454	0.4377	0.2158	0.2287	0.1442	0.8277	0.0174	0.1626	0.4913	0.1979	0.1563	0.0302

GA = Group analysis, *Cp* = Compound parameter, *Lc* = Length of main channel, Lb^2 = Length of watershed, *Li* = Length of contour lines, *Lu* = Length of channels, *Hmin* = Minimum height, *Hmax* = Maximum height, *Cc* = Gravelius compactness coefficient, *Re* = Elongation elation, *Rf* = Form factor, *Ia* = Elongation index, *J* = Mean slope of watershed, *tgα* = Mass coefficient, *Dd* = Drainage density, *j* = mean slope of main channel, *a* = medium distance, *TcK* = Kirpich concentration time, *Scp* = Sinuosity of main channel, *Qp* = Average peak flow, *T* = Texture ratio, *Ru* = Unitary shape factor, *Fu* = River frequency, *Rn* = Resistance number, *Rh* = Relief ratio, *Rr* = Relative relief, *Lo* = Length of general flow, *Di* = Drainage intensity.

4. Discussion

The prioritization of watersheds, based on susceptibility to erosion, has been carried out in different regions of the world [72,73], using different prioritization methods [74]. This study contributes to the lack of knowledge regarding the susceptibility to erosion in northern Mexico. This was assessed by implementing two methods for prioritization based on the analysis of a set of 33 parameters, which differ from other studies [5,7,75]. The inclusion of several geomorphometric parameters and their relationships within several connected watersheds enriched the study of their erosion susceptibility. In this sense, multivariate techniques have proved to be appropriate methods for establishing priorities, reducing the dimensionality of the dataset by losing the least amount of information [76].

This study integrated a multivariate analysis of several geomorphometric parameters that served to identify those watersheds, which may be prone to erosion. That was possible by evaluating the behavior of such geomorphometric parameters and representing them in a geospatial basis [77,78]. Their relationships provided significant information about the main sources of variability among the studied watersheds [74]. Previous research studies have reported that topography, geomorphology, and land use/land cover are the most important factors in the watershed susceptibility to erosion [79–82]. In this study, the factors with the greatest influence on the hydrological behavior of watersheds and their erosion susceptibility were the average peak flow and the all channel lengths, as it has also been found in previous studies [6,83].

The PCA is considered a statistical exploratory technique, whose results have helped explain the distribution of environmental attributes [69]. Results from the PCA were useful to identify the sources of variance, which were mainly represented by the dominant parameters influencing the data structure.

Then, the basin's hydrological configuration was explained by those geomorphometric parameters explaining the greatest portion of the variance among the watersheds. The PCA results from this study are consistent with the observations made by Meshram and Sharma [32] and Farhan et al. [33].

From the PCA analysis, PC1 and PC2 are mainly influenced by linear geomorphometric parameters. Some of the linear parameters with an influence on PC1 are the average peak flow. This is shown in Figure 3, where the lowest PC1 coefficients correspond to the watersheds with the lowest mean slope values of the small channels. Regarding PC2, drainage density is one of the linear parameters with an influence. Watersheds with low drainage density indicate the presence of permeable surface material, good vegetation cover, and low relief [84,85]. The map of PC2 (Figure 3) was highly influenced by drainage density, since the watersheds with low values of this parameter are located in the south–central part of the study area and grouped in such a map.

PC3 and PC4 are influenced by linear parameters such as mean distance and shape parameters such as elongation ratio, as well as topographic parameters such as minimum height and mean slope. These factors are associated with the main channel, relief, and slopes, among others. In Figure 3, the watersheds with the greatest heights and slopes correspond to the watersheds located in a mountainous zone, while the watersheds with the lowest elevations and slopes correspond to the arid and semi-arid zones of the state of Chihuahua [86]. Regarding PC5, it is mainly influenced by the elongation index (shape parameter) and the general flow length (linear parameter). The high values of the elongation index correspond to enlarged watersheds, which are related to high drainage densities. A watershed with a high drainage density implies a quick hydrological response to rainfall events, while non-enlarged watersheds correspond to fan-shaped watersheds, which are characterized by short channels [87].

Group analysis (GA) was one of the methodologies used in this study to group and then prioritize the watersheds. It was useful to relate watersheds that share the same characteristics based on their geomorphometric parameters. The groups delineated by the analysis have a unique combination in terms of their geomorphometric attributes [40]. The groups of watersheds follow a territorial pattern. Group 1 includes watersheds located in the zones with the least slopes, where the predominant economic activity is agriculture. Group 2 and 3 belong to watersheds located in transition zones because of their moderate slopes. Groups 4 and 5 are watersheds with rugged topography, with vegetation of shrublands and oak forests.

The compound parameter (*Cp*) was the second methodology employed to prioritize the watersheds. The value of *Cp* was calculated for each of the 31 watersheds composing the CRB (Figure 5a). Based on the value of *Cp*, watershed RH24Hf received the highest priority (1), followed by the watershed RH24Le (2). By comparing the results from GA and *Cp*, Group 4 was identically integrated by the two methodologies. This group is characterized by watersheds having the highest average values of maximum and minimum height, elongation ratio, elongation index, mean watershed slope, slope of the main channel and unit shape factor. The high values of these parameters correspond to watersheds with a high erosion susceptibility [88]. Conversely, Group 5 was formed by watersheds having the highest values of main channel length, watershed length, contour length, all channel lengths, Gravelius compactness coefficient, shape factor, massivity coefficient, mean distance, Kirpich concentration time, sinuosity of the main channel, average peak flow, texture ratio, river frequency and resistance number. This coincide with high values of *Cp*, which correspond to watersheds distributed in the southwestern zone of the study area and may have a low erosion susceptibility [61].

The two prioritization schemes used in this study gave similar results according to the spatial distribution of watershed groups. The prioritization of watersheds, obtained through GA and *Cp*, highlighted those watersheds with potential for the implementation of soil and water conservation practices. Based on the ANOVA analyses performed to statistically compare the GA and *Cp* methodologies, the former resulted in more effectively classifying the watersheds, since it permitted better differentiating the watershed groups.

Results from the GA show that erosion susceptibility is strongly related to linear parameters (surface hydrology) for southwestern watersheds, where steep slopes of both the watershed and the main channel influence soil erosion [89]. Watersheds RH24Lg, RH24Le, RH24Lf, RH24Mc, RH24Lb, RH24Nc, and RH24Ne have the steepest slopes, making them more prone to erosion [29].

One of the advantages of using the watershed as a territorial unit is the analysis of multiple geomorphometric parameters, which are related to the watershed's hydrological configuration, topography, and shape. Most of the watershed surface attributes depend on local topographic conditions [5]. In this study, the Basin's altitudinal gradient, a surface attribute, assists in exhibiting the contrasts among watersheds groups, while showing a homogeneous geographic distribution within them. The linear and shape-type parameters are important because of their influence on soil erosion.

The description and spatial grouping of the 31 watersheds through their 26 parameters using multivariate techniques proved to be useful to understand the main factors that control the variance in the CRB. Prioritization through the two types of grouping was also effective in detecting those watersheds susceptible to erosion. The proposed methodology for prioritizing watersheds on a geospatial basis is a feasible approach for identifying watersheds that are susceptible to erosion. However, prioritization with parameters that are based on shape, linear, and relief of the watersheds may not be sufficient. Thus, the incorporation of information regarding the activities on the territory of the CRB would help to improve the efficacy of the classification of watersheds based on their erosion susceptibility. Therefore, future research could include socioeconomic attributes that contribute to soil loss, such as agriculture [39]. Despite the limitations of this study, the contribution of this work represents an advance in the identification of the watersheds that are most susceptible to erosion in the CRB. This in turn contributes to land-use planning, which may help mitigate soil degradation processes.

5. Conclusions

The application of GA and *Cp* methodologies allowed integrating a large set of geomorphometric parameters, which served to classify watersheds according to their characteristics.

GA more effectively clustered the watersheds of the Conchos River Basin than *Cp*, since the groups formed by GA were more differentiated based on the analysis of the watersheds' geomorphometric parameters. The results of GA show that watersheds RH24Lf, RH24Lb, RH24Nc, and RH24Jb might be subjected to strong erosion processes, and are potential candidates to be subjected to soil conservation practices.

The present study demonstrates the usefulness of integrating GIS and multivariate techniques to prioritize watersheds based on their erosion susceptibility. Such an integration approach showed the spatial relationships of the different geomorphometric parameters analyzed. Although the present study permitted a definition of watershed groups according to the values of their geomorphometric parameters and their relation with erosion susceptibility, the integration of additional variables in the analysis may provide a more insightful classification and thus a more reliable watershed prioritization. Such variables could include land use/land cover, soil type, lithology, geomorphology, and socioeconomic activities, among others.

Author Contributions: Conceptualization, A.P.-A. and F.V.-G.; Data curation, G.V.-Q.; Formal analysis, J.A.P.-A., A.P.-A., M.C.V.-A., A.E.R.-R., and M.M.-S.; Investigation, M.C.V.-A.; Methodology, M.M.-S.; Project administration, A.P.-A.; Software, J.A.P.-A. and G.V.-Q.; Supervision, F.V.-G.; Writing—original draft, J.A.P.-A.; Writing—review and editing, M.C.V.-A., A.E.R.-R., and F.V.-G.

Funding: This research received no external funding.

Acknowledgments: We deeply thank Consejo Nacional de Ciencia y Tecnología (CONACYT-Mexico) for supporting the doctoral graduate studies of the first author of this manuscript.

Conflicts of Interest: The authors declare no conflict of interest.

Appendix A

Table A1. Values of 33 geomorphometric parameters used for the analysis of the 31 watersheds composing the Conchos River Basin.

W_{id}	*A*	*P*	Lb^2	*Lc*	*Li*	*Lu*	*Nu*	No_1	H_{min}	H_{max}	H_{med}	*Cc*	*Re*	*Rf*	*Ia*	*J*	*tgα*
RH24Hf	183.19	71.99	12.18	12.66	123.87	505.43	649	336	940	1100	780	3.00	1.25	14.47	0.87	6.76	0.19
RH24Ia	78.62	50.60	24.19	11.17	69.21	189.88	252	111	920	1040	800	3.22	0.41	7.04	1.59	8.80	0.09
RH24Ib	1209.37	259.75	140.20	34.15	2261.86	3915.19	5039	2367	1470	2140	800	4.21	0.28	35.41	0.96	18.70	0.82
RH24Ja	2344.14	297.44	132.14	61.84	3197.22	6430.37	7716	3854	1470	2140	800	3.47	0.41	37.91	1.63	13.64	1.59
RH24Jb	4527.91	373.79	121.30	56.88	5415.42	14,899.51	28,074	13,371	1580	2260	900	3.13	0.63	79.60	0.71	11.96	2.87
RH24Jc	1880.25	290.99	40.60	82.93	1877.94	5304.81	8784	4197	1610	2240	980	3.79	1.21	22.67	3.66	9.99	1.17
RH24Ka	2345.09	283.55	104.99	45.81	3530.78	6993.14	8771	4431	1620	2240	1000	3.30	0.52	51.19	0.89	15.06	1.45
RH24Kb	3447.12	546.62	5.47	106.05	2200.40	7188.10	10,601	5123	1645	2210	1080	5.25	12.12	32.51	3.26	6.38	2.10
RH24Kc	2928.21	433.41	141.09	83.92	3015.00	6069.72	9153	4357	1700	2320	1080	4.52	0.43	34.89	2.41	10.30	1.72
RH24Kd	1076.82	190.50	64.88	37.64	2111.47	2426.44	4431	2100	2020	2640	1400	3.28	0.57	28.61	1.32	19.61	0.53
RH24Ke	380.77	120.27	33.60	22.50	872.80	1023.82	1812	893	2000	2460	1540	3.48	0.66	16.92	1.33	22.92	0.19
RH24Kf	920.06	232.60	93.65	64.15	1030.60	1673.31	2311	1123	1690	2260	1120	4.33	0.37	14.34	4.47	11.20	0.54
RH24Kg	1167.54	193.88	43.29	30.35	717.78	1864.60	2331	1120	1540	1800	1280	3.20	0.89	38.47	0.79	6.15	0.76
RH24La	953.97	169.49	65.07	31.48	781.19	2467.01	3747	1843	1530	1760	1300	3.10	0.54	30.30	1.04	8.19	0.62
RH24Lb	5428.30	640.78	283.10	133.17	10,026.32	13,139.53	19,455	9786	2070	2820	1320	4.91	0.29	40.76	3.27	18.47	2.62
RH24Lc	1760.18	228.02	93.55	51.28	3884.50	2710.82	2872	1428	2320	2820	1820	3.07	0.51	34.32	1.49	22.07	0.76
RH24Ld	2017.67	324.63	16.10	75.21	4868.58	3104.04	3952	1912	2400	2980	1820	4.08	3.15	26.83	2.80	24.13	0.84
RH24Le	2267.52	286.04	145.51	73.49	5722.48	5511.96	12,338	6122	2200	2780	1620	3.39	0.37	30.85	2.38	25.24	1.03
RH24Lf	4296.09	482.75	158.28	97.01	9577.97	9829.22	22,334	11,021	2150	2860	1440	4.16	0.47	44.28	2.19	22.29	2.00
RH24Lg	2618.69	322.00	156.71	85.60	7070.69	6394.05	17,221	8428	2420	3280	1560	3.55	0.37	30.59	2.80	27.00	1.08
RH24Lh	1457.89	220.72	111.39	65.27	946.94	2997.72	5510	2692	1840	2360	1320	3.26	0.39	22.34	2.92	6.50	0.79
RH24Ma	3778.37	465.35	215.32	130.55	1552.45	5018.08	7351	3374	1680	2140	1220	4.27	0.32	28.94	4.51	4.11	2.25
RH24Mb	3808.53	347.58	126.67	92.62	3022.41	8366.91	15,137	7443	1970	2560	1380	3.18	0.55	41.12	2.25	7.94	1.93
RH24Mc	2854.62	435.90	117.91	78.48	5141.29	7178.68	12,804	6264	2330	3020	1640	4.60	0.51	36.37	2.16	18.01	1.23
RH24Md	2485.07	400.59	20.76	122.53	1646.85	3661.01	5738	2822	2070	2880	1260	4.53	2.71	20.28	6.04	6.63	1.20
RH24Me	804.95	211.92	95.40	68.67	1032.98	1770.22	2756	1355	2170	2820	1520	4.21	0.34	11.72	5.86	12.83	0.37
RH24Na	1430.41	232.19	104.82	71.05	574.84	1594.10	2164	1023	1580	2040	1120	3.46	0.41	20.13	3.53	4.02	0.91
RH24Nb	1115.90	186.37	72.02	51.87	1298.61	2417.57	3364	1666	1750	2260	1240	3.15	0.52	21.51	2.41	11.64	0.64
RH24Nc	4655.43	494.37	207.40	110.90	7684.79	10,752.04	23,253	11,600	2020	2780	1260	4.09	0.37	41.98	2.64	16.51	2.30
RH24Nd	2761.21	363.46	166.11	118.17	4289.87	5719.63	8250	4146	2080	2800	1360	3.90	0.36	23.37	5.06	15.54	1.33
RH24Ne	2077.04	315.23	128.60	93.97	3020.71	5385.38	9996	4964	1930	2500	1360	3.90	0.40	22.10	4.25	14.54	1.08

Table A1. *Cont.*

W_{id}	*A*	*P*	Lb^2	*Lc*	*Li*	*Lu*	*Nu*	No_1	H_{min}	H_{max}	H_{med}	*Cc*	*Re*	*Rf*	*Ia*	*J*	*tgα*
W_{id}	*Co*	*Dd*	*j*	*a*	*TcK*	*Scp*	*Qp*	*T*	*Ru*	*Fu*	*Rn*	*Rh*	*Rr*	*Lo*	*Rc*	*Di*	
RH24Hf	183.19	2.76	2.63	0.90	3.12	0.96	108.12	9.01	0.94	3.54	3034.96	86.90	15.28	1.38	0.44	1.28	
RH24Ia	78.62	2.42	0.99	2.73	7.69	2.17	69.53	4.98	1.26	3.21	2511.63	93.10	20.55	1.21	0.39	1.33	
RH24Ib	1209.37	3.24	0.96	4.03	30.21	4.11	289.58	19.40	0.98	4.17	6927.97	62.67	8.24	1.62	0.23	1.29	
RH24Ja	2344.14	2.74	1.01	2.73	28.21	2.14	409.07	25.94	1.28	3.29	5870.37	34.61	7.19	1.37	0.33	1.20	
RH24Jb	4527.91	3.29	1.12	1.80	25.41	2.13	576.83	75.11	0.85	6.20	7436.74	39.73	6.05	1.65	0.41	1.88	
RH24Jc	1880.25	2.82	3.10	0.94	7.39	0.49	364.60	30.19	1.91	4.67	6319.78	27.01	7.70	1.41	0.28	1.66	
RH24Ka	2345.09	2.98	1.18	2.17	22.28	2.29	409.16	30.93	0.95	3.74	6679.75	48.89	7.90	1.49	0.37	1.25	
RH24Kb	3447.12	2.09	20.67	0.09	0.76	0.05	500.29	19.39	1.81	3.08	4608.40	20.84	4.04	1.04	0.14	1.47	
RH24Kc	2928.21	2.07	0.88	2.61	31.35	1.68	459.45	21.12	1.55	3.13	4808.99	27.64	5.35	1.04	0.20	1.51	
RH24Kd	1076.82	2.25	1.91	1.98	12.78	1.72	272.55	23.26	1.15	4.11	5948.79	70.15	13.86	1.13	0.37	1.83	
RH24Ke	380.77	2.69	2.74	1.72	6.70	1.49	158.41	15.07	1.15	4.76	6614.51	109.32	20.45	1.34	0.33	1.77	
RH24Kf	920.06	1.82	1.22	3.09	20.17	1.46	251.06	9.94	2.11	2.51	4110.25	35.23	9.72	0.91	0.21	1.38	
RH24Kg	1167.54	1.60	1.20	1.27	11.19	1.43	284.31	12.02	0.89	2.00	2874.65	59.31	9.28	0.80	0.39	1.25	
RH24La	953.97	2.59	0.71	2.11	18.79	2.07	255.85	22.11	1.02	3.93	4551.47	55.91	10.38	1.29	0.42	1.52	
RH24Lb	5428.30	2.42	0.53	3.84	65.12	2.13	634.11	30.36	1.81	3.58	6825.98	21.18	4.40	1.21	0.17	1.48	
RH24Lc	1760.18	1.54	1.07	2.23	21.19	1.82	352.25	12.60	1.22	1.63	4343.02	54.99	12.37	0.77	0.43	1.06	
RH24Ld	2017.67	1.54	7.20	0.36	2.62	0.21	378.27	12.17	1.67	1.96	4584.51	39.62	9.18	0.77	0.24	1.27	
RH24Le	2267.52	2.43	0.80	3.06	33.33	1.98	402.04	43.13	1.54	5.44	6757.71	37.83	9.72	1.22	0.35	2.24	
RH24Lf	4296.09	2.29	0.90	2.41	33.98	1.63	561.22	46.26	1.48	5.20	6543.52	29.48	5.92	1.14	0.23	2.27	
RH24Lg	2618.69	2.44	1.10	3.06	31.20	1.83	433.42	53.48	1.67	6.58	8008.77	38.32	10.19	1.22	0.32	2.69	
RH24Lh	1457.89	2.06	0.93	2.92	25.53	1.71	319.25	24.96	1.71	3.78	4852.66	36.16	10.69	1.03	0.38	1.84	
RH24Ma	3778.37	1.33	0.43	3.50	57.31	1.65	524.84	15.80	2.12	1.95	2842.15	16.39	4.60	0.66	0.22	1.46	
RH24Mb	3808.53	2.20	0.93	2.05	28.21	1.37	527.02	43.55	1.50	3.97	5624.03	27.64	7.37	1.10	0.40	1.81	
RH24Mc	2854.62	2.51	1.17	2.21	24.45	1.50	453.38	29.37	1.47	4.49	7594.57	38.48	6.93	1.26	0.19	1.78	
RH24Md	2485.07	1.47	7.80	0.42	3.09	0.17	421.73	14.32	2.46	2.31	4242.82	23.50	7.19	0.74	0.19	1.57	
RH24Me	804.95	2.20	1.36	3.36	19.59	1.39	234.14	13.00	2.42	3.42	6201.64	41.06	13.31	1.10	0.23	1.56	
RH24Na	1430.41	1.11	0.88	2.77	24.95	1.48	316.10	9.32	1.88	1.51	2273.45	28.71	8.79	0.56	0.33	1.36	
RH24Nb	1115.90	2.17	1.42	2.16	15.54	1.39	277.67	18.05	1.55	3.01	4896.23	43.57	12.13	1.08	0.40	1.39	
RH24Nc	4655.43	2.31	0.73	3.04	45.23	1.87	585.26	47.04	1.63	4.99	6420.61	25.07	5.62	1.15	0.24	2.16	
RH24Nd	2761.21	2.07	0.87	3.16	35.74	1.41	445.58	22.70	2.25	2.99	5799.98	23.70	7.70	1.04	0.26	1.44	
RH24Ne	2077.04	2.59	0.89	2.82	29.09	1.37	384.04	31.71	2.06	4.81	6482.05	26.60	7.93	1.30	0.26	1.86	

W_{id} = Watershed identification, *A* = Area, *P* = Perimeter, Lb^2 = Length of watershed, *Lc* = Main channel length, *Li* = Length of contour lines, *Lu* = Length of channels, *Nu* = Number of channels, No_1 = Total number of channels of order 1 of the watershed, *Hmin* = Minimum height, *Hmax* = Maximum height, *Hmed* = Medium height, *Cc* = Gravelius compactness coefficient, *Re* = Elongation elation, *Rf* = Form factor, *Ia* = Elongation index, *J* = Medium slope of the watershed, *tgα* = Massive coefficient, *Co* = Orographic coefficient, *Dd* = Drainage density, *j* = Mean slope of main channel, *a* = Medium distance, *TcK* = Kirpich concentration time, *Scp* = Sinuosity of main channel, *Qp* = Average peak flow, *T* = Texture ratio, *Ru* = Unitary shape factor, *Fu* = River frequency, *Rn* = Resistance number, *Rh* = Relief ratio, *Rr* = Relative relief, *Lo* = Length of general flow, *Rc* = Circularity ratio, *Di* = Drainage intensity.

Table 2. Eigenvectors of the correlation matrix.

GP	PC1	PC2	PC3	PC4	PC5
Lc	0.2678	0.0223	−0.2753	−0.1330	0.0095
Lb^2	0.2345	−0.2721	−0.0327	−0.0059	0.0880
Li	0.2872	0.0343	0.0149	0.1182	−0.1631
Lu	**0.3014**	0.0344	0.1560	−0.1236	−0.0398
H_{max}	0.2170	−0.1034	−0.0543	0.3656	−0.1169
H_{min}	0.0664	−0.0941	−0.1183	**0.4424**	−0.3506
Cg	0.1308	−0.2425	0.0677	0.0028	0.2681
Re	−0.0340	−0.1992	**0.4039**	0.0072	0.0603
Rf	0.2180	0.1309	0.2002	−0.1898	−0.3142
Ia	0.0510	−0.3141	−0.1533	0.1096	**0.3442**
J	0.1269	0.1467	−0.0457	**0.4034**	−0.1466
tgα	0.2759	−0.0721	0.1407	−0.2389	−0.0850
Dd	0.0720	**0.3174**	0.1766	−0.0248	**0.3345**
j	−0.0484	−0.2118	0.4035	0.0593	0.0789
a	0.1271	0.1028	**−0.4225**	−0.0973	0.1850
TcK	0.2483	0.0091	−0.2925	−0.1728	−0.0094
Scp	0.0694	0.2821	−0.2272	−0.1822	0.0258
Qp	**0.3031**	−0.1104	0.0908	−0.0989	−0.1442
T	0.2642	0.1539	0.1430	0.0413	−0.0230
Ru	0.0641	**−0.3193**	−0.1582	0.1261	**0.3286**
Fu	0.1737	0.2478	0.1291	0.1961	0.2488
Rn	0.2305	0.1724	0.0819	0.2424	0.1779
Rh	−0.2207	0.2512	0.0133	0.1126	−0.0453
Rr	−0.2415	0.1661	−0.0986	0.2273	0.0458
Lo	0.0720	**0.3174**	0.1766	−0.0248	**0.3345**
Di	0.1911	0.0795	0.0241	0.2992	0.1179

GP = Geomorphometric parameter, PC1 = Principal component 1, PC2 = Principal component 2, PC3 = Principal component 3, PC4 = Principal component 4, PC5 = Principal component 5, *Lc* = Length of main channel, Lb^2 = Length of watershed, *Li* = Length of contour lines, *Lu* = Length of channels, *Hmin* = Minimum height, *Hmax* = Maximum height, *Cc* = Gravelius compactness coefficient, *Re* = Elongation ratio, *Rf* = Form factor, *Ia* = Elongation index, *J* = Mean slope of watershed, *tgα* = Mass coefficient, *Dd* = Drainage density, *j* = mean slope of the main channel, *a* = Medium distance, *TcK* = Kirpich concentration time, *Scp* = Sinuosity of the main channel, *Qp* = Average peak flow, *T* = Texture ratio, *Ru* = Unit shape factor, *Fu* = River frequency, *Rn* = Resistance number, *Rh* = Relief ratio, *Rr* = Relative relief, *Lo* = General flow length, *Di* = Drainage intensity. Bold letters indicate the dominant coefficient.

B.

Table 3. Abbreviations.

Abbreviation	Meaning
ANOVA	Analysis of variance
Cp	Compound parameter
CRB	Conchos River Basin
DTM	Digital terrain model
GIS	Geographic information systems
GA	Group analysis
INEGI *	National Institute of Statistics, Geography and Informatics
MANOVA	Multivariate analysis of variance
CONABIO *	National Commission for the Knowledge and Use of Biodiversity
PCA	Principal component analysis
PCs	Principal components

* Acronyms in Spanish.

References

1. Alexakis, D.D.; Hadjimitsis, D.G.; Agapiou, A. Integrated use of remote sensing, GIS and precipitation data for the assessment of soil erosion rate in the catchment area of "Yialias" in Cyprus. *Atmos. Res.* **2013**, *131*, 108–124. [CrossRef]
2. Ganasri, B.; Ramesh, H. Assessment of soil erosion by RUSLE model using remote sensing and GIS—A case study of Nethravathi Basin. *Geosci. Front.* **2016**, *7*, 953–961. [CrossRef]
3. Gajbhiye, S.; Sharma, S.K.; Awasthi, M.K. Application of Principal Components Analysis for Interpretation and Grouping of Water Quality Parameters. *Int. J. Hybrid Inf. Technol.* **2015**, *8*, 89–96. [CrossRef]
4. Gutiérrez, M.; Carreón-Hernández, E. Salinidad en el bajo Rio Conchos: Aportes y tendencias. *Terra Latinoam.* **2004**, *22*, 499–506.
5. Biswas, S.; Sudhakar, S.; Desai, V.R. Prioritisation of subwatersheds based on morphometric analysis of drainage basin: A remote sensing and gis approach. *J. Indian Soc. Remote Sens.* **1999**, *27*, 155–166. [CrossRef]
6. Adhami, M.; Sadeghi, S.H. Sub-watershed prioritization based on sediment yield using game theory. *J. Hydrol.* **2016**, *541*, 977–987. [CrossRef]
7. Malik, A.; Kumar, A.; Kushwaha, D.P.; Kisi, O.; Salih, S.Q.; Al-Ansari, N.; Yaseen, Z.M. The Implementation of a Hybrid Model for Hilly Sub-Watershed Prioritization Using Morphometric Variables: Case Study in India. *Water* **2019**, *11*, 1138. [CrossRef]
8. Robinne, F.-N.; Bladon, K.D.; Silins, U.; Emelko, M.B.; Flannigan, M.D.; Parisien, M.-A.; Wang, X.; Kienzle, S.W.; Dupont, D.P. A Regional-Scale Index for Assessing the Exposure of Drinking-Water Sources to Wildfires. *Forests* **2019**, *10*, 384. [CrossRef]
9. Khosravi, K.; Pham, B.T.; Chapi, K.; Shirzadi, A.; Shahabi, H.; Revhaug, I.; Prakash, I.; Bui, D.T. A comparative assessment of decision trees algorithms for flash flood susceptibility modeling at Haraz watershed, northern Iran. *Sci. Total Environ.* **2018**, *627*, 744–755. [CrossRef]
10. Chen, T.; Niu, R.Q.; Li, P.X.; Zhang, L.P.; Du, B. Regional soil erosion risk mapping using RUSLE, GIS, and remote sensing: A case study in Miyun Watershed, North China. *Environ. Earth Sci.* **2011**, *63*, 533–541. [CrossRef]
11. Rahaman, S.A.; Ajeez, S.A.; Aruchamy, S.; Jegankumar, R. Prioritization of Sub Watershed Based on Morphometric Characteristics Using Fuzzy Analytical Hierarchy Process and Geographical Information System—A Study of Kallar Watershed, Tamil Nadu. *Aquat. Procedia* **2015**, *4*, 1322–1330. [CrossRef]
12. Sharma, S.K.; Gajbhiye, S.; Nema, R.K.; Tignath, S. Assessing vulnerability to soil erosion of a watershed of tons River basin In Madhya Pradesh using Remote sensing and GIS. *Int. J. Environ. Res. Dev.* **2014**, *4*, 153–164.
13. Keesstra, S.; Pereira, P.; Novara, A.; Brevik, E.C.; Azorin-Molina, C.; Parras-Alcántara, L.; Jordán, A.; Cerdà, A. Effects of soil management techniques on soil water erosion in apricot orchards. *Sci. Total Environ.* **2016**, *551*, 357–366. [CrossRef] [PubMed]
14. Clarke, J.I. *Morphometry from Maps. Essays in Geomorphology*; Elsevier Publishing Co.: New York, NY, USA, 1966; pp. 235–274.
15. Horton, R.E. Erosional development of streams and their drainage basins: A hydrophysical approach to quantitative morphology. *Geol. Soc. Am. Bull.* **1945**, *56*, 275–370. [CrossRef]
16. Strahler, A.N. Quantitative geomorphology of drainage basins and channel networks. In *Handbook of Applied Hydrology*, 1st ed.; Chow, V.T., Ed.; McGraw-Hill: New York, NY, USA, 1964; Volume 1, pp. 4–39.
17. Singh, P.; Gupta, A.; Singh, M. Hydrological inferences from watershed analysis for water resource management using remote sensing and GIS techniques. *Egypt. J. Remote Sens. Space Sci.* **2014**, *17*, 111–121. [CrossRef]
18. Nautiyal, M.D. Morphometric analysis of a drainage basin using aerial photographs: A case study of Khairkuli basin District Deharadun. *J. Indian Soc. Remote Sens.* **1994**, *22*, 251–262. [CrossRef]
19. Shrimali, S.; Aggarwal, S.; Samra, J. Prioritizing erosion-prone areas in hills using remote sensing and GIS—A case study of the Sukhna Lake catchment, Northern India. *Int. J. Appl. Earth Obs. Geoinf.* **2001**, *3*, 54–60. [CrossRef]
20. Thakker, A.K.; Dhiman, S.P. Morphometric analysis and prioritization of miniwatersheds in Mohr watershed, Gujarat using Remote sensing and GIS techniques. *J. Indian Soc. Remote Sens.* **2007**, *35*, 313–321. [CrossRef]
21. Sharma, S.K.; Rajput, G.S.; Tignath, S.; Pandey, R.P. Morphometric analysis of a watershed using GIS. *J. Indian Water Res. Soc.* **2010**, *30*, 33–39.

22. Viramontes-Olivas, O.A.; Reyes-Gómez, V.M.; Escoboza-García, L.F.; Román-Calleros, J.A.; Pérez-Márquez, A.; Pinedo-Álvarez, C.; Sánchez-Fernández, P.B.; Miranda Ojeda, N. Hidrología de los suelos de las cuencas media y baja del Río Conchos, Chihuahua, México. *Rev. Latinoam. Recur. Nat.* **2008**, *4*, 31–45.
23. Tilahun, S.A.; Guzman, C.D.; Zegeye, A.D.; Ayana, E.K.; Collick, A.S.; Yitaferu, B.; Steenhuis, T.S. Spatial and temporal patterns of soil erosion in the semi-humid Ethiopian highlands: A case study of Debre Mawi watershed. In *Nile River Basin*, 1st ed.; Assefa, M., Melesse, W., Abtew Shimelis, G., Eds.; Springer: Cham, Switzerland, 2014; pp. 149–163.
24. Zavoianu, I. *Morphometry of Drainage Bassins*, 2nd ed.; Editura Academiei: Bucharest, Romania; Elsevier Science Publishers: Amsterdam, The Netherlands, 1985; pp. 9–235.
25. Schumm, S.A. Evolution of drainage systems and slopes in badlands at perth amboy, new jersey. *GSA Bull.* **1956**, *67*, 597. [CrossRef]
26. Saha, S.; Mukhopadhyay, S. Determination of Significant Geomorphic Parameters of Kunur River Basin through Principal Component Analysis. *J. Eng. Comput. Appl. Sci.* **2012**, *4*, 232–243.
27. Sharma, S.K.; Tignath, S.; Gajbhiye, S.; Patil, R. Application of principal component analysis in grouping geomorphic parameters of Uttela watershed for hydrological modeling. *Int. J. Remote Sens. Geosci.* **2013**, *2*, 63–70.
28. Gavit, B.K.; Purohit, R.C.; Singh, P.K.; Kothari, M.K.; Jain, H.K. Grouping of Significant Geomorphic Parameters using Multivariate Parameters using Multivariate. *Res. J. Recent Sci.* **2016**, *5*, 32–38.
29. Youssef, A.M.; Pradhan, B.; Hassan, A.M. Flash flood risk estimation along the St. Katherine road, southern Sinai, Egypt using GIS based morphometry and satellite imagery. *Environ. Earth Sci.* **2011**, *62*, 611–623. [CrossRef]
30. Makwana, J.; Tiwari, M.K. Prioritization of agricultural sub-watersheds in semiarid middle region of Gujarat using remote sensing and GIS. *Environ. Earth Sci.* **2016**, *75*, 137. [CrossRef]
31. Sharma, S.K.; Gajbhiye, S.; Tignath, S. Application of principal component analysis in grouping geomorphic parameters of a watershed for hydrological modeling. *Appl. Water Sci.* **2014**, *5*, 89–96. [CrossRef]
32. Meshram, S.G.; Sharma, S.K. Prioritization of watershed through morphometric parameters: A PCA-based approach. *Appl. Water. Sci.* **2017**, *7*, 1505–1519. [CrossRef]
33. Farhan, Y.; Anbar, A.; Al-Shaikh, N.; Mousa, R. Prioritization of semi-arid agricultural watershed using morphometric and principal component analysis, remote sensing, and GIS techniques, the Zerqa River Watershed, Northern Jordan. *Agric. Sci.* **2017**, *8*, 113–148. [CrossRef]
34. Bach, K.; Schawe, M.; Beck, S.; Gerold, G.; Gradstein, S.R.; Moraes, R. Vegetación, suelos y clima en los diferentes pisos altitudinales de un bosque montano de Yungas, Bolivia: Primeros resultados. *Ecol. Boliv.* **2003**, *38*, 3–14.
35. Sanderson, M.; Santini, M.; Valentini, R.; Pope, E. *Relationships between Forests and Weather. EC Directorate General of the Environment*; Met Office: Devon, UK, 2012; Available online: http://ec.europa.eu/environment/forests/pdf/EU_Forests_annex1.pdf (accessed on 29 July 2019).
36. Batayneh, A.; Zumlot, T. Multivariate statistical approach to geochemical methods in water quality factor identification; application to the shallow aquifer system of the Yarmouk basin of north Jordan. *Res. J. Environ. Earth Sci.* **2012**, *4*, 756–768.
37. Oketola, A.A.; Adekolurejo, S.M.; Osibanjo, O. Water Quality Assessment of River Ogun Using Multivariate Statistical Techniques. *J. Environ. Prot.* **2013**, *4*, 466–479. [CrossRef]
38. Castillo-Rodríguez, M.; López-Blanco, J.; Muñoz-Salinas, E. A geomorphologic GIS-multivariate analysis approach to delineate environmental units, a case study of La Malinche volcano (central México). *Appl. Geogr.* **2010**, *30*, 629–638. [CrossRef]
39. Miranda, J.; Andrade, E.; López-Suárez, A.; Ledesma, R.; Cahill, T.; Wakabayashi, P. A receptor model for atmospheric aerosols from a southwestern site in Mexico City. *Atmos. Environ.* **1996**, *30*, 3471–3479. [CrossRef]
40. Tritsch, I.; Sist, P.; Narvaes, I.D.S.; Mazzei, L.; Blanc, L.; Bourgoin, C.; Cornu, G.; Gond, V. Multiple Patterns of Forest Disturbance and Logging Shape Forest Landscapes in Paragominas, Brazil. *Forests* **2016**, *7*, 315. [CrossRef]
41. Prieto-Amparán, J.A.; Santellano-Estrada, E.; Villarreal-Guerrero, F.; Martinez-Salvador, M.; Pinedo-Alvarez, A.; Vázquez-Quintero, G.; Valles-Aragón, M.C.; Manjarrez-Domínguez, C. Spatial Analysis of Temperate Forest Structure: A Geostatistical Approach to Natural Forest Potential. *Forests* **2019**, *10*, 168. [CrossRef]

42. Altaf, S.; Meraj, G.; Ahmad Romshoo, S. Morphometry and land cover based multicriteria analysis for assessing the soil erosion susceptibility of the western Himalayan watershed. *Environ. Monit. Assess.* **2014**, *186*, 8391–8412. [CrossRef]
43. Montero-Martínez, M.J.; Ibáñez-Hernández, Ó.F. *La cuenca del río Conchos: Una mirada desde las ciencias ante el cambio climático*, 1st ed.; Instituto Mexicano de la Tecnología del Agua: Morelos, Mexico, 2017; p. 10.
44. Ordoñez, R. Modelación de la erosión hídrica en la Cuenca Alta del Río Conchos, Chihuahua. Master's Thesis, Universidad Autónoma de Chihuahua, Chih, Mexico, June 2017.
45. Mahlknecht, J.; Horst, A.; Aravena, R.; Hernández-Limón, G.; Hernández-Limón, G. Groundwater geochemistry of the Chihuahua City region in the Rio Conchos Basin (northern Mexico) and implications for water resources management. *Hydrol. Process.* **2008**, *22*, 4736–4751. [CrossRef]
46. Sánchez, A. 1944 Water Treaty between Mexico and the United States: Present situation and future potential. *Front. Norte* **2006**, *18*, 125–144.
47. CONAGUA (Comisión Nacional del Agua). Comisión de cuenca del río Conchos. In *Diagnóstico Versión 1.0*; Gerencia Estatal Chihuahua: Chihuahua, México, 2001.
48. Núñez López, D.; Treviño Garza, E.J.; Reyes Gómez, V.M.; Muñoz Robles, C.A.; Aguirre Calderón, O.A.; Jiménez Pérez, J. Uso de modelos de regresión para interpolar espacialmente la precipitación media mensual en la cuenca del río Conchos. *Rev. Mex. Cienc. Agrícolas* **2014**, *5*, 201–213. [CrossRef]
49. Viramontes-Olivas, O.A.; Escoboza-García, L.F.; Pinedo-Alvarez, C.; Pinedo-Alvarez, A.; Reyes Gómez, V.; Román Calleros, J.; Pérez-Márquez, A. Morfometría de la cuenca del río San Pedro, Conchos, Chihuahua. *Tecnociencia Chihuah.* **2008**, *1*, 21–31.
50. Aboites-Aguilar, L. *Hacia una historia del río Conchos*; CIESAS, Juárez: Chihuahua, México, 2002.
51. INEGI (Instituto Nacional de Estadística, Geografía e Informática). *Guía para la Interpretación de la Cartografía uso del Suelo y Vegetación, Escala 1:250,000 Serie V*; INEGI: Aguascalientes, México, 2014; p. 30. Available online: https://www.inegi.org.mx/contenidos/temas/mapas/usosuelo/metadatos/guia_interusosuelov.pdf (accessed on 31 July 2019).
52. CONABIO (Comisión Nacional para el Conocimiento y Uso de la Biodiversidad). Portal de Geoinformación. Available online: http://www.conabio.gob.mx/informacion/gis/ (accessed on 29 July 2019).
53. SIATL. Simulador de Flujos de Agua de Cuencas Hidrográficas. Available online: http://antares.inegi.org.mx/analisis/red_hidro/siatl/ (accessed on 29 July 2019).
54. INEGI (Instituto Nacional de Estadística, Geografía e Informática). Sistema de descarga del Continuo de Elevaciones Mexicano. Available online: https://www.inegi.org.mx/app/geo2/elevacionesmex/ (accessed on 29 July 2019).
55. Hamdy, O.; Zhao, S.; Salheen, M.A.; Eid, Y.Y. Identifying the Risk Areas and Urban Growth by ArcGIS-Tools. *Geosciences* **2016**, *6*, 47. [CrossRef]
56. Miller, V.C. *A Quantitative Geomorphic Study of Drainage Basin Characteristics in the Clinch Mountain Area, Virginia and Tennessee*; Office of Naval Research (USA): Arlington, VA, USA, 1953; pp. 1–51.
57. Mueller, J.E. An introduction to the hydraulic and topographic sinuosity indexes 1. *Ann. Assoc. Am. Geogr.* **1968**, *58*, 371–385. [CrossRef]
58. Kirpich, Z.P. Time of concentration of small agricultural watersheds. *Civ. Eng.* **1940**, *10*, 362.
59. Şen, Z. Modified hydrograph method for arid regions. *Hydrol. Process.* **2008**, *22*, 356–365. [CrossRef]
60. Faniran, A. The Index of Drainage Intensity—A Provisional New Drainage Factor. *Aust. J. Sci.* **1968**, *31*, 328–330.
61. Patel, D.P.; Gajjar, C.A.; Srivastava, P.K. Prioritization of Malesari mini-watersheds through morphometric analysis: A remote sensing and GIS perspective. *Environ. Earth Sci.* **2013**, *69*, 2643–2656. [CrossRef]
62. Agarwal, K.K.; Bali, R.; Kumar, M.G.; Srivastava, P.; Singh, P.V. Active tectonics in and around Kimin-Ziro area, Lower Subansiri District, Arunachal Pradesh, NE India. *Z. Geomorphol.* **2009**, *53*, 109–120. [CrossRef]
63. Dillon, W.R.; Goldstein, M. *Multivariate Analysis. Methods and Applications*; John Wiley & Sons Inc.: New York, NY, USA, 1984; pp. 1–608.
64. Jolliffe, I.T. *Principal Component Analysis*, 2nd ed.; John Wiley & Sons Inc.: Charlottesville, VA, USA, 2002; pp. 150–166.
65. Yidana, S.M.; Banoeng-Yakubo, B.; Akabzaa, T.M. Analysis of groundwater quality using multivariate an spatial analyses in the Keta basin, Ghana. *J. Afr. Earth Sci.* **2010**, *58*, 220–234. [CrossRef]
66. Bro, R.; Smilde, A.K. Principal component analysis. *Anal. Methods* **2014**, *6*, 2812. [CrossRef]

67. Ward, J.H. Hierarchical Grouping to Optimize an Objective Function. *J. Am. Stat. Assoc.* **1963**, *58*, 236–244. [CrossRef]
68. Eder, B.K.; Davis, J.M.; Bloomfield, P. An Automated Classification Scheme Designed to Better Elucidate the Dependence of Ozone on Meteorology. *J. Appl. Meteorol.* **1994**, *33*, 1182–1199. [CrossRef]
69. Johnson, R.A.; Wichern, D.W. *Applied Multivariate Statistical Analysis*, 6th ed.; Pearson Prentice-Hall Inc.: Upper Saddle River, NJ, USA, 2007; pp. 296–301.
70. Nooka Ratnam, K. Check Dam positioning by prioritization ofmicro watershed using SYI model and morphometric analysis—Remote sensing and GIS perspective. *J. Indian Soc. Remote Sens.* **2005**, *33*, 25–38. [CrossRef]
71. Fajardo, J.; Lessmann, J.; Bonaccorso, E.; Devenish, C.; Munoz, J. Combined Use of Systematic Conservation Planning, Species Distribution Modelling, and Connectivity Analysis Reveals Severe Conservation Gaps in a Megadiverse Country (Peru). *PLoS ONE* **2014**, *9*, e114367. [CrossRef] [PubMed]
72. Paul, I.I.; Bayode, E.N. Watershed Characteristics and Their Implication for Hydrologic Response in the Upper Sokoto Basin, Nigeria. *J. Geogr. Geol.* **2012**, *4*, 147–155. [CrossRef]
73. Darabi, H.; Shahedi, K.; Solaimani, K.; Miryaghoubzadeh, M. Prioritization of subwatersheds based on flooding conditions using hydrological model, multivariate analysis and remote sensing technique. *Water Environ. J.* **2014**, *28*, 382–392. [CrossRef]
74. Rawat, K.S.; Tripathi, V.K.; Mishra, A.K. Sediment yield index mapping and prioritization of Madia subwatershed, Sagar District of Madhya Pradesh (India). *Arab. J. Geosci.* **2014**, *7*, 3131–3145. [CrossRef]
75. Suji, V.R.; Sheeja, R.V.; Karuppasamy, S.; Suji, V.R.; Sheeja, R.V.; Karuppasamy, S. Prioritization using Morphometric Analysis and Land use/Land Cover Parameters for Vazhichal Watershed using Remote Sensing and GIS Techniques. *Int. J. Innov. Res. Sci. Tech.* **2015**, *1*, 2349–6010.
76. Prieto-Amparán, J.A.; Rocha-Gutiérrez, B.A.; Ballinas-Casarrubias, M.D.L.; Valles-Aragón, M.C.; Peralta-Pérez, M.D.R.; Pinedo-Alvarez, A. Multivariate and Spatial Analysis of Physicochemical Parameters in an Irrigation District, Chihuahua, Mexico. *Water* **2018**, *10*, 1037. [CrossRef]
77. Singh, P.K.; Kumar, V.; Purohit, R.C.; Kothari, M.; Dashora, P.K. Application of principal component analysis in grouping geomorphic parameters for hydrologic modeling. *Water Resour. Manag.* **2009**, *23*, 325. [CrossRef]
78. Yunus, A.P.; Oguchi, T.; Hayakawa, Y.S. Morphometric Analysis of Drainage Basins in the Western Arabian Peninsula Using Multivariate Statistics. *Int. J. Geosci.* **2014**, *5*, 527–539. [CrossRef]
79. Adediran, A.O.; Parcharidis, I.; Poscolieri, M.; Pavlopoulos, K. Computer-assisted discrimination of morphological units on north-central Crete (Greece) by applying multivariate statistics to local relief gradients. *Geomorphology* **2004**, *58*, 357–370. [CrossRef]
80. Javed, A.; Khanday, M.Y.; Ahmed, R. Prioritization of sub-watersheds based on morphometric and land use analysis using remote sensing and GIS techniques. *J. Indian Soc. Remot* **2009**, *37*, 261–274. [CrossRef]
81. Kompani-Zare, M.; Soufi, M.; Hamzehzarghani, H.; Dehghani, M. The effect of some watershed, soil characteristics and morphometric factors on the relationship between the gully volume and length in Fars Province, Iran. *Catena* **2011**, *86*, 150–159. [CrossRef]
82. Welde, K. Identification and prioritization of subwatersheds for land and water management in Tekeze dam watershed, Northern Ethiopia. *Int. Soil Water Conserv. Res.* **2016**, *4*, 30–38. [CrossRef]
83. Subyani, A.M.; Qari, M.H.; Matsah, M.I. Digital elevation model and multivariate statistical analysis of morphometric parameters of some wadis, western Saudi Arabia. *Arab J. Geosci.* **2012**, *5*, 147–157. [CrossRef]
84. Luo, W. Quantifying groundwater-sapping landforms with a hypsometric technique. *J. Geophys. Res. Space Phys.* **2000**, *105*, 1685–1694. [CrossRef]
85. Harlin, J.M.; Wijeyawickrema, C. Irrigation and groundwater depletion in caddo county, oklahoma. *JAWRA J. Am. Water Resour. Assoc.* **1985**, *21*, 15–22. [CrossRef]
86. CONABIO (Comisión Nacional para el Conocimiento y Uso de la Biodiversidad). *La biodiversidad en Chihuahua: Estudio de Estado*, 1st ed.; Comisión Nacional para el Conocimiento y Uso de la Biodiversidad: Méx, México, 2014; pp. 23–25.
87. Rai, P.K.; Mishra, V.N.; Mohan, K. A study of morphometric evaluation of the Son basin, India using geospatial approach. *Remote Sens. Appl. Soc. Environ.* **2017**, *7*, 9–20.

88. Rai, P.K.; Chaubey, P.K.; Mohan, K.; Singh, P. Geoinformatics for assessing the inferences of quantitative drainage morphometry of the Narmada Basin in India. *Appl. Geomat.* **2017**, *9*, 167–189. [CrossRef]
89. Tucker, G.E.; Bras, R.L. Hillslope processes, drainage density, and landscape morphology. *Water Resour. Res.* **1998**, *34*, 2751–2764. [CrossRef]

sustainability

MDPI

Article

Water Quality Sustainability Evaluation under Uncertainty: A Multi-Scenario Analysis Based on Bayesian Networks

Anna Sperotto [1,2], Josè Luis Molina [3], Silvia Torresan [1], Andrea Critto [1,2,*], Manuel Pulido-Velazquez [4] and Antonio Marcomini [1,2]

1 Fondazione Centro Euro-Mediterraneo sui Cambiamenti Climatici (Fondazione CMCC), c/o via Augusto Imperatore 16, 73100 Lecce, Italy
2 Department of Environmental Sciences, Informatics and Statistics, University Ca' Foscari Venice, 30123 Venezia, Italy
3 High Polytechnic School of Engineering, University of Salamanca, Av. de los Hornos Caleros, 50, 05003 Ávila, Spain
4 Research Institute of Water and Environmental Engineering (IIAMA), Universitat Politècnica de València, 46022 València, Spain
* Correspondence: critto@unive.it; Tel.: +39-041-234-8975

Received: 1 August 2019; Accepted: 23 August 2019; Published: 31 August 2019

Abstract: With increasing evidence of climate change affecting the quality of water resources, there is the need to assess the potential impacts of future climate change scenarios on water systems to ensure their long-term sustainability. The study assesses the uncertainty in the hydrological responses of the Zero river basin (northern Italy) generated by the adoption of an ensemble of climate projections from 10 different combinations of a global climate model (GCM)–regional climate model (RCM) under two emission scenarios (representative concentration pathways (RCPs) 4.5 and 8.5). Bayesian networks (BNs) are used to analyze the projected changes in nutrient loadings (NO_3, NH_4, PO_4) in mid- (2041–2070) and long-term (2071–2100) periods with respect to the baseline (1983–2012). BN outputs show good confidence that, across considered scenarios and periods, nutrient loadings will increase, especially during autumn and winter seasons. Most models agree in projecting a high probability of an increase in nutrient loadings with respect to current conditions. In summer and spring, instead, the large variability between different GCM–RCM results makes it impossible to identify a univocal direction of change. Results suggest that adaptive water resource planning should be based on multi-model ensemble approaches as they are particularly useful for narrowing the spectrum of plausible impacts and uncertainties on water resources.

Keywords: water quality; climate change; Bayesian networks; uncertainty; multi-models

1. Introduction

The maintenance of good water quality resources is essential to protect both ecosystems and human health, and they represent one of the main targets of both the European Water Framework Directive (2000/60/CE) and Sustainable Development Goals (i.e., SDG6) [1].

Changes in the global climate system are expected to have major consequences on the qualitative aspect of available water resources [2–7]; thus, assessing the impacts of future climate change scenarios on water systems is necessary to ensure a sustainable management of water supply for multiple purposes.

The inherent complexity, variability, and randomness of water systems, their interaction with socio-economic factors including the land use and population growth, and the high degree of uncertainty stemming from climate change make the assessment of climate change impacts on water resources particularly challenging [8].

Uncertainty plays a prominent role in climate change science and climate change impact science, with hydrology and water resources research in particular [8–11]. According to Parker et al. [12] and Hawkins et al. [13], it can be attributed to a number of reasons including (i) scenario uncertainty, arising from our limited understanding about the path of greenhouse gasses emissions and socio-economic development; (ii) internal climate variability, due to the inherent variability of the climate system components, processes, and their interaction; (iii) model uncertainty, caused by the different formulations used to represent climatic processes in climate and impact models.

A proper understanding of the type, sources, and effects of uncertainty is needed to achieve the goals of reliability and sustainability in water system management and planning under changing conditions [14,15]. Uncertainty quantification is vital to facilitate a risk-based approach to decision-making, where the range of possible futures is considered [16,17], and costs and benefits of adaptation are estimated accordingly. For this reason, uncertainties should be communicated as an inevitable component of each climate impact assessment study in a form which is also understandable by a non-scientific community to avoid misjudged information and to prevent overconfidence in impact projections [18].

A promising way to evaluate and deal with uncertainty is represented by ensemble modeling approaches [19]. Multi-model ensembles are commonly used to investigate structural model uncertainty, employing more than one climate model to perform multiple simulations and analyzing how climate change projections differ. The development of ensembles in both climate and impact studies is strongly encouraged by the Intergovernmental Panel on Climate Change (IPCC) since the Fourth Assessment Report (AR4, 2007) [20], which suggests the use of multiple climate models and scenarios to cover different sources of uncertainty [20]. The variability among ensemble components, in fact, can be used as a measure of the state of the knowledge. Furthermore, it could be useful to describe the confidence about the impact of climate change on the system modeled, supporting more robust decisions. In other words, if most ensemble members give comparable results, high confidence in projected climate change impacts is obtained, while, by contrast, if a large spread between components exist, there is less confidence in the forecasted impacts. Furthermore, it was shown that ensembles often give a more accurate prediction of future climate impacts than even the best individual model [21–23].

Relying on the extensive experience acquired in climate modeling, the use of ensembles was also transferred to the water resources field where attempts to build ensembles of impact models and scenarios (i.e., hydrological, water quality) are becoming increasingly common [24–26] to support water system management and adaptation.

In this respect, the paper proposes a Bayesian network (BN)-based approach to develop an ensemble of impact scenarios simulating the effect of different climate change projections on the quality of water of the Zero river basin (Italy). Accordingly, BNs are used as a modeling framework to evaluate the uncertainty due to global climate model (GCM)–regional climate model (RCM) structure and representative concentration pathways (RCPs), helping in determining and communicating the level of confidence of projected water quality alterations between baseline and future climate regimes.

BN outcomes (i.e., multiple impact scenarios) can be used to inform the spectrum of plausible effects of expected climate change on the Zero river basin and, thus, support the choice of effective adaptation strategies for a sustainable management of water resource quality at the local scale.

After a brief introduction to the study area (Section 1), this paper describes the methodology and input data employed (Section 2) and, finally, discusses the scenarios developed for the Zero river basin case study (Section 3), together with their uncertainty analysis.

Study Area

The Zero river basin (ZRB) (latitudes 45°28′ north (N)–45°48 N, longitudes 11°54′ east (E)–12°25′ E) (Figure 1) is located within the Venetian floodplain (northern Italy) and it is a sub-basin of the Venice lagoon watershed (Figure 1a), covering an area of 140 km^2. The Zero river (Figure 1b), which is 47 km long, together with the Dese rivers, provides the greatest contribution of freshwater (21% of the total) to the lagoon of Venice [27].

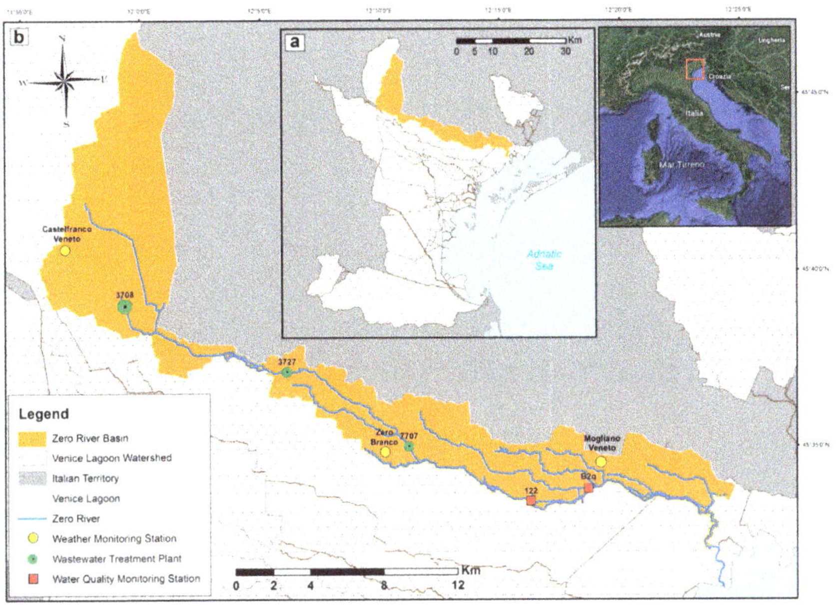

Figure 1. The Zero river basin case study (**a**) and input data location (**b**).

The basin features a Mediterranean climate but typical traits of more continental climates [28], with an average annual precipitation of around 1000 mm (period 2007–2012) and an average annual temperature of 14 °C (period 2004–2013). It is characterized by marked inter-annual climate variability, which can originate years climatologically very different from each other. The land use of the ZRB is mainly characterized by agricultural areas, representing 73% of the total surface, while the remaining surface of the basin is covered by artificial (24%), semi-natural, and forested areas (4%).

Agricultural areas are dominated by industrial crop typologies, including corn (45%) (i.e., *Zea mays* L.), soy (9%) (i.e., *Glycine max* L.), and autumn–winter cereals (13%) such as winter wheat (i.e., *Triticum aestivum* L.) and barley (i.e., *Hordeum vulgare* L.). A negligible percentage of the agricultural land is also used for the cultivation of beets and other permanent horticultural crops.

Artificial surfaces are mainly represented by housing areas (54%), industrial businesses (32%), and transportation and services (14%). Accordingly, several industrial and residential activities exist on the basin. Three wastewater treatment plans (i.e., Morgano, Zero-Branco, and Castelfranco Veneto) (Figure 1b) with capacities ranging from 2500 to 32,000 of population equivalents (PE) directly discharge into the Zero river.

The intensive agriculture, characterized by an elevated level of fertilization, and the dense urbanization represent significant pollution sources for the area and the main factor responsible for the excessive nutrient loadings in the reaching bodies of the Venice lagoon. Nutrient pollution is a major concern in the area considering the risk of eutrophication and toxic algae blooms which can threaten the good qualitative status of waters with consequent implications for environmental and human health [29]. Climate change, inducing extreme changes in temperature and precipitation trends, could exacerbate such nutrient pollution, altering those hydrological processes (e.g., runoff, river flow, water retention time, evapotranspiration) that regulate the mobilization of nutrients from land to inland and coastal water bodies.

2. Materials and Methods

To understand the complexity of the interactions between climate, hydrology and water quality parameters in the Zero river basin, we adopted the BN-based integrated approach proposed in Figure 2.

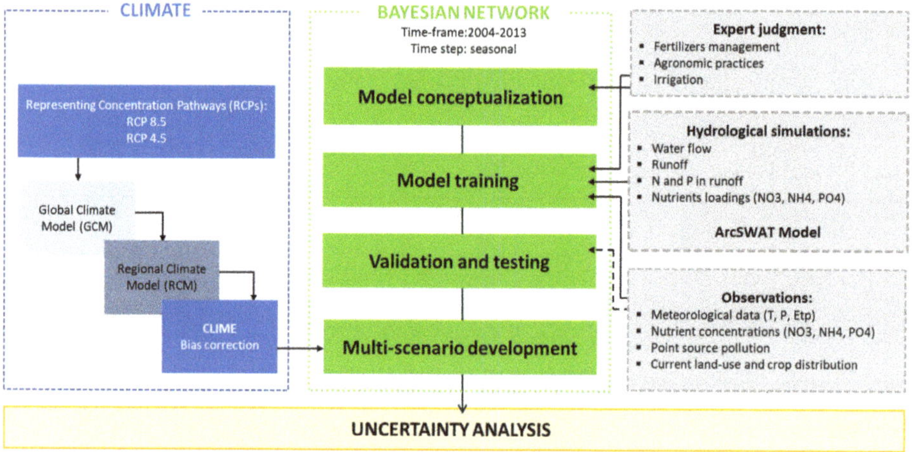

Figure 2. The Bayesian network (BN)-based integrated approach.

The core of the proposed approach is represented by a Bayesian network, which is employed as a modeling tool for the simulation of multiple nutrient loadings scenarios, and for the analysis of their uncertainty. Different data coming from hydrological models, historical observations, and expert judgment (Table A1, Appendix A) were structured and combined into a probabilistic form to develop and train the BN at different levels of implementation.

Qualitative information elicited from experts was used to develop the conceptual model of the network and to train socio-economic and agronomic variables (i.e., irrigation, fertilizer application) of the model for which quantitative data were not available. Observations regarding the main climatic parameters (i.e., precipitation, temperature, and evapotranspiration) and point-source pollution sources (i.e., wastewater treatment plants (WWTPs) and industrial discharges), together with Soil and Water Assessment Tool (SWAT) model simulations (i.e., runoff, river discharge (Q)) for the current conditions developed by Reference [30], were used for the training of the network for the period 2004–2013. Additional observed hydrologic data (i.e., river discharge (Q), nutrient concentrations (i.e., NO_3^-, NH_4^+, PO_4^{3-})) coming from the water quality monitoring station were used to evaluate the performance of the model under current conditions.

After the training and validation, an ensemble of climate change projections generated by coupling different combinations of Global Climate Models (GCMs) with regional climate models (RCMs) was used as input for scenarios analysis to assess the effect (and uncertainty) of future climate change on nutrient loadings.

2.1. Climate Change Projections

To assess the effect of climate change on nutrient loadings (i.e., NO_3^-, NH_4^+, PO_4^{3-}), changes in temperature and precipitation over future scenarios were selected as climate change indicators and used as input for the development of alternative nutrient loading scenarios using the BN model. The main aim of the study, however, was to capture uncertainties across a range of available GCM–RCM structures and representative concentration pathways (RCPs); thus, in order to represent the widest range of temperature and precipitation changes projected for the case study area, different climate change model outputs were considered (Table 1). This allowed including both the "worst" and "best" case scenarios in the BN, giving the users substantial flexibility in exploring and understanding the possible implications of climate change in the future. GCM–RCM combinations were selected among those available considering different kinds of features, including (i) their representativeness for the case study area and for the selected time periods (i.e., 2041–2070 and 2071–2100); (ii) their ability to perform at high spatial resolution; (iii) their availability in an open-source format.

Table 1. Global climate model (GCM)–regional climate model (RCM) projections selected and implemented in the Bayesian network (BN). SMHI—Swedish Meteorological and Hydrological Institute; DMI—Danish Meteorological Institute; CMCC—Centro Euro-Mediterraneo sui Cambiamenti Climatici.

No.	Global Climate Model (GCM)	Regional Climate Model (RCM)	Representative Concentration Pathways (RCPs)	Resolution	Time Range	Institute
1	HadGEM2-ES	RCA4	4.5, 8.5	12 km	1970–2099	SMHI
2	IPSL-CM5A-MR	RCA5	4.5, 8.6	12 km	1970–2100	SMHI
3	CNRM-CM5	RCA6	4.5, 8.7	12 km	1970–2100	SMHI
4	EC-EARTH	RCA7	4.5, 8.8	12 km	1970–2100	SMHI
5	MPI-ESM-LR	RCA8	4.5, 8.9	12 km	1970–2100	SMHI
6	CNRM-CM5	CCLM	4.5, 8.10	12 km	1950–2100	CLMcom
7	CMCC-CM	COSMO-CLM	4.5, 8.11	8 km	1976–2100	CMCC
8	HadGEM2-ES	RACMO22E	4.5, 8.12	12 km	1950–2099	KNMI
9	EC-EARTH	HIRHAM5	4.5, 8.13	12 km	1951–2100	DMI
10	EC-EARTH	RACMO22E	4.5, 8.14	12 km	1950–2100	KNMI

Ensembles of 10 climate change models were selected (Table 1), including the CMCC-CM/COSMO-CLM GCM–RCM and nine GCM–RCM model combinations from the EURO-CORDEX project [31].

The CMCC-CM global model [32] is a coupled atmosphere–ocean general circulation model, while the COSMO-CLM (CCLM) [33] is a high-resolution (between 1 and 50 km) climate regional model; both were developed by the Centro Euro-Mediterraneo sui Cambiamenti Climatici (CMCC), and, when coupled, they allow a spatial resolution of 8 km for the selected region.

EURO-CORDEX is the European branch of the CORDEX initiative sponsored by the World Climate Research Program (WRCP) with the aim of organizing an internationally coordinated framework to produce improved regional climate change projections for all land regions worldwide based on dynamical statistical downscaling models forced by multiple GCMs. CORDEX results were used as input for climate change impact and adaptation studies within the Fifth Assessment Report (AR5) of the Intergovernmental Panel on Climate Change (IPCC). In this study, nine climate change scenarios resulting from different combinations of GCMs and RCMs at 12-km spatial resolution were selected (Table 1). Different GCMs and RCMs were developed by different research groups including the Danish Meteorological Institute (DMI), the Swedish Meteorological and Hydrological Institute (SMHI), and the Met Office Hadley Centre (MOHC) (Table 1).

Based on the outputs of the selected GCM–RCMs (Table 1), different climate change scenarios were developed for the Zero river basin case study by extrapolating the mean temperature (°C) and the cumulative precipitation (mm) calculated on a monthly basis. Specifically, for each GCM–RCM, five different 30-year scenarios were developed for a control period (i.e., 1983–2012), a mid-term period (i.e., 2041–2070) and a long-term period (i.e., 2071–2100) under two different representative concentration pathways (i.e., RCP4.5 and RCP8.5). RCP4.5 represents the moderate emission scenario which predicts an increase in radiative forcing up to 4.5 $W \cdot m^{-2}$ by 2100 and a stabilization of the emissions (i.e., 650 ppm) shortly after 2100 [34], while RCP8.5 was chosen as representative of the extreme emission scenario, in which the greenhouse gas (GHG) emissions and concentrations increase considerably over the 21st century, leading to a radiative forcing of 8.5 $W \cdot m^{-2}$ by 2100 [35], thus describing a future without any specific climate mitigation target.

To make the outputs of GCM–RCMs suitable to be implemented at the spatial scale of impact assessment models, a bias correction was applied [30]. GCMs, in fact, have a spatial resolution too coarse for local-scale assessments and, for this reason, they are generally coupled with RCMs to consider the effects of orography, land–sea surface contrast, and land surface characteristics. However, RCMs also often show significant biases due to imperfect conceptualization, discretization, and spatial averaging within grid cells; therefore, a bias correction is required.

For the data used in this study, the linear scaling (LS) method was applied to correct the biases in the monthly values of temperature and precipitation based on observed ones. The LS method was applied using the software CLIME, a geographic information system (GIS) software for climate data analysis developed by the Regional Models and Geo-Hydrogeological Impacts (REMHI) division of CMCC, as extensively described in Reference [30]. Specifically, the method was implemented for all 10 climate scenarios for every weather station of the case study (Figure 1) using the rainfall and temperature observations for the period 1993–2012 as a correction factor. Once corrected, outputs of the GCM–RCMs for each of the 10 climate scenarios and for each of the three weather stations of the case study (i.e. Castelfranco Veneto, Zero-Branco, Mogliano Veneto) were elaborated to obtain suitable inputs for the BN model.

2.2. Bayesian Network Model

A BN model was employed to assess and compare the impacts of different climate change scenarios on nutrient loadings (i.e., NO_3^-, NH_4^+, PO_4^{3-}) in the transitional waters of the Zero river basin, thus generating an ensemble of impact scenarios supporting the identification of climate change effect on water quality.

The BN was implemented by building on a BN model previously developed and validated in a case study [36] which was extended to allow the incorporation of multiple GCM–RCM inputs. The BN for the Zero river basin was developed and run using the software HUGIN Expert, version 8 [37,38]. For additional details about the methodology and data used to develop the BN, please refer to Reference [36].

2.2.1. BN Development and Training

The BN structure was designed following the DPSIR (Drivers-Pressures-State-Impacts-Responses) framework, starting from the conceptual model described in Reference [36]. An influence (i.e., "box and arrow") diagram was developed including the most relevant systems variables (i.e., nodes), as well as the links between them (i.e., directed arcs), allowing the identification of the main cause–effect pathways between input variables, represented by climatic changes and land use, and output variables, represented by the increase in nutrient loadings (i.e., NO_3^-, NH_4^+, PO_4^{3-}) discharged by the Zero river basin into the Venice lagoon. Successively, the BN was trained, assigning states, prior information and conditional probabilities to all nodes of the network, translating the conceptual model into an operative probabilistic form.

The training was performed using a heterogeneous set of information for the period 2004–2013 (Tables A1 and A2, Appendix A) at seasonal time steps including historical observations, hydrological model simulations (i.e., SWAT), and expert opinions. Specifically, for nodes associated with climatic variables (i.e., temperature, precipitation, evapotranspiration), probabilities were learned directly from the frequency of observations of weather monitoring stations available in the case study (Figure 1). Nodes related with point pollution sources were trained using the nutrient loadings measured in the outflow from three different WWTPs (i.e., Morgano, Zero-Branco, Castelfranco Veneto) (Figure 1) located in the basin, summing up their respective contribution.

Probability distributions of hydrological variables (i.e., runoff, river flow, nutrients loadings, N and P in the runoff) were instead calculated based on the frequencies of results of hydrological simulations performed with the Soil and Water Assessment Tool (SWAT) [39]. Finally, nodes describing agronomic practices (i.e., water needs, irrigation, P and N fertilizer application) were trained through expert elicitation or by applying empirical equations due to the lack of quantitative information and experiences in the case study. An exhaustive description of assumptions and information used to train the BN can be found in Reference [36]. Figure 3 shows the configuration of the BN for the Zero river basin once states, prior information, and conditional probabilities of each node were parametrized.

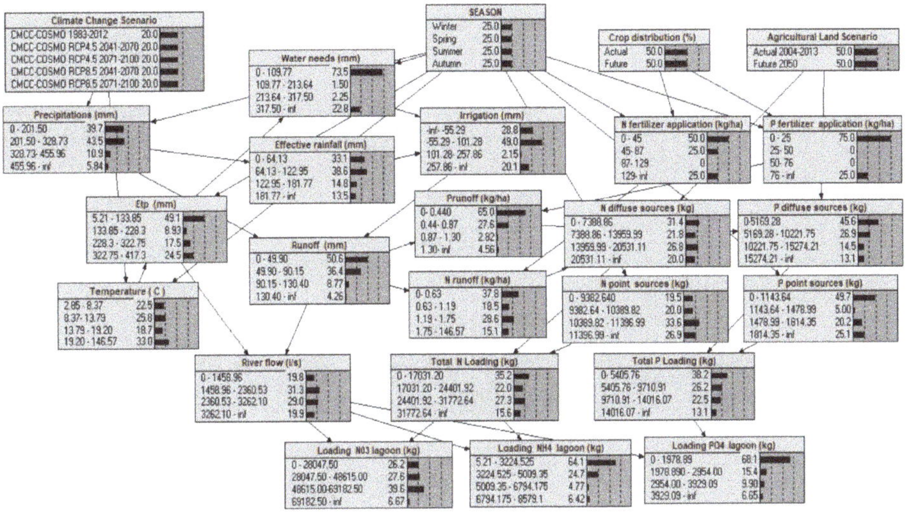

Figure 3. Configuration of the Bayesian network for the Zero river basin trained with the information for the period 2004–2013.

2.2.2. BN Evaluation

The developed BN was evaluated with the main aim to assess if it was able to correctly represent nutrient loadings and dynamics in the case study. Specifically, two main forms of model evaluation were performed, namely, model predictive accuracy and sensitivity analysis [36].

Predictive Accuracy

Predictive accuracy assessment was performed by comparing BN simulations with observations from water quality monitoring stations available for the case study. Specifically, observed nutrient loadings were calculated, multiplying the observed water flow (Q) and nutrient concentrations measured at two different water quality monitoring stations located in the case study (Figure 1) managed by ARPAV Servizio Acque Interne and the former MAV (Magistrato alle Acque di Venezia). The manual station 122 (45°33′ N and longitude 12°15′ E), provided seasonal data, and the automatic station B2q (45°34′ N and longitude 12°17′ E) provided daily data. Both stations were specifically identified and they are routinely used for (i) the assessment of the good environmental status of the Zero river according to the requirements of the Water Framework Directive (WFD); (ii) the assessment of the compliance with the maximum admissible load of nutrients discharged into the Venice lagoon from the drainage basin fixed by the national competent law (DM 09/02/1999).

For this reason, the two stations were considered particularly representative to measure the condition of the Zero river and its basin, thus providing reliable data for the evaluation of BN performance.

Specifically, loadings of nitrogen nitrate (NO_3^-) and ammonium (NH_4^+) were obtained using data from station B2q, while loadings of phosphate (PO_4^{3-}) were calculated using data from station 122. A consistent set of observations was available only from year 2007 to 2012 and, therefore, the evaluation was conducted only for this period.

For each output node (i.e., NO_3^-, NH_4^+, and PO_4^{3-} loadings), correctly classified instances (CCIs) were assessed as the percentage of cases correctly predicted divided by the total number of cases, providing the measure of how many instances the model predicted correctly when tested against known case outcomes (i.e., observations). Error rates, used as evaluation criteria, were then computed and depicted in confusion matrices as suggested in Reference [40].

Sensitivity Analysis

Another form of evaluating the developed model entails sensitivity analysis, which allows testing the sensitivity of model outcomes to variations of model parameters [41]. Sensitivity analysis in BNs can measure the sensitivity of outcome probabilities to changes in input nodes or other model parameters, such as changes in node type of state and their coarseness; therefore, it useful to detect the most relevant variables within the network. Sensitivity analysis was performed using two types of measures: entropy and Shannon's measure of mutual information [42].

The entropy (H(x)) of the probability distribution of a variable (x) expresses the measure of the associated uncertainty of the random process with a particular probability distribution (P(x)) [42]; it is calculated using the following function:

$$H(x) = -\sum P(x) \log P(x). \tag{1}$$

Reducing entropy by collecting information, in addition to the current knowledge about the variable x, is interpreted as reducing the uncertainty about the true state of x. Accordingly, the entropy function enables an assessment of the level of uncertainty/certainty about the state of the output node and of the additional information required to specify a particular alternative.

Entropy can be seen as a score of a variable richness (i.e., how much information is within the data for that particular variable) [43,44], and it was used to rank nodes from the most uncertain to the least uncertain, where the most uncertain variables are the least informative within the network.

In addition, the sensitivity of one node to multiple other nodes was evaluated using Shannon's measure of mutual information (MI) as follows:

$$MI\ (Y,X) = H(Y) - H(Y|X). \tag{2}$$

MI enables assessing the effect of collecting information about one variable (Y) on reducing the total uncertainty about variable X. When MI is equal to zero, the condition of one node does not affect the state of the other and, therefore, the nodes can be defined as mutually independent [43].

2.2.3. Scenarios and Uncertainty Analysis

The model developed as above was used in this study to perform scenario analysis, allowing the assessment of the relative change in outcome probabilities of nutrients under different climate change conditions (Section 2.1), thus obtaining an ensemble of multiple impact scenarios. For each GCM–RCM combination (Table 1) and climate change scenario (i.e., 2041–2070 and 2071–2100 under two different representative concentration pathways, RCP4.5 and RCP8.5), the probability distributions of temperature and precipitation were calculated based on the frequency in the respective model simulations (Section 2.1). The BN was then run, alternatively fixing the evidence of being in a particular scenario by assigning 100% probability to the related state in the "climate change scenario node", letting the information propagate through nodes linked by conditional probability (Figure 3) and calculating the change in the posteriori probabilities of output variables (i.e., NO_3^-, NH_4^+, PO_4^{3-} loadings).

Moreover, uncertainties in projected loadings due to the application of the ensemble of 10 GCM–RCM couples was performed by comparing outputs obtained with each of the different GCM–RCM combinations across scenarios and seasons. Specifically, the changes in the probability of each loading class in the mid-term (i.e., 2041–2070) and long-term (i.e., 2071–2100) simulated periods were compared against the respective baseline scenario (i.e., 1983–2012) for each combination of GCM–RCM.

3. Results

3.1. BN Evaluation

3.1.1. Accuracy

As described above, a data-based evaluation was performed to assess the ability of the model to correctly predict instances in an independent dataset. Accordingly, BN predictions were tested against observations from water quality monitoring stations available from ARPAV for the case study (Figure 1), generating confusion matrices representing the percentage of CCIs and, consequently, the error rates (Figure 4). Observations were available only for 2007–2012 and, therefore, the evaluation was conducted only for this period.

Loading NO3		Actual			
		State 1	State 2	State 3	State 4
Predicted	State 1	80.0	20.0	0.0	0.0
	State 2	20.0	80.0	0.0	0.0
	State 3	0.0	45.5	54.5	0.0
	State 4	0.0	0.0	0.0	100.0

Loading NH4		Actual			
		State 1	State 2	State 3	State 4
Predicted	State 1	66.7	5.6	22.2	5.6
	State 2	25.0	50.0	0.0	25.0
	State 3	0.0	0.0	100.0	0.0
	State 4	0.0	0.0	0.0	100.0

Loading PO4		Actual			
		State 1	State 2	State 3	State 4
Predicted	State 1	90.0	10.0	0.0	0.0
	State 2	33.3	66.7	0.0	0.0
	State 3	0.0	0.0	100.0	0.0
	State 4	0.0	0.0	0.0	100.0

(%) 0 20 40 60 80 100

Figure 4. Confusion matrices for output nodes of the BN model tested against the observed dataset (2007–2012). The cells lying on the leading diagonal of the matrices represent the correctly predicted instances, while those off the diagonal are incorrect predictions. Adapted from Sperotto et al., 2019 [35].

In addition, the expected values of the probability distributions of nutrient loadings (i.e., NO_3^-, NH_4^+, PO_4^{3-}) for observed data were compared with those obtained through the Bayesian network outputs (Figure 5).

Overall, the BN was able to reproduce the observed nutrient dynamics with loadings closely replicated for most seasons. The evaluation produced very good results for phosphate (PO_4^{3-}), while, for ammonium (NH_4^+) and nitrate (NO_3^-), the correlation between observed and predicted nutrient loadings was slightly worse. Indeed, overall, the BN was able to correctly classify 87.50% of instances for PO_4^{3-}, 63.64% for NH_4^+, and the 66.67% for NO_3^-, when tested against the observed dataset (Figure 4). The BN overpredicted the decrease in ammonium and nitrate loadings between spring and summer, while it underestimated the autumn loadings (Figure 5) for all three nutrient species (i.e., PO_4^{3-}, NH_4^+, NO_3^-) and the winter loadings of NH_4^+ and NO_3^-.

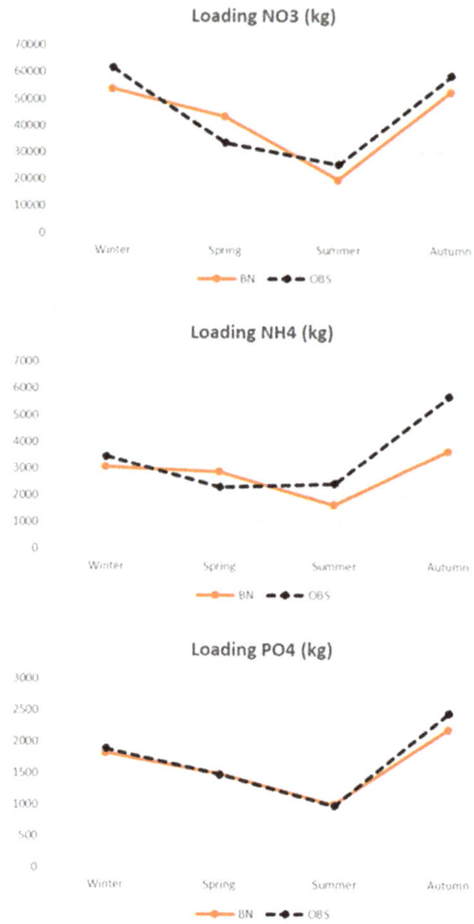

Figure 5. Expected values of the probability distributions of nutrient loadings (NO_3^-, NH_4^+, $P0_4^{3-}$) from observed data (black) and Bayesian network outputs (red) for the period 2007–2012.

3.1.2. Sensitivity Analysis

Entropy was calculated for each node of the network (Table 2), allowing us to rank all variables from the most uncertain to the least uncertain, where the most uncertain variables were those characterized by high entropy and were, thus, the least informative probability distributions.

Results showed that the variable characterized by the least informative probability distribution and a particularly high value of entropy was "effective rainfall" (3.26); however, variables characterized by intermediate values of uncertainty were nodes directly influencing "loading NO_3^-", including "N point sources" (1.36), "river flow" (1.34), "N runoff"(1.33), "N diffuse sources" (1.31), and "total N Loading" (1.31). These uncertainties propagated through the network and, as a result, "loading NO_3^-" was the output node characterized by the highest value of entropy (1.24), while, for others, the uncertainty was moderate (0.98 and 0.97).

Table 2. Node ranking according to entropy score.

Variable	Entropy H(*x*)
Effective rainfall	3.26
Season	1.39
Temperature	1.38
N point sources	1.36
River flow	1.34
N runoff	1.33
N diffuse sources	1.33
Total N Loading	1.31
Evapotranspiration	1.28
Loading NO_3^-	1.24
Irrigation	1.24
P point sources	1.17
Runoff	1.12
Precipitation	1.06
N fertilizer application	1.04
Loading NH_4^+	0.98
Loading PO_4^{3-}	0.97
P runoff	0.96
P diffuse sources	0.96
Water needs	0.56
P fertilizer application	0.56

Table 3 provides a ranking of the top five most influential variables on output nodes based on the mutual information analysis. The output nodes "NO_3^- loadings" and "NH_4^+ loadings" were both highly sensitive to "river flow" (MI = 0.54 and 0.37, respectively). "PO_4^{3-} loadings" resulted highly sensitive to "total P Loading" (M = 0.53), "P runoff" (M = 0.39), and "P diffuse sources" (M = 0.39).

Table 3. Summary of the mutual information (MI) analysis presenting the top five most influential variables on output nodes.

Sensitive Node	Node Affecting Sensitivity	MI
Loading NO_3^-	River flow	0.54
	Total N loading	0.30
	N diffuse sources	0.23
	N runoff	0.23
	Evapotranspiration	0.23
Loading NH_4^+	River flow	0.37
	Loading NO_3^-	0.18
	Total N loading	0.17
	Runoff	0.15
	N diffuse sources	0.11
Loading PO_4^{3-}	Total P loading	0.53
	P runoff	0.39
	P diffuse sources	0.39
	River flow	0.34
	Runoff	0.31

In general, hydrological variables, which in turn were strongly influenced by climatic ones (i.e., precipitation, temperature), were those most influential on other network variables. By contrast, variables related to agronomic practices and land use had a mild effect (MI < 0.1) on other variables with the exception of "N fertilizer application", which moderately affected "NO_3^- loadings" (MI = 0.22). In particular, point sources had negligible effects on all output nodes (MI < 0.04) with respect to diffuse sources.

3.2. Climate Change Scenarios for the Zero River Basin

Different climate change scenarios were developed for the Zero river basin case study by extrapolating the mean temperature (°C) and the cumulative precipitation (mm) calculated on a monthly basis, based on the outputs of the selected GCM–RCMs (Table 1). Specifically, for each GCM–RCM, five different 30-year scenarios were developed for a control period (i.e., 1983–2012), a mid-term period (i.e., 2041–2070), and a long-term period (i.e., 2071–2100) under two different representative concentration pathways (i.e., RCP4.5 and RCP8.5).

Figure 6 shows the variability of temperature for different time periods and RCPs across different climate change scenarios used to inform the BN. It is possible to observe that the temperature variability across the future projection was quite narrow.

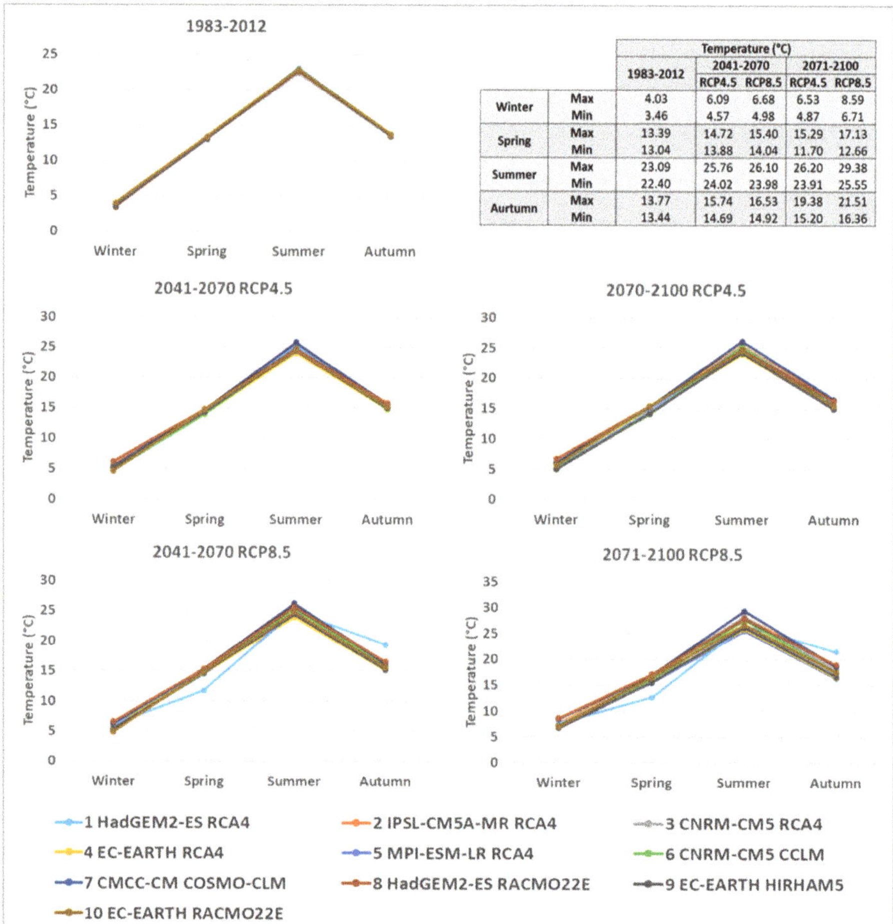

		Temperature (°C)				
		1983-2012	2041-2070		2071-2100	
			RCP4.5	RCP8.5	RCP4.5	RCP8.5
Winter	Max	4.03	6.09	6.68	6.53	8.59
	Min	3.46	4.57	4.98	4.87	6.71
Spring	Max	13.39	14.72	15.40	15.29	17.13
	Min	13.04	13.88	14.04	11.70	12.66
Summer	Max	23.09	25.76	26.10	26.20	29.38
	Min	22.40	24.02	23.98	23.91	25.55
Aurtumn	Max	13.77	15.74	16.53	19.38	21.51
	Min	13.44	14.69	14.92	15.20	16.36

Figure 6. Variability in mean seasonal temperature within the global climate model (GCM)–regional climate model (RCM) ensembles for the Zero river basin.

All climate scenarios agreed on projected temperature during the control period (i.e., 1983–2012). Greater variability, instead, was depicted for RCP8.5 where one model in particular (i.e., MPI-ESM-LR/RCA4, Model 5) of the ensemble projected lower temperatures in spring and higher temperatures in autumn. In general, all models predicted an increase in mean seasonal temperature with respect to baseline across the different considered scenarios (Figure A3, Appendix A). MPI-ESM-LR/RCA4 (Model 5) represented the only exception, predicting a decrease in temperature in spring for RCP8.5 (Figure A3, Appendix A). A greater increase in temperature with respect to baseline was predicted by RCP8.5 for the period 2071–2100.

By contrast, precipitation featured marked variability in all scenarios, as shown in Figure 7. All 10 GCM/RCMs of the ensemble generated quite similar statistics for the control period (i.e., 1983–2012) with a narrow range between maximum and minimum values for all seasons. By contrast, the variability increased consistently along the century, especially for RCP8.5. Greater variability can be seen in summer, autumn, and winter, where the range between maximum and minimum values projected by different GCM/RCMs became quite wide. However, while for winter and autumn most models agreed on an increase in the cumulative precipitation (Figure A4, Appendix A), for spring and summer, models gave the opposite results, making it impossible to agree on the direction of change (i.e., decrease or increase).

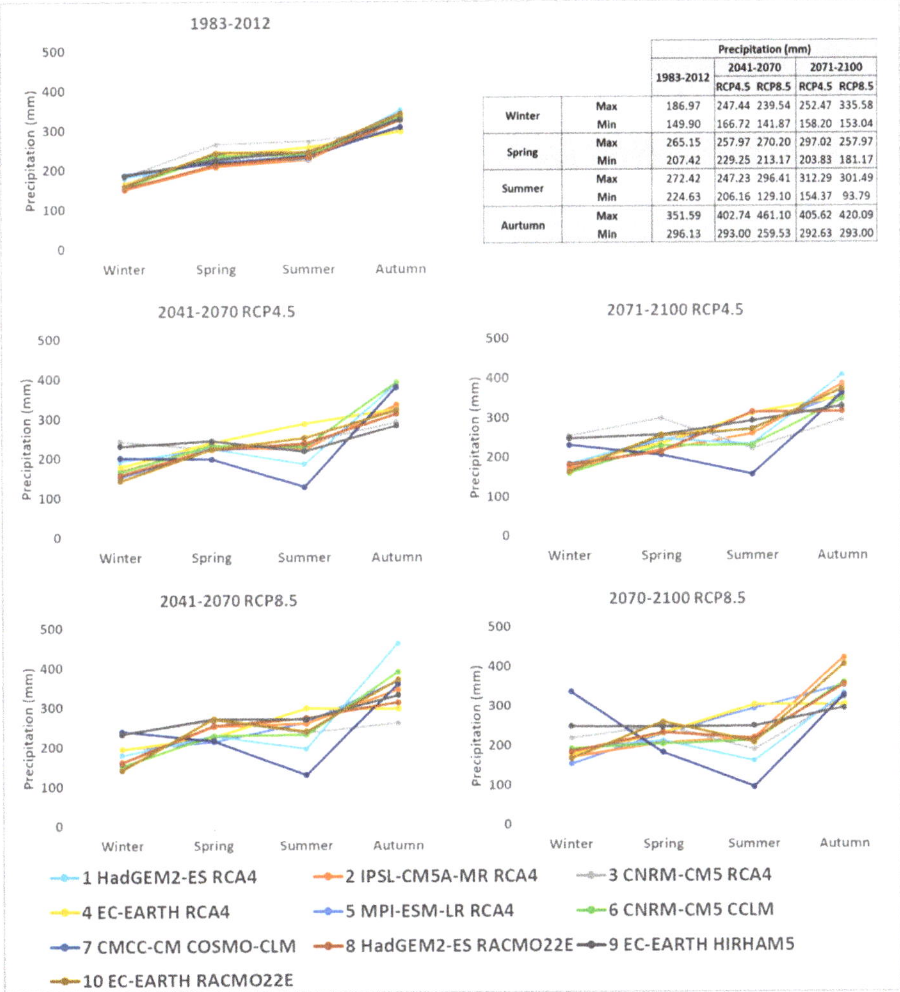

		Precipitation (mm)				
		1983-2012	2041-2070		2071-2100	
			RCP4.5	RCP8.5	RCP4.5	RCP8.5
Winter	Max	186.97	247.44	239.54	252.47	335.58
	Min	149.90	166.72	141.87	158.20	153.04
Spring	Max	265.15	257.97	270.20	297.02	257.97
	Min	207.42	229.25	213.17	203.83	181.17
Summer	Max	272.42	247.23	296.41	312.29	301.49
	Min	224.63	206.16	129.10	154.37	93.79
Aurtumn	Max	351.59	402.74	461.10	405.62	420.09
	Min	296.13	293.00	259.53	292.63	293.00

Figure 7. Variability in cumulative seasonal precipitation within the GCM/RCM ensembles for the Zero river basin.

3.3. Hydrological Responses to Climate Change

The BN model was run, alternatively fixing the probability distribution of precipitation and temperature according to the medium- and long-term projections (i.e., 2041–2070, 2071–2100) provided by the different available combinations of GCM–RCMs under two different representative concentration pathways (i.e., RCP4.5 and RCP8.5). Accordingly, the network was used to develop multiple impact scenarios linking the effect of future climate change projections on nutrient loadings. The developed scenarios represent the probability of different classes of nutrient loadings (i.e., low, medium, high, very high) calculated by the BN model as a result of changes in the probability distribution of input variables (i.e., temperature and precipitation).

Figure 8 gives a concise overview of the probabilistic results obtained through the BN for each season and scenario across the different GCM–RCM models considered (Table 1). Specifically, each triangular portion of the graph represents one of the different climate change scenarios considered (i.e., RCP4.5 2041–2070, RCP8.5 2041–2070, RCP4.5 2071–2100, RCP8.5 2071–2100), while, inside them, each slice represents the results of different GCM–RCMs arranged in a clockwise direction (i.e., from 1–10, in Table 1). Each slice, in turn, is divided into the four different classes of loadings with an amplitude corresponding to the value of the associate probability (i.e., from 0–100).

With regard to NO_3^- (Figure 8a), the impact scenarios reported that higher loadings will take place in autumn and winter, while the lowest loadings are predicted for summer. Across different models, in fact, in autumn, higher probabilities were associated with high (i.e., 48,615-69,182 kg/season, orange) and very high loading classes (i.e., >69,182 kg/season, red). The highest loading was predicted by the MPI-ESM-LR/RCA4 (Model 5) under the RCP8.5 2071–2100 scenario with 70% probability associated with the high loading class (Figure A5, Appendix A).

In summer, by contrast, a higher probability was associated with low (i.e., 0–28,047 kg/season, green) loading classes, with the CMCC-CM/COSMO-CLM (Model 7) predicting the highest probability (77%) under the long-term RCP8.5 scenario (Figure A5, Appendix A).

For ammonium (i.e., NH_4^+), results across different models predicted high probabilities of low loading during summer and spring (Figure 8b). The lowest loading was predicted by the CMCC-CM/COSMO-CLM (Model 7) under RCP8.5 2041–2070 with a 97% probability of the low loading class (i.e., 0–3224 kg/season, green) (Figure A6, Appendix A). In autumn, the probability of low loading states decreased gradually across the scenarios, followed by an increase in the probability of medium (i.e., 3224–5009 kg/season, yellow) and very high loadings (i.e., >6794 kg/season, red), respectively reaching 38% and 24% under RCP8.5 2071–2100 in the simulation with IPSL-CM5A-MR/RCA4 (Model 2).

Results for PO_4^{3-} showed a marked seasonality with high autumn loads and low loads in spring and summer across different scenarios (Figure 8c). In summer, in fact, higher probabilities were associated with the low loading state (i.e., 0–1978 kg/season, green). Specifically, the lowest loadings were predicted by the CMCC-CM/COSMO-CLM (Model 7) under the medium- and long-term RCP8.5 scenarios with a probability of 98% (Figure A5, Appendix A). High loadings were instead predicted for autumn with probabilities of high (i.e., 2954–3929 kg/season, orange) and very high classes (i.e., >3929 kg/season, red) increasing across scenarios. The IPSL-CM5A-MR/RCA4 (Model 2), which described the most extreme loadings for the season, predicted probabilities of 34% and 16% of being in very high and high classes under the long-term RCP8.5 scenario (Figure A5, Appendix A).

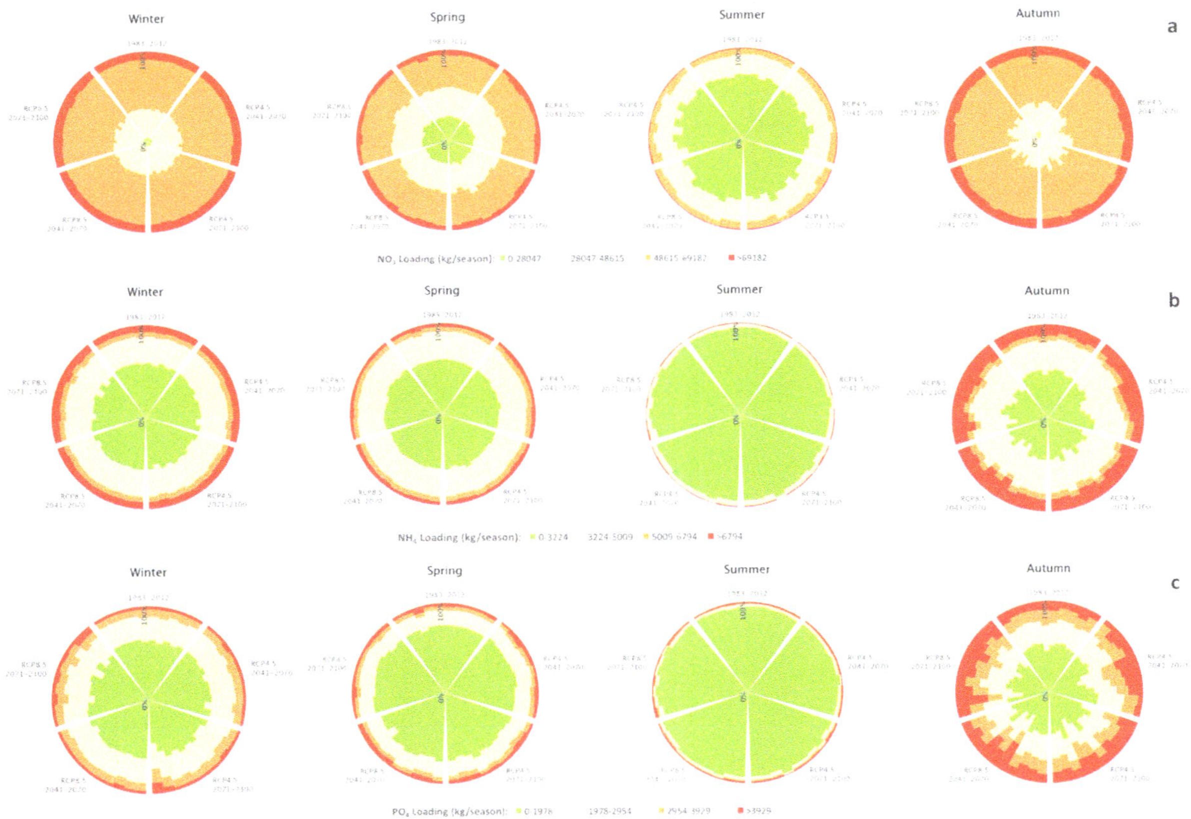

Figure 8. Probability of different classes of NO_3^- (**a**), NH_4^+ (**b**), and PO_4^{3-} (**c**) loadings associated with different seasons and scenarios across the GCM–RCM combinations considered.

Uncertainty Analysis

The variability of results was also analyzed by comparing outputs obtained with each of the 10 GCM–RCM combinations across scenarios and seasons. To make the results comparable, the change in the probability of each loading class with respect to the baseline scenario (i.e., 1983–2012) was calculated for each combination of GCM–RCM.

Accordingly, in Figure 9, which provides an example for PO_4^{3-} loadings, negative values describe a decrease in probability of specific loading classes (i.e., colored bars) with respect to baseline, while positive values indicate an increase. Orthophosphate (i.e., PO_4^{3-}) loadings showed clear variability during spring and summer. During these periods, in fact, half of considered models predicted an increase in loading, while others predicted a strong decrease. Less marked variability, however, was depicted under RCP8.5 2071–2100, where most models agreed on a reduction in loadings in the summer–spring period and an increase in probabilities associated with the low class. A good agreement among models, instead, can be depicted in autumn especially under RCP8.5 2071–2100, where most models predicted an increase in probabilities of very high and high loadings. Despite the good agreement on the increase in loading, a moderate variability in the magnitude of the change with respect to baseline remained. For RCP8.5 2041–2070, for instance, the maximum variation was related to MPI-ESM-LR/RCA4 (Model 5) (i.e., +20%), while EC-EARTH/RCA4 (Model 4) predicted an increase of +1.5%. In RCP8.5 2071–2100, the increase in probability ranged from +28% for IPSL-CM5A-MR/RCA4 (Model 2) to 2% for EC-EARTH/RCA4 and EC-EARTH/HIRHAM5 (Models 4 and 9). Also, in winter, a general increase in loading was predicted with an increase in probabilities associated with higher classes and a consequent decrease in probabilities of lower classes. The maximum increase (i.e., +10%) was depicted with EC-EARTH/RACMO22E (Model 10) under RCP4.5 2071–2100.

Results for NO_3^- and NH_4^+ loadings presented a similar tendency (Figure A6, Appendix A). The best agreement among models resulted for the autumn season, where an increase in loading was predicted across all scenarios and for all GCM–RCM combinations. Specifically, for NO_3^-, an increase in probability of the high loading class was depicted, while, for NH_4^+, the increase was associated with the highest loading class (i.e., very high). Also, for winter, the variability of results was quite low, with most models agreeing on an increase in probability of high and very high classes across different scenarios. By contrast, two models (i.e., CNRM-CM5/CCLM and CMCC-CM/COSMO-CLM (Model 6 and 7)) predicted a decrease in loadings for both NO_3^- and NH_4^+. Large variability resulted for both summer and spring seasons; hence, it was not possible to identify a clear direction of change.

Overall, the results for different nutrient species highlighted that, in general, the best agreement between models resulted for autumn and winter, especially for RCP8.5 scenarios. In summer and spring, instead, variability was high and, thus, there was less confidence in the changes projected. This seasonal pattern of variability to some extent reflects that of precipitation (Section 3.1, Figures 6 and 7), suggesting that this variable could play a major role in the model in determining both the direction and the magnitude of changes in nutrient loadings. Comparing the results (Figures 9, A3 and A4, Appendix A) with the changes in precipitation across the different models (Figure 9 and Tables A1 and A2, Appendix A), a strong correlation between the increase in precipitation and increase in the probability of high loading can be found.

In summer and spring, in fact, those models which predicted the highest increase in probability of high loadings were also those showing a positive variation (i.e., increase) in precipitation with respect to baseline.

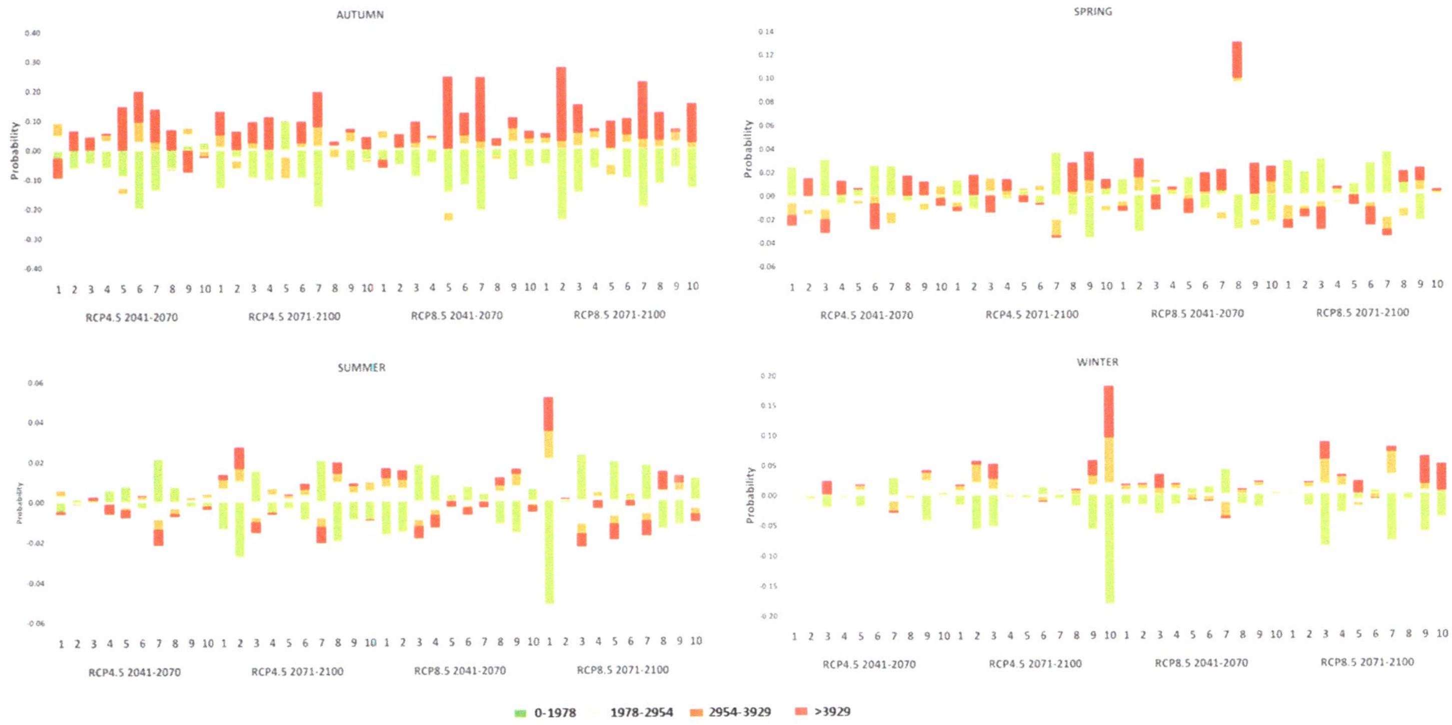

Figure 9. Variations in the probability of each PO_4^{3-} loading class with respect to baseline (i.e., 1983–2012) under different scenarios and GCM–RCM combinations. Note: GCM–RCM combinations are numbered as follows: 1. HadGEM2-ES/RCA4; 2. IPSL-CM5A-MR/RCA4; 3. CNRM-CM5/RCA4; 4. EC-EARTH/RCA4; 5. MPI-ESM-LR/RCA4; 6. CNRM-CM5/CCLM; 7. CMCC-CM/COSMO-CLM; 8. HadGEM2-ES/RACMO22E; 9. EC-EARTH/HIRHAM5; 10.EC-EARTH/RACMO22E.

In this context, it is interesting to notice how, in spring (Figures A6 and A7, Appendix A) some models (e.g., Models 8–10) contemporarily predicted an increase in probability of the two most extreme classes (i.e., low and very high). The same models were also those predicting the highest increase in precipitation with respect to baseline (Figure A7, Appendix A), suggesting that the unexpected high probabilities of very high loading classes could be related to the projections of extreme precipitation events in the considered scenarios.

4. Discussion and Conclusions

A BN was used to develop an ensemble of impact scenarios to investigate the potential mid- and long-term impacts of climate change on the nutrient loadings in the waters of the Zero river basin in northern Italy, one of the main tributaries of the Venice lagoon. Moreover, the uncertainty related to the implementation of 10 GCM–RCM combinations forced with RCP4.5 and RCP8.5 emission scenarios was analyzed.

The BN used was implemented by building on a model previously developed and tested in a case study [36], integrating a heterogeneous set of data coming from multiple information sources (i.e., observations, hydrological model simulations, climate change projections). The BN was evaluated through a cross comparison between predicted and observed loadings, providing satisfactory results on a seasonal time step; therefore, it is considered suitable for projecting future climate change scenarios.

According to the analysis of future climate for the Zero river basin, all projections agreed on an increase in the mean seasonal temperature with respect to baseline for both RCPs.

By contrast, precipitation featured marked variability across scenarios; while for winter and autumn most models agreed on an increase in the cumulative precipitation, for spring and summer, some models gave opposite results, making it impossible to agree on the direction of change (i.e., decrease or increase). The variability increased consistently along the century, especially for RCP8.5.

The impact scenarios developed showed that seasonal changes in precipitation and temperature are likely to affect nutrient loadings and, consequently, the water quality of the Zero river. Results suggest with good confidence that, across the considered scenarios, nutrient loadings will increase, especially during the autumn and winter seasons. Most models, in fact, agreed in projecting a high probability of an increase in nutrient loadings with respect to the current conditions. In summer and spring, instead, the large variability between different GCM–RCM results made it impossible to identify a clear direction of change.

The results were consistent with those obtained by Reference [45] applying the SWAT model for simulating the effect of climate change on hydrological and ecological parameters in the same case study. However, while conclusions for autumn are similar to those reached by other authors [26,46,47] for similar catchments in Europe and the United States (US), for spring and summer, the results differ. Xu et al., in particular, found that, in the Lake Erie region, spring loading of P will increase under RCP8.5 scenarios driven by an anticipation of snow-melting processes. Such discrepancies can be attributed to local and regional climatic characteristics which should, therefore, be taken into account carefully, as they can have a significant role in governing nutrient transport dynamics.

In the Zero river basin, nutrient loadings were found to be particularly sensitive to hydrological variables (i.e., river flow, runoff, N and P in runoff) directly correlated with climate variables (i.e., precipitation, temperature) and diffuse pollution, especially considering that most dramatic changes (e.g., increases in precipitation and runoff) will happen during seasons characterized by intensive agricultural activities (e.g., manure application, irrigation).

In spring and summer, in fact, NO_3^- and NH_4^+ are commonly applied as fertilizers. In dry and warm conditions, NH_4^+ is readily adsorbed to clay mineral and is, therefore, scarcely prone to movements; however, it becomes easily available in autumn, driven by runoff and extreme precipitation events. NO_3^-, on the other hand, is highly soluble and, thus, suitable to be transported by hydrological flow. In autumn, the elevated temperature and wet conditions projected will enhance the nitrification process, making NO_3^- highly available. This, combined with the seasonal increase in the river flow, could explain the great increase in NO_3^- loading during autumn. In the soil, the soluble form of phosphorus (PO_4^{3-}) is mobile, and it can be transported by diffusion or by surface water flow. At elevated temperatures and in dry conditions, however, PO_4^{3-} is easily adsorbed to clay particles or immobilized by organic matter accumulating in the upper soil layer. In autumn, an increase in runoff, following the enrichment of the topsoil of phosphorus occurring during the summer, increases PO_4^{3-} transport and, thus, its loading in the river. In addition, the projected increase of dry prolonged conditions in summer might speed up soil erosion phenomena and, consequently, enhance the runoff of adsorbed mineral forms of phosphorus through the basin, leading to peaks of PO_4^{3-} loading in autumn as soon as the drought breaks.

At the same time, the large uncertainty in spring and summer loadings makes it difficult to predict the possible implications for the trophic state of the Venice lagoon. A significant increase in spring and summer nutrients delivered by the Zero river, during the season of growth for most phytoplankton species, would significantly increase the risk of harmful algae blooms and eutrophication phenomena.

The BN was revealed to be a suitable tool to characterize and communicate uncertainty on the effect of climate change and land use on water quality attributes in a policy-relevant manner; however, it is important to also acknowledge some limitations. Some uncertainty exists mainly due to the availability and quality of input data, especially regarding agronomic practices. Due to data constraints, in fact, fertilizer application and irrigation were considered uniform across the whole catchment, while they could vary considerably, both spatially and temporally. Furthermore, due to scarce information regarding point pollution sources, nutrient (N and P) loading was considered while neglecting to take into account the type of WWTPs and how they work in cases with a large amount of inflow water, for instance, during extreme precipitation events.

Improving the accuracy of input data throughout the catchment and involving a higher number of experts in the model development would improve its calibration, validation, and results.

Finally, changes in land use (i.e., agricultural land extension, crop typology distribution) and agricultural management practices (i.e., amount of fertilizer application), which were kept constant over future scenarios in this BN version, should be accounted for in future model improvements to provide a realistic picture of future risks threatening water quality sustainability. Accordingly, further improvements of the proposed approach will consider the implementation of a dynamic version of the BN [48] to better handle temporal dynamics over future scenarios, while also integrating land-use change projections.

Overall, the results obtained from this study show that the selection of climate change information to feed impact studies should be considered carefully as it strongly affects the outcome and the conclusions of the assessment. Studies based on only one GCM–RCM combination should be interpreted with caution, as results are highly dependent on the assumptions of the selected combination. Adaptation and management decisions are taken based on this information with the consequence that societies may underprepare for real risks affecting water systems, increasing the likelihood of severe impacts, or, by contrast, they may overreact, wasting resources and efforts targeting irrelevant threats.

Accordingly, an adaptive water resource planning method should be based on ensembles and multi-model probabilistic approaches rather than on an individual scenario and a single-value projection for the future. Through the identification of worst- or best-case scenarios, it is possible to bound the spectrum of plausible climate change impacts into an uncertainty space, inside which a set of optimal adaptation strategies can be defined and tested for the sustainable and climate-proof management of the water system.

Author Contributions: Conceptualization, A.S., J.L.M., A.C., M.P.-V. and A.M.; Data curation, A.S.; Formal analysis, A.S. and J.L.M.; Investigation, A.S.; Methodology, A.S., J.L.M. and S.T.; Supervision, A.C. and A.M.; Validation, A.S., J.L.M. and M.P.-V.; Visualization, A.S., J.L.M., S.T. and A.C.; Writing – original draft, A.S., J.L.M. and S.T.; Writing – review & editing, S.T., A.C., M.P.-V. and A.M.

Funding: This research was funded.

Acknowledgments: The authors would like to thank all the public authorities and local experts that provided territorial data and information supporting the implementation of the methodology; we would also like to thank Giampietro Basei who helped in the graphical design of the results.

Conflicts of Interest: The authors declare no conflicts of interest. The funders had no role in the design of the study; in the collection, analyses, or interpretation of data; in the writing of the manuscript, or in the decision to publish the results.

Appendix A

Table A1. List of input data used for the implementation of the BN in the Zero river basin.

Data Type	Description	Time Scale	Resolution	Source
		Observations		
Land cover map	■ Land-use map of the Veneto region	2006	1:10,000	Regione del Veneto Infrastruttura dati territoriali (http://idt.regione.veneto.it/app/metacatalog/)
Climatic data	■ Daily precipitation ■ Maximum/minimum daily temperature ■ Daily evapotranspiration	2004–2013	3 stations (i.e., Castelfranco, Veneto, Zero-Branco, Mogliano Veneto) (Figure 1)	ARPAV Servizio Meteorologico
Water quantity and quality data	■ Observed daily river discharge ■ Observed nutrient (NO_3^-, NH_4^+, PO_4^{3-}) concentrations in the lagoon	2007–2012	2 stations (i.e., manual station (Code 122), automatic station (Code: B2q) (Figure 1)	ARPAV Servizio Acque Interne MAV (Magistrato Acque Venezia)
Point-source pollution	■ Monthly N and P loadings from WWTP and industrial discharges	2004–2013	3 stations (i.e., Morgano, Zero-Branco, Castelfranco Veneto) (Figure 1)	ARPAV Servizio Acque Interne
		Hydrological simulations		
Water quantity and quality data	■ Simulated runoff ■ Simulated N and P load in the runoff	2004–2013	River basin	SWAT (Soil Water Assessment Tool) simulations [30]
	■ Simulated river discharge Simulated nutrient loadings (NO_3^-, NH_4^+, PO_4^{3-}) in the lagoon	2004–2013	1 station (i.e., manual station (Code 122)	

Table A2. Overview of nodes and states in the Bayesian network model for the Zero river basin.

Node	Description	Type	States	Parametrization Method
Season	Alternative seasons	Labeled	Winter	Expert judgement
			Spring	
			Summer	
			Autumn	
Climate change scenario	Alternative climate change scenarios	Labeled	Baseline 1983–2012	CMCC-CM/COSMO-CLM simulations
			RCP 4.5 2041–2070	
			RCP 4.5 2071–2100	
			RCP 8.5 2041–2070	
			RCP 8.5 2071–2100	
Agricultural land scenario	Extension of land (ha) occupied by agricultural activities under different scenarios	Labeled	Actual 2004–2013;	Observations, LUISA simulations
			Future 2050	
Temperature	Seasonal average temperature (°C)	Numeric interval	0–8.37	Observations
			8.37–13.79	
			13.79–19.21	
			>19.21	
Precipitation	Seasonal cumulative precipitation (mm)	Numeric interval	0–201.50;	Observations
			201.50–328.73	
			328.73–455.96	
			>455.96	
Potential Evapotranspiration	Seasonal cumulative potential evapotranspiration (mm)	Numeric interval	0–133.85	Observations
			133.85–228.3	
			228.3–322.75	
			>322.75	

Table A2. *Cont.*

Node	Description	Type	States	Parametrization Method
Effective rainfall	Seasonal cumulative effective rainfall reaching the soil (mm)	Numeric interval	0–64.13	SWAT simulations
			64.13–122.95	
			122.95–181.77	
			>181.77	
Crop water needs	Seasonal water demand for different crop typology (mm)	Numeric interval	0–109.77	Equation [36]
			109.77–213.64	
			213.64–317.50	
			>317.50	
Irrigation	Seasonal amount of water applied as irrigation	Numeric interval	<−55.29	Equation [36]
			−55.29–101.28	
			101.28–257.86	
			>257.86	
N fertilizer application	Nitrogen fertilizer applied for each season according to different crop typology (kg/ha)	Numeric interval	0–45.74	Expert judgment
			45.74–87.52	
			87.52–129.30	
			>129.30	
P fertilizer application	Phosphorus fertilizer applied for each season according to different crop typology (kg/ha)	Numeric interval	0–25.41	Expert judgment
			25.41–50.83	
			50.83–76.25	
			>76.25	
N diffuse sources	Seasonal amount of nitrogen coming from agricultural practices (kg)	Numeric interval	0–7388.86	Equation [36]
			7388.86–13,959.99	
			13,959.99–20,531.11	
			>20,531.11	

Table A2. *Cont.*

Node	Description	Type	States	Parametrization Method
P diffuse sources	Seasonal amount of phosphorus coming from agricultural practices (kg)	Numeric interval	0–5169.28	Equation [36]
			5169.28–10,221.75	
			10,221.75–15,274.21	
			>15,274.21	
N point sources	Seasonal amount of nitrogen coming from point sources (i.e., wastewater treatment plants (WWTPs) and industrial discharges) (kg)	Numeric interval	0–9382.64	Observations
			9382.64–10,389.82	
			10,389.82–11,396.99	
			>11,396.99	
P point sources	Seasonal amount of phosphorus coming from point sources (i.e., WWTPs and industrial discharges) (kg)	Numeric interval	0–1143.64	Observations
			1143.64–1478.99	
			1478.99–1814.35	
			>1814.35	
River discharge	Seasonal average river discharge (L/s)	Numeric interval	0–1458.96	SWAT simulations
			1458.96–2360.53	
			2360.535–3262.102	
			>3262.10	
Runoff	Seasonal cumulative runoff (mm)	Numeric interval	0–49.90	SWAT simulations
			49.90–90.15	
			90.15–130.40	
			>130.40	
N in runoff	Seasonal amount of nitrogen loaded in the runoff (kg/ha)	Numeric interval	0–0.63	SWAT simulations
			0.63–1.19	
			1.19–1.75	
			>1.75	

Table A2. *Cont.*

Node	Description	Type	States	Parametrization Method
P in runoff	Seasonal amount of phosphorus loaded in the runoff (kg/ha)	Numeric interval	0–0.44	SWAT simulations
			0.44–0.87	
			0.87–1.30	
			>1.30	
Total N loading	Seasonal nitrogen load in the river (kg)	Numeric interval	0–17,031.20	Equation [36]
			17,031.20–24,401.92	
			24,401.92–31,772.64	
			> 31,772.64	
Total P loading	Seasonal phosphorus load in the river (kg)	Numeric interval	0–5405.76	Equation [36]
			5405.76–9710.91	
			9710.91–14,016.07	
			>14,016.07	
Loading NO_3^- lagoon	Seasonal loading of NO_3^- reaching the lagoon (kg)	Numeric interval	0–28,047.50	SWAT simulations
			28,047.50–48,615.00	
			48,615.00–69,182.50	
			>69,182.50	
Loading NH_4^+ lagoon	Seasonal loading of NH_4^+ reaching the lagoon (kg)	Numeric interval	0–3224.52	SWAT simulations
			3224.52–5009.3	
			5009.3–6794.17	
			>6794.17	
Loading PO_4^{3-} lagoon (kg)	Seasonal loading of PO_4^{3-} reaching the lagoon (kg)	Numeric interval	0–1978.90	SWAT simulations
			1978.90–2954.00	
			2954.00–3929.10	
			>3929.10	

Season	RCP4.5 2041-2070									
	1 HadGEM2-ES RCA4	2 IPSL-CM5A-MR RCA4	3 CNRM-CM5 RCA4	4 EC-EARTH RCA4	5 MPI-ESM-LR RCA4	6 CNRM-CM5 CCLM	7 CMCC-CM COSMO-CLM	8 HadGEM2-ES RACMO22E	9 EC-EARTH HIRHAM5	10 EC-EARTH RACMO22E
Winter	1.02	2.06	1.16	1.20	1.28	0.99	1.69	2.63	1.54	0.54
Spring	0.74	1.55	1.58	0.74	1.17	0.79	1.47	1.63	1.09	1.23
Summer	1.61	2.06	2.30	1.34	2.07	1.92	2.91	1.96	1.49	1.41
Autumn	0.98	2.06	1.44	0.99	1.59	1.01	1.99	2.09	1.34	1.35

Season	RCP4.5 2071-2100									
	1 HadGEM2-ES RCA4	2 IPSL-CM5A-MR RCA4	3 CNRM-CM5 RCA4	4 EC-EARTH RCA4	5 MPI-ESM-LR RCA4	6 CNRM-CM5 CCLM	7 CMCC-CM COSMO-CLM	8 HadGEM2-ES RACMO22E	9 EC-EARTH HIRHAM5	10 EC-EARTH RACMO22E
Winter	1.89	2.43	1.94	1.83	1.59	1.36	2.44	3.22	1.37	1.62
Spring	1.31	2.19	1.70	1.40	1.43	0.95	2.16	2.23	1.01	2.02
Summer	1.46	2.72	2.88	1.30	2.07	2.43	3.25	2.03	1.41	1.67
Autumn	1.45	2.56	2.56	1.47	1.86	1.71	2.99	2.76	1.34	1.70

Season	RCP8.5 2041-2070									
	1 HadGEM2-ES RCA4	2 IPSL-CM5A-MR RCA4	3 CNRM-CM5 RCA4	4 EC-EARTH RCA4	5 MPI-ESM-LR RCA4	6 CNRM-CM5 CCLM	7 CMCC-CM COSMO-CLM	8 HadGEM2-ES RACMO22E	9 EC-EARTH HIRHAM5	10 EC-EARTH RACMO22E
Winter	1.81	2.17	1.97	1.80	2.48	1.45	2.65	3.07	1.73	0.83
Spring	1.21	2.13	2.02	1.37	-1.33	1.46	2.10	2.21	1.47	1.43
Summer	1.35	2.80	3.42	1.23	1.76	2.62	3.35	2.93	1.55	1.81
Autumn	1.51	2.83	2.78	1.49	5.67	2.21	2.73	3.02	1.67	1.82

Season	RCP8.5 2071-2100									
	1 HadGEM2-ES RCA4	2 IPSL-CM5A-MR RCA4	3 CNRM-CM5 RCA4	4 EC-EARTH RCA4	5 MPI-ESM-LR RCA4	6 CNRM-CM5 CCLM	7 CMCC-CM COSMO-CLM	8 HadGEM2-ES RACMO22E	9 EC-EARTH HIRHAM5	10 EC-EARTH RACMO22E
Winter	3.31	4.00	3.41	3.21	3.97	3.35	4.86	5.12	3.09	2.97
Spring	2.25	3.73	3.68	2.58	-0.37	2.81	3.87	4.04	2.41	3.16
Summer	2.81	5.60	5.94	3.20	3.42	4.93	6.54	5.27	3.39	3.80
Autumn	2.63	5.02	4.54	2.88	7.80	3.64	5.14	5.47	3.15	3.75

Figure A1. Variation in mean seasonal temperature with respect to baseline (i.e., 1983–2012) within GCM–RCM ensemble.

Season	RCP4.5 2041-2070									
	1 HadGEM2-ES RCA4	2 IPSL-CM5A-MR RCA4	3 CNRM-CM5 RCA4	4 EC-EARTH RCA4	5 MPI-ESM-LR RCA4	6 CNRM-CM5 CCLM	7 CMCC-CM COSMO-CLM	8 HadGEM2-ES RACMO22E	9 EC-EARTH HIRHAM5	10 EC-EARTH RACMO22E
Winter	-14.07	17.83	59.87	13.30	11.76	7.73	13.42	4.34	44.78	-16.16
Spring	-15.77	21.13	-43.52	4.08	3.20	-2.22	-19.95	9.58	18.00	-21.32
Summer	-17.69	-2.35	-35.25	30.18	-43.19	-10.00	-106.05	4.01	-28.97	7.56
Autumn	24.20	-2.67	-9.31	26.26	34.31	56.90	69.32	-14.42	-46.75	-20.44

Season	RCP4.5 2071-2100									
	1 HadGEM2-ES RCA4	2 IPSL-CM5A-MR RCA4	3 CNRM-CM5 RCA4	4 EC-EARTH RCA4	5 MPI-ESM-LR RCA4	6 CNRM-CM5 CCLM	7 CMCC-CM COSMO-CLM	8 HadGEM2-ES RACMO22E	9 EC-EARTH HIRHAM5	10 EC-EARTH RACMO22E
Winter	6.00	20.48	69.77	15.60	2.03	1.39	41.65	27.77	60.34	3.61
Spring	2.59	8.39	31.87	-1.20	24.41	-7.73	-13.09	-0.42	29.90	10.32
Summer	28.65	31.93	-53.24	52.86	2.41	-9.47	-78.80	80.84	44.51	25.57
Autumn	50.91	45.40	-7.46	53.85	54.03	10.85	50.80	-12.56	-1.08	29.17

Season	RCP8.5 2041-2070									
	1 HadGEM2-ES RCA4	2 IPSL-CM5A-MR RCA4	3 CNRM-CM5 RCA4	4 EC-EARTH RCA4	5 MPI-ESM-LR RCA4	6 CNRM-CM5 CCLM	7 CMCC-CM COSMO-CLM	8 HadGEM2-ES RACMO22E	9 EC-EARTH HIRHAM5	10 EC-EARTH RACMO22E
Winter	29.04	11.61	51.32	30.23	-1.24	-4.56	52.57	8.42	46.99	-16.95
Spring	-25.92	44.28	2.61	-9.45	7.13	-4.80	-1.12	41.46	45.36	27.64
Summer	19.06	33.65	-37.51	40.32	-32.79	-6.85	-104.07	41.10	23.26	-4.43
Autumn	60.83	7.90	-40.56	-0.45	109.50	55.55	50.35	-13.13	2.48	28.75

Season	RCP8.5 2071-2100									
	1 HadGEM2-ES RCA4	2 IPSL-CM5A-MR RCA4	3 CNRM-CM5 RCA4	4 EC-EARTH RCA4	5 MPI-ESM-LR RCA4	6 CNRM-CM5 CCLM	7 CMCC-CM COSMO-CLM	8 HadGEM2-ES RACMO22E	9 EC-EARTH HIRHAM5	10 EC-EARTH RACMO22E
Winter	-12.24	20.87	35.42	13.45	9.43	34.49	148.61	31.06	62.16	7.90
Spring	-10.40	-2.52	-17.16	-3.72	-9.90	-30.93	-35.75	19.41	20.65	15.50
Summer	49.25	-7.07	-84.18	45.40	-67.79	-26.06	-139.39	-14.79	1.71	-36.75
Autumn	45.43	83.05	19.28	5.85	-21.26	23.80	15.97	25.87	-34.98	61.53

Figure A2. Variation in cumulative seasonal precipitation with respect to baseline (i.e., 1983–2012) within GCM–RCM ensemble.

Winter

Classes	1983-2012										RCP4.5 2041-2070										RCP4.5 2071-2100										RCP8.5 2041-2070										RCP8.5 2071-2100									
	1	2	3	4	5	6	7	8	9	10	1	2	3	4	5	6	7	8	9	10	1	2	3	4	5	6	7	8	9	10	1	2	3	4	5	6	7	8	9	10	1	2	3	4	5	6	7	8	9	10
0-28047	0.05	0.05	0.05	0.05	0.05	0.05	0.04	0.05	0.05	0.05	0.05	0.05	0.05	0.05	0.04	0.05	0.05	0.05	0.04	0.05	0.05	0.04	0.04	0.05	0.05	0.05	0.04	0.05	0.04	0.05	0.05	0.05	0.05	0.05	0.05	0.05	0.05	0.05	0.04	0.05	0.05	0.05	0.04	0.05	0.05	0.05	0.04	0.05	0.04	0.05
28047-48615	0.33	0.32	0.32	0.32	0.31	0.32	0.30	0.32	0.31	0.32	0.32	0.32	0.31	0.32	0.30	0.32	0.32	0.32	0.29	0.32	0.32	0.30	0.30	0.32	0.31	0.32	0.30	0.32	0.29	0.31	0.31	0.31	0.31	0.31	0.32	0.32	0.32	0.32	0.30	0.32	0.32	0.31	0.28	0.31	0.31	0.32	0.27	0.32	0.28	0.30
48615-69182	0.55	0.55	0.55	0.55	0.56	0.55	0.56	0.55	0.56	0.55	0.55	0.55	0.56	0.55	0.56	0.55	0.55	0.55	0.57	0.55	0.55	0.57	0.57	0.55	0.56	0.55	0.56	0.55	0.57	0.56	0.55	0.55	0.56	0.55	0.55	0.55	0.55	0.55	0.56	0.55	0.55	0.55	0.58	0.56	0.56	0.55	0.58	0.55	0.58	0.56
>69182	0.08	0.08	0.08	0.08	0.09	0.08	0.09	0.08	0.09	0.08	0.08	0.08	0.09	0.08	0.09	0.08	0.08	0.08	0.10	0.08	0.08	0.09	0.09	0.08	0.09	0.08	0.09	0.09	0.10	0.09	0.09	0.09	0.09	0.09	0.09	0.08	0.08	0.08	0.09	0.08	0.08	0.09	0.10	0.09	0.09	0.08	0.11	0.08	0.10	0.09

Spring

Classes	1983-2012										RCP4.5 2041-2070										RCP4.5 2071-2100										RCP8.5 2041-2070										RCP8.5 2071-2100									
	1	2	3	4	5	6	7	8	9	10	1	2	3	4	5	6	7	8	9	10	1	2	3	4	5	6	7	8	9	10	1	2	3	4	5	6	7	8	9	10	1	2	3	4	5	6	7	8	9	10
0-28047	0.26	0.26	0.26	0.25	0.26	0.27	0.26	0.26	0.26	0.26	0.27	0.28	0.28	0.26	0.27	0.28	0.29	0.27	0.27	0.27	0.27	0.27	0.26	0.27	0.27	0.27	0.31	0.27	0.25	0.27	0.27	0.26	0.27	0.27	0.26	0.27	0.28	0.26	0.27	0.26	0.29	0.29	0.28	0.27	0.26	0.30	0.30	0.29	0.26	0.28
28047-48615	0.31	0.32	0.30	0.32	0.32	0.32	0.32	0.32	0.31	0.32	0.33	0.32	0.32	0.31	0.33	0.33	0.33	0.31	0.32	0.31	0.32	0.31	0.29	0.31	0.32	0.32	0.34	0.31	0.29	0.32	0.32	0.30	0.31	0.32	0.33	0.31	0.32	0.31	0.31	0.30	0.33	0.33	0.32	0.32	0.33	0.33	0.34	0.33	0.31	0.32
48615-69182	0.38	0.37	0.35	0.38	0.37	0.36	0.38	0.38	0.38	0.36	0.36	0.34	0.32	0.37	0.35	0.35	0.33	0.35	0.35	0.37	0.37	0.35	0.37	0.37	0.37	0.36	0.32	0.36	0.39	0.35	0.37	0.38	0.36	0.37	0.36	0.34	0.34	0.36	0.35	0.37	0.34	0.34	0.33	0.36	0.37	0.33	0.32	0.34	0.38	0.34
>69182	0.05	0.05	0.09	0.05	0.05	0.06	0.05	0.05	0.05	0.06	0.04	0.07	0.08	0.06	0.05	0.04	0.04	0.07	0.06	0.05	0.05	0.07	0.07	0.05	0.05	0.05	0.04	0.07	0.07	0.05	0.05	0.06	0.07	0.05	0.04	0.07	0.06	0.07	0.07	0.07	0.04	0.04	0.06	0.04	0.04	0.04	0.03	0.05	0.05	0.05

Summer

Classes	1983-2012										RCP4.5 2041-2070										RCP4.5 2071-2100										RCP8.5 2041-2070										RCP8.5 2071-2100									
	1	2	3	4	5	6	7	8	9	10	1	2	3	4	5	6	7	8	9	10	1	2	3	4	5	6	7	8	9	10	1	2	3	4	5	6	7	8	9	10	1	2	3	4	5	6	7	8	9	10
0-28047	0.72	0.70	0.69	0.68	0.71	0.70	0.72	0.68	0.71	0.69	0.68	0.71	0.70	0.67	0.72	0.69	0.76	0.70	0.70	0.66	0.67	0.65	0.71	0.64	0.70	0.69	0.76	0.64	0.68	0.64	0.67	0.66	0.73	0.70	0.70	0.71	0.74	0.65	0.65	0.69	0.60	0.70	0.74	0.64	0.75	0.69	0.77	0.69	0.68	0.71
28047-48615	0.21	0.22	0.22	0.23	0.21	0.22	0.22	0.24	0.21	0.23	0.24	0.21	0.22	0.25	0.21	0.23	0.20	0.23	0.22	0.25	0.25	0.24	0.22	0.26	0.22	0.23	0.20	0.25	0.24	0.27	0.24	0.24	0.21	0.23	0.23	0.22	0.20	0.26	0.26	0.23	0.28	0.22	0.20	0.27	0.20	0.24	0.18	0.21	0.23	0.22
48615-69182	0.06	0.07	0.08	0.08	0.07	0.07	0.06	0.07	0.07	0.07	0.07	0.07	0.07	0.07	0.06	0.07	0.04	0.07	0.07	0.08	0.07	0.10	0.06	0.08	0.07	0.07	0.04	0.09	0.08	0.08	0.08	0.08	0.06	0.07	0.07	0.06	0.06	0.08	0.08	0.07	0.11	0.07	0.05	0.08	0.05	0.06	0.04	0.09	0.08	0.06
>69182	0.01	0.01	0.01	0.01	0.01	0.01	0.01	0.01	0.01	0.01	0.01	0.01	0.01	0.01	0.01	0.01	0.00	0.01	0.01	0.01	0.01	0.01	0.01	0.01	0.01	0.01	0.00	0.01	0.01	0.01	0.01	0.01	0.00	0.01	0.01	0.01	0.01	0.01	0.01	0.01	0.02	0.01	0.00	0.01	0.00	0.01	0.00	0.01	0.01	0.01

Autumn

Classes	1983-2012										RCP4.5 2041-2070										RCP4.5 2071-2100										RCP8.5 2041-2070										RCP8.5 2071-2100									
	1	2	3	4	5	6	7	8	9	10	1	2	3	4	5	6	7	8	9	10	1	2	3	4	5	6	7	8	9	10	1	2	3	4	5	6	7	8	9	10	1	2	3	4	5	6	7	8	9	10
0-28047	0.06	0.06	0.05	0.06	0.04	0.07	0.06	0.06	0.04	0.05	0.02	0.02	0.03	0.02	0.01	0.02	0.01	0.02	0.02	0.02	0.02	0.02	0.02	0.02	0.02	0.01	0.01	0.02	0.02	0.02	0.02	0.02	0.02	0.02	0.02	0.01	0.02	0.02	0.02	0.02	0.02	0.02	0.02	0.02	0.06	0.02	0.02	0.03	0.02	0.02
28047-48615	0.31	0.32	0.33	0.32	0.28	0.31	0.34	0.33	0.32	0.31	0.32	0.30	0.32	0.31	0.24	0.21	0.26	0.31	0.34	0.34	0.25	0.33	0.30	0.28	0.36	0.27	0.23	0.36	0.29	0.30	0.31	0.31	0.30	0.32	0.20	0.26	0.21	0.34	0.27	0.29	0.30	0.19	0.26	0.31	0.28	0.27	0.22	0.28	0.29	0.24
48615-69182	0.54	0.53	0.52	0.52	0.56	0.54	0.52	0.52	0.55	0.54	0.55	0.58	0.56	0.56	0.65	0.65	0.62	0.58	0.53	0.54	0.63	0.57	0.59	0.61	0.54	0.61	0.64	0.54	0.59	0.57	0.56	0.57	0.59	0.56	0.69	0.62	0.67	0.55	0.60	0.58	0.58	0.70	0.62	0.56	0.58	0.61	0.66	0.60	0.59	0.63
>69182	0.09	0.09	0.09	0.09	0.11	0.09	0.09	0.09	0.09	0.11	0.12	0.09	0.09	0.11	0.10	0.12	0.11	0.09	0.10	0.10	0.11	0.08	0.10	0.09	0.08	0.10	0.12	0.08	0.11	0.10	0.11	0.10	0.10	0.10	0.09	0.11	0.10	0.09	0.11	0.12	0.10	0.09	0.11	0.11	0.08	0.11	0.10	0.09	0.10	0.11

Note: GCM–RCM combinations are numbered as follows: 1. HadGEM2-ES/RCA4; 2. IPSL-CM5A-MR/RCA4; 3. CNRM-CM5/RCA4; 4. EC-EARTH/RCA4; 5. MPI-ESM-LR/RCA4; 6. CNRM-CM5/CCLM; 7. CMCC-CM/COSMO-CLM; 8. HadGEM2-ES/RACMO22E; 9. EC-EARTH/HIRHAM5; 10. EC-EARTH/RACMO22E.

Figure A3. Probability of different classes of NO_3^- loadings associated with different seasons and scenarios across the GCM–RCM combinations considered.

Winter

Classes	1983-2012										RCP4.5 2041-2070										RCP4.5 2071-2100										RCP8.5 2041-2070										RCP8.5 2071-2100									
	1	2	3	4	5	6	7	8	9	10	1	2	3	4	5	6	7	8	9	10	1	2	3	4	5	6	7	8	9	10	1	2	3	4	5	6	7	8	9	10	1	2	3	4	5	6	7	8	9	10
0-3224	0.59	0.59	0.59	0.59	0.57	0.58	0.56	0.59	0.57	0.59	0.59	0.59	0.57	0.59	0.56	0.58	0.58	0.59	0.54	0.59	0.59	0.55	0.55	0.59	0.57	0.59	0.56	0.58	0.53	0.57	0.58	0.58	0.56	0.57	0.58	0.59	0.59	0.58	0.55	0.59	0.59	0.58	0.53	0.57	0.56	0.58	0.51	0.58	0.52	0.56
3224-5009	0.28	0.28	0.29	0.28	0.29	0.29	0.30	0.28	0.29	0.28	0.28	0.29	0.29	0.29	0.30	0.29	0.29	0.28	0.31	0.28	0.28	0.30	0.31	0.29	0.30	0.28	0.30	0.29	0.32	0.29	0.29	0.29	0.30	0.29	0.29	0.28	0.28	0.29	0.30	0.28	0.28	0.29	0.32	0.30	0.30	0.29	0.33	0.29	0.32	0.30
5009-6794	0.06	0.06	0.06	0.06	0.06	0.06	0.06	0.06	0.06	0.06	0.06	0.06	0.06	0.06	0.07	0.06	0.06	0.06	0.07	0.06	0.06	0.06	0.06	0.06	0.06	0.06	0.06	0.06	0.07	0.06	0.06	0.06	0.06	0.06	0.06	0.06	0.06	0.06	0.06	0.06	0.06	0.06	0.06	0.06	0.06	0.06	0.07	0.06	0.07	0.06
>6794	0.07	0.07	0.07	0.07	0.07	0.07	0.08	0.07	0.07	0.07	0.07	0.07	0.07	0.07	0.08	0.07	0.07	0.07	0.08	0.07	0.07	0.08	0.08	0.07	0.08	0.07	0.08	0.07	0.09	0.07	0.07	0.07	0.08	0.07	0.07	0.07	0.07	0.07	0.08	0.07	0.07	0.07	0.09	0.08	0.08	0.07	0.10	0.07	0.09	0.08

Spring

Classes	1983-2012										RCP4.5 2041-2070										RCP4.5 2071-2100										RCP8.5 2041-2070										RCP8.5 2071-2100									
	1	2	3	4	5	6	7	8	9	10	1	2	3	4	5	6	7	8	9	10	1	2	3	4	5	6	7	8	9	10	1	2	3	4	5	6	7	8	9	10	1	2	3	4	5	6	7	8	9	10
0-3224	0.59	0.60	0.59	0.59	0.60	0.60	0.60	0.60	0.60	0.59	0.61	0.61	0.61	0.60	0.61	0.61	0.63	0.60	0.60	0.61	0.60	0.61	0.59	0.60	0.61	0.60	0.64	0.61	0.58	0.61	0.60	0.60	0.61	0.60	0.60	0.61	0.62	0.61	0.60	0.60	0.63	0.63	0.63	0.62	0.60	0.64	0.65	0.63	0.61	0.62
3224-5009	0.31	0.31	0.29	0.32	0.31	0.30	0.31	0.31	0.31	0.31	0.31	0.29	0.28	0.31	0.30	0.30	0.29	0.30	0.30	0.30	0.31	0.29	0.30	0.31	0.31	0.30	0.28	0.29	0.31	0.30	0.31	0.30	0.29	0.31	0.31	0.29	0.29	0.29	0.29	0.30	0.29	0.28	0.28	0.30	0.31	0.28	0.28	0.28	0.30	0.29
5009-6794	0.06	0.06	0.06	0.06	0.06	0.06	0.06	0.06	0.06	0.06	0.06	0.06	0.06	0.06	0.06	0.06	0.06	0.06	0.06	0.05	0.06	0.06	0.06	0.05	0.06	0.06	0.05	0.06	0.06	0.06	0.06	0.06	0.06	0.06	0.06	0.06	0.06	0.06	0.06	0.06	0.05	0.05	0.05	0.05	0.06	0.05	0.05	0.05	0.05	0.05
>6794	0.04	0.04	0.05	0.03	0.04	0.04	0.03	0.03	0.04	0.04	0.03	0.04	0.05	0.04	0.03	0.03	0.03	0.04	0.04	0.03	0.03	0.04	0.05	0.04	0.03	0.04	0.03	0.04	0.05	0.04	0.03	0.04	0.04	0.03	0.03	0.05	0.04	0.04	0.04	0.05	0.02	0.03	0.04	0.03	0.03	0.03	0.02	0.03	0.04	0.03

Summer

Classes	1983-2012										RCP4.5 2041-2070										RCP4.5 2071-2100										RCP8.5 2041-2070										RCP8.5 2071-2100									
	1	2	3	4	5	6	7	8	9	10	1	2	3	4	5	6	7	8	9	10	1	2	3	4	5	6	7	8	9	10	1	2	3	4	5	6	7	8	9	10	1	2	3	4	5	6	7	8	9	10
0-3224	0.95	0.94	0.94	0.93	0.94	0.95	0.95	0.94	0.94	0.94	0.95	0.94	0.94	0.94	0.95	0.94	0.97	0.95	0.94	0.94	0.94	0.92	0.95	0.93	0.94	0.94	0.97	0.92	0.94	0.93	0.94	0.93	0.96	0.95	0.95	0.95	0.96	0.93	0.93	0.95	0.90	0.94	0.96	0.93	0.96	0.95	0.97	0.93	0.93	0.95
3224-5009	0.04	0.05	0.05	0.05	0.05	0.04	0.04	0.05	0.05	0.05	0.04	0.05	0.05	0.05	0.04	0.05	0.02	0.04	0.05	0.05	0.05	0.06	0.04	0.06	0.05	0.05	0.02	0.06	0.05	0.06	0.05	0.06	0.04	0.05	0.04	0.04	0.04	0.06	0.06	0.05	0.08	0.05	0.03	0.05	0.03	0.04	0.03	0.06	0.05	0.04
5009-6794	0.00	0.00	0.00	0.01	0.00	0.00	0.00	0.00	0.00	0.00	0.00	0.00	0.00	0.00	0.00	0.00	0.00	0.00	0.00	0.00	0.00	0.01	0.00	0.01	0.00	0.00	0.00	0.01	0.00	0.00	0.00	0.01	0.00	0.00	0.00	0.00	0.00	0.01	0.01	0.00	0.01	0.00	0.00	0.00	0.00	0.00	0.00	0.01	0.01	0.00
>6794	0.01	0.01	0.01	0.01	0.01	0.01	0.01	0.01	0.01	0.01	0.01	0.01	0.01	0.01	0.00	0.01	0.00	0.01	0.01	0.01	0.01	0.01	0.00	0.01	0.01	0.01	0.00	0.01	0.01	0.01	0.01	0.01	0.00	0.01	0.01	0.00	0.00	0.01	0.01	0.01	0.01	0.01	0.00	0.01	0.00	0.01	0.00	0.01	0.01	0.01

Autumn

Classes	1983-2012										RCP4.5 2041-2070										RCP4.5 2071-2100										RCP8.5 2041-2070										RCP8.5 2071-2100									
	1	2	3	4	5	6	7	8	9	10	1	2	3	4	5	6	7	8	9	10	1	2	3	4	5	6	7	8	9	10	1	2	3	4	5	6	7	8	9	10	1	2	3	4	5	6	7	8	9	10
0-3224	0.49	0.51	0.51	0.51	0.46	0.50	0.52	0.52	0.48	0.49	0.47	0.44	0.47	0.45	0.37	0.35	0.39	0.45	0.49	0.49	0.39	0.46	0.43	0.42	0.49	0.41	0.37	0.49	0.43	0.45	0.45	0.45	0.43	0.46	0.34	0.39	0.34	0.47	0.41	0.43	0.44	0.32	0.40	0.54	0.46	0.40	0.36	0.43	0.43	0.38
3224-5009	0.32	0.32	0.32	0.32	0.35	0.32	0.32	0.32	0.33	0.34	0.35	0.35	0.33	0.35	0.37	0.39	0.37	0.34	0.34	0.34	0.37	0.34	0.35	0.35	0.33	0.36	0.38	0.33	0.36	0.35	0.35	0.35	0.35	0.35	0.37	0.37	0.38	0.34	0.36	0.36	0.35	0.38	0.36	0.35	0.33	0.36	0.37	0.35	0.36	0.37
5009-6794	0.06	0.06	0.06	0.06	0.07	0.06	0.07	0.06	0.06	0.07	0.07	0.06	0.06	0.07	0.06	0.07	0.07	0.06	0.07	0.07	0.07	0.06	0.06	0.06	0.06	0.07	0.07	0.06	0.07	0.07	0.07	0.06	0.06	0.07	0.06	0.07	0.06	0.06	0.07	0.07	0.07	0.06	0.07	0.07	0.06	0.07	0.06	0.06	0.07	0.07
>6794	0.13	0.11	0.11	0.11	0.12	0.12	0.09	0.11	0.13	0.11	0.11	0.15	0.13	0.13	0.20	0.19	0.17	0.15	0.10	0.11	0.18	0.14	0.15	0.17	0.12	0.17	0.18	0.12	0.15	0.13	0.13	0.14	0.15	0.12	0.24	0.17	0.22	0.13	0.16	0.13	0.14	0.24	0.17	0.12	0.15	0.16	0.20	0.15	0.14	0.18

Note: GCM–RCM combinations are numbered as follows: 1. HadGEM2-ES/RCA4; 2. IPSL-CM5A-MR/RCA4; 3. CNRM-CM5/RCA4; 4. EC-EARTH/RCA4; 5. MPI-ESM-LR/RCA4; 6. CNRM-CM5/CCLM; 7. CMCC-CM/COSMO-CLM; 8. HadGEM2-ES/RACMO22E; 9. EC-EARTH/HIRHAM5; 10. EC-EARTH/RACMO22E.

Figure A4. Probability of different classes of NH_4^+ loadings associated with different seasons and scenarios across the GCM–RCM combinations considered.

Winter																																																		
Classes	1983-2012										RCP4.5 2041-2070										RCP4.5 2071-2100										RCP8.5 2041-2070										RCP8.5 2071-2100									
	1	2	3	4	5	6	7	8	9	10	1	2	3	4	5	6	7	8	9	10	1	2	3	4	5	6	7	8	9	10	1	2	3	4	5	6	7	8	9	10	1	2	3	4	5	6	7	8	9	10
0-1978	0.65	0.65	0.64	0.65	0.62	0.64	0.61	0.65	0.62	0.65	0.65	0.64	0.62	0.64	0.60	0.63	0.63	0.65	0.58	0.65	0.63	0.59	0.59	0.64	0.62	0.65	0.60	0.63	0.56	0.46	0.63	0.63	0.61	0.63	0.63	0.65	0.65	0.63	0.60	0.65	0.65	0.63	0.56	0.61	0.61	0.64	0.53	0.64	0.56	0.61
1978-2954	0.25	0.25	0.25	0.25	0.26	0.25	0.26	0.26	0.26	0.25	0.25	0.25	0.25	0.25	0.27	0.25	0.25	0.25	0.28	0.25	0.26	0.27	0.26	0.25	0.26	0.25	0.27	0.26	0.27	0.27	0.26	0.26	0.25	0.26	0.26	0.25	0.25	0.25	0.27	0.25	0.25	0.26	0.27	0.26	0.25	0.26	0.30	0.26	0.26	0.24
2954-3929	0.08	0.08	0.08	0.08	0.09	0.08	0.10	0.08	0.09	0.08	0.08	0.08	0.08	0.08	0.10	0.08	0.08	0.08	0.11	0.08	0.08	0.11	0.09	0.08	0.09	0.08	0.10	0.08	0.11	0.15	0.08	0.08	0.09	0.08	0.08	0.08	0.08	0.08	0.10	0.08	0.08	0.08	0.12	0.09	0.09	0.08	0.14	0.08	0.10	0.08
>3929	0.03	0.03	0.03	0.03	0.03	0.03	0.03	0.03	0.03	0.03	0.03	0.03	0.05	0.03	0.03	0.03	0.03	0.03	0.04	0.03	0.03	0.03	0.05	0.03	0.03	0.03	0.03	0.03	0.06	0.11	0.03	0.03	0.05	0.03	0.03	0.03	0.03	0.03	0.03	0.03	0.03	0.03	0.06	0.03	0.05	0.03	0.04	0.03	0.08	0.07

Spring																																																		
Classes	1983-2012										RCP4.5 2041-2070										RCP4.5 2071-2100										RCP8.5 2041-2070										RCP8.5 2071-2100									
	1	2	3	4	5	6	7	8	9	10	1	2	3	4	5	6	7	8	9	10	1	2	3	4	5	6	7	8	9	10	1	2	3	4	5	6	7	8	9	10	1	2	3	4	5	6	7	8	9	10
0-1978	0.75	0.75	0.72	0.75	0.75	0.75	0.75	0.75	0.75	0.75	0.77	0.75	0.75	0.74	0.76	0.78	0.78	0.75	0.75	0.75	0.76	0.74	0.72	0.75	0.76	0.74	0.79	0.73	0.71	0.75	0.76	0.72	0.72	0.75	0.77	0.74	0.75	0.72	0.74	0.73	0.78	0.77	0.75	0.76	0.76	0.78	0.79	0.76	0.73	0.75
1978-2954	0.17	0.17	0.16	0.18	0.17	0.16	0.17	0.17	0.17	0.17	0.17	0.16	0.15	0.17	0.16	0.17	0.16	0.16	0.16	0.17	0.17	0.16	0.16	0.17	0.17	0.17	0.15	0.16	0.17	0.16	0.17	0.17	0.17	0.17	0.17	0.16	0.16	0.27	0.16	0.17	0.16	0.16	0.16	0.17	0.17	0.16	0.15	0.16	0.17	0.16
2954-3929	0.05	0.05	0.05	0.05	0.05	0.05	0.05	0.05	0.05	0.04	0.04	0.04	0.04	0.05	0.04	0.04	0.04	0.05	0.05	0.05	0.05	0.05	0.06	0.05	0.05	0.05	0.04	0.05	0.06	0.05	0.05	0.06	0.05	0.05	0.04	0.05	0.04	0.05	0.05	0.06	0.04	0.04	0.05	0.05	0.05	0.04	0.04	0.04	0.06	0.05
>3929	0.03	0.03	0.07	0.03	0.03	0.04	0.03	0.03	0.03	0.04	0.02	0.05	0.06	0.04	0.03	0.02	0.03	0.04	0.04	0.03	0.03	0.05	0.06	0.04	0.03	0.04	0.03	0.05	0.05	0.04	0.02	0.05	0.06	0.03	0.02	0.06	0.05	0.06	0.06	0.05	0.02	0.03	0.05	0.03	0.02	0.02	0.02	0.04	0.04	0.04

Summer																																																		
Classes	1983-2012										RCP4.5 2041-2070										RCP4.5 2071-2100										RCP8.5 2041-2070										RCP8.5 2071-2100									
	1	2	3	4	5	6	7	8	9	10	1	2	3	4	5	6	7	8	9	10	1	2	3	4	5	6	7	8	9	10	1	2	3	4	5	6	7	8	9	10	1	2	3	4	5	6	7	8	9	10
0-1978	0.95	0.94	0.94	0.93	0.95	0.95	0.96	0.94	0.95	0.94	0.95	0.95	0.94	0.94	0.95	0.95	0.98	0.95	0.94	0.94	0.94	0.92	0.96	0.93	0.94	0.94	0.98	0.92	0.94	0.93	0.94	0.93	0.96	0.95	0.95	0.96	0.96	0.93	0.93	0.95	0.90	0.94	0.96	0.93	0.97	0.95	0.97	0.93	0.93	0.95
1978-2954	0.02	0.03	0.03	0.03	0.03	0.03	0.02	0.03	0.03	0.03	0.03	0.03	0.03	0.03	0.02	0.03	0.01	0.03	0.03	0.03	0.03	0.04	0.02	0.04	0.03	0.03	0.01	0.04	0.03	0.04	0.03	0.03	0.02	0.03	0.03	0.02	0.02	0.03	0.03	0.03	0.05	0.03	0.02	0.03	0.02	0.03	0.02	0.03	0.03	0.03
2954-3929	0.01	0.01	0.01	0.02	0.01	0.01	0.01	0.01	0.01	0.01	0.01	0.01	0.01	0.02	0.01	0.01	0.00	0.01	0.01	0.02	0.02	0.02	0.01	0.02	0.01	0.01	0.01	0.02	0.01	0.02	0.01	0.02	0.01	0.01	0.01	0.01	0.01	0.02	0.02	0.01	0.02	0.01	0.01	0.02	0.01	0.01	0.01	0.02	0.02	0.01
>3929	0.01	0.01	0.01	0.02	0.01	0.01	0.01	0.01	0.01	0.02	0.01	0.01	0.02	0.01	0.01	0.01	0.00	0.01	0.01	0.01	0.01	0.03	0.01	0.02	0.01	0.02	0.00	0.02	0.02	0.01	0.02	0.02	0.01	0.01	0.01	0.01	0.01	0.02	0.02	0.01	0.03	0.02	0.01	0.01	0.01	0.01	0.01	0.02	0.02	0.01

Autumn																																																		
Classes	1983-2012										RCP4.5 2041-2070										RCP4.5 2071-2100										RCP8.5 2041-2070										RCP8.5 2071-2100									
	1	2	3	4	5	6	7	8	9	10	1	2	3	4	5	6	7	8	9	10	1	2	3	4	5	6	7	8	9	10	1	2	3	4	5	6	7	8	9	10	1	2	3	4	5	6	7	8	9	10
0-1978	0.51	0.53	0.54	0.53	0.45	0.51	0.53	0.54	0.51	0.50	0.48	0.47	0.50	0.48	0.36	0.31	0.40	0.48	0.52	0.52	0.38	0.50	0.45	0.43	0.55	0.41	0.34	0.55	0.44	0.46	0.47	0.48	0.45	0.49	0.30	0.39	0.33	0.52	0.40	0.44	0.46	0.29	0.40	0.47	0.44	0.41	0.33	0.42	0.44	0.36
1978-2954	0.24	0.25	0.25	0.25	0.28	0.25	0.26	0.24	0.24	0.27	0.29	0.24	0.24	0.28	0.24	0.27	0.26	0.24	0.28	0.27	0.25	0.24	0.25	0.24	0.25	0.25	0.28	0.25	0.27	0.27	0.28	0.25	0.25	0.27	0.21	0.26	0.23	0.25	0.27	0.29	0.26	0.22	0.26	0.28	0.23	0.27	0.24	0.25	0.26	0.26
2954-3929	0.13	0.14	0.13	0.14	0.18	0.14	0.14	0.13	0.13	0.16	0.17	0.14	0.13	0.15	0.16	0.21	0.16	0.13	0.15	0.14	0.17	0.11	0.14	0.14	0.10	0.15	0.20	0.10	0.16	0.15	0.15	0.14	0.14	0.15	0.15	0.17	0.16	0.12	0.17	0.18	0.15	0.16	0.17	0.16	0.14	0.17	0.17	0.15	0.16	0.18
>3929	0.12	0.09	0.08	0.09	0.10	0.11	0.07	0.08	0.12	0.07	0.05	0.15	0.13	0.09	0.24	0.21	0.18	0.15	0.05	0.07	0.20	0.15	0.16	0.19	0.10	0.18	0.19	0.10	0.14	0.11	0.09	0.13	0.15	0.09	0.34	0.18	0.29	0.11	0.16	0.10	0.14	0.34	0.18	0.10	0.19	0.16	0.26	0.18	0.14	0.20

Note: GCM–RCM combinations are numbered as follows: 1. HadGEM2-ES/RCA4; 2. IPSL-CM5A-MR/RCA4; 3. CNRM-CM5/RCA4; 4. EC-EARTH/RCA4; 5. MPI-ESM-LR/RCA4; 6. CNRM-CM5/CCLM; 7. CMCC-CM/COSMO-CLM; 8. HadGEM2-ES/RACMO22E; 9. EC-EARTH/HIRHAM5; 10. EC-EARTH/RACMO22E.

Figure A5. Probability of different classes of PO_4^{3-} loadings associated with different seasons and scenarios across the GCM–RCM combinations considered.

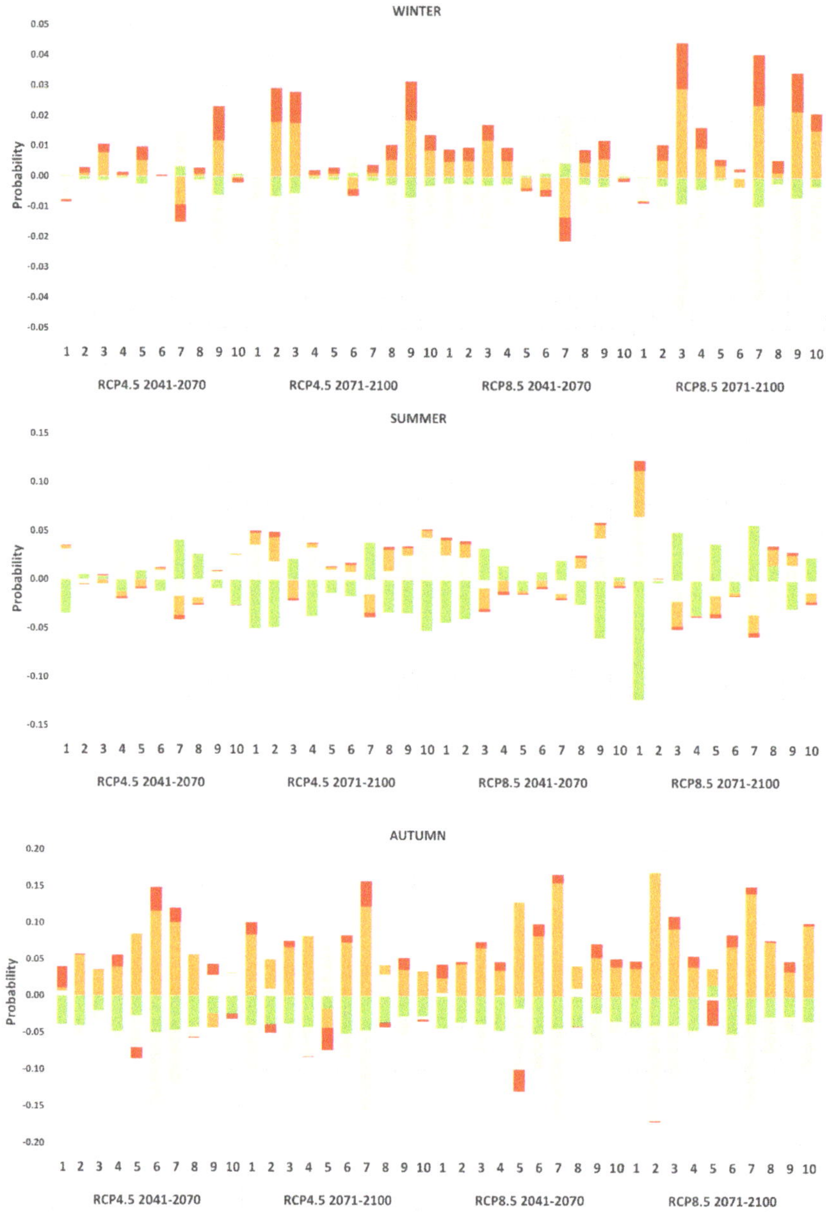

Figure A6. *Cont.*

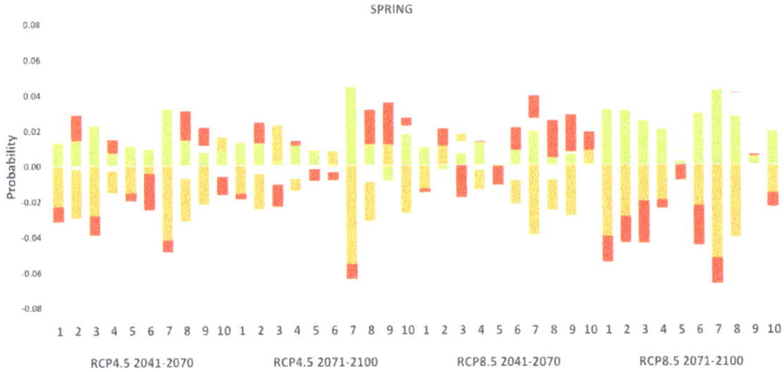

Figure A6. Variations in the probability of each NO_3^- loading class with respect to baseline (i.e., 1983–2012) under different scenarios and GCM–RCM combinations.

Figure A7. *Cont.*

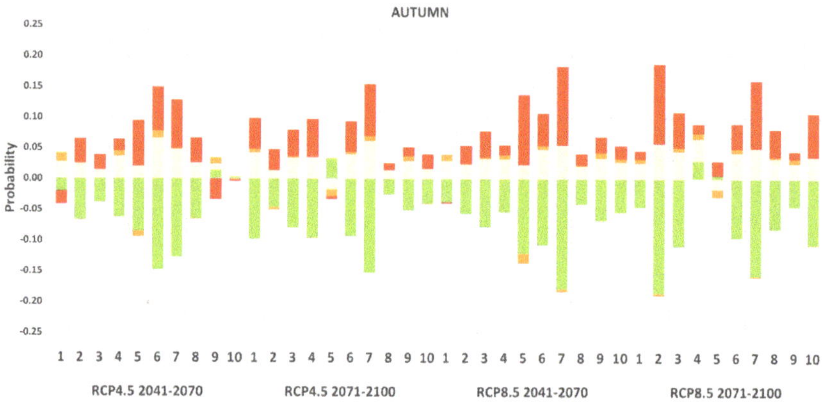

Figure A7. Variations in the probability of each NH_4^+ loading class with respect to baseline (i.e., 1983–2012) under different scenarios and GCM–RCM combinations.

References

1. Resolution, A. RES/70/1. Transforming our World: The 2030 Agenda for Sustainable Development. Available online: https://sustainabledevelopment.un.org/post2015/transformingourworld (accessed on 31 August 2019).
2. Pasini, S.; Torresan, S.; Rizzi, J.; Zabeo, A.; Critto, A.; Marcomini, A. Climate change impact assessment in Veneto and Friuli Plain groundwater. Part II: a spatially resolved regional risk assessment. *Sci. Total Environ.* **2012**, *440*, 219–235. [CrossRef] [PubMed]
3. Iyalomhe, F.; Rizzi, J.; Pasini, S.; Torresan, S.; Critto, A.; Marcomini, A. Regional Risk Assessment for climate change impacts on coastal aquifers. *Sci. Total Environ.* **2015**, *537*, 100–114. [CrossRef] [PubMed]
4. Bussi, G.; Whitehead, P.G.; Bowes, M.J.; Read, D.S.; Prudhomme, C.; Dadson, S.J. Impacts of climate change, land-use change and phosphorus reduction on phytoplankton in the River Thames (UK). *Sci. Total Environ.* **2016**, *572*, 1507–1519. [CrossRef] [PubMed]
5. Huttunen, I.; Lehtonen, H.; Huttunen, M.; Piirainen, V.; Korppoo, M.; Veijalainen, N.; Viitasalo, M.; Vehviläinen, B. Effects of climate change and agricultural adaptation on nutrient loading from Finnish catchments to the Baltic Sea. *Sci. Total Environ.* **2015**, *529*, 168–181. [CrossRef] [PubMed]
6. Whitehead, P.; Butterfield, D.; Wade, D. *Potential Impacts of Climate Change on River Water Quality*; Environment Agency: Bristol, UK, 2008; ISBN 9781844329069.
7. Carrasco, G.; Molina, J.-L.; Patino-Alonso, M.-C.; Castillo, M.D.C.; Vicente-Galindo, M.-P.; Galindo-Villardón, M.-P. Water quality evaluation through a multivariate statistical HJ-Biplot approach. *J. Hydrol.* **2019**, 123993. [CrossRef]
8. Molina, J.-L.; Zazo, S.; Martín, A.-M. Causal Reasoning: Towards Dynamic Predictive Models for Runoff Temporal Behavior of High Dependence Rivers. *Water* **2019**, *11*, 877. [CrossRef]
9. Beck, M.; Krueger, T. The epistemic, ethical, and political dimensions of uncertainty in integrated assessment modeling. *Wiley Interdiscip. Rev. Clim. Chang.* **2016**, *7*, 627–645. [CrossRef]
10. Carter, T.R.; Kenkyū, K.K.K.C.K. *IPCC Technical Guidelines for Assessing Climate Change Impacts and Adaptations: Part of the IPCC Special Report to the First Session of the Conference of the Parties to the UN Framework Convention on Climate Change*; IPCC: Geneva, Switzerland, 1994.
11. Kundzewicz, Z.W.; Krysanova, V.; Benestad, R.E.; Hov, Ø.; Piniewski, M.; Otto, I.M. Uncertainty in climate change impacts on water resources. *Environ. Sci. Policy* **2018**, *79*, 1–8. [CrossRef]
12. Parker, W.S. Ensemble modeling, uncertainty and robust predictions. *Wiley Interdiscip. Rev. Clim. Chang.* **2013**, *4*, 213–223. [CrossRef]
13. Hawkins, E.; Sutton, R. The potential to narrow uncertainty in regional climate predictions. *Bull. Am. Meteorol. Soc.* **2009**, *90*, 1095–1108. [CrossRef]

14. Ajami, N.K.; Hornberger, G.M.; Sunding, D.L. Sustainable water resource management under hydrological uncertainty. *Water Resour. Res.* **2008**, *44*. [CrossRef]
15. Larson, K.; White, D.; Gober, P.; Wutich, A. Decision-making under uncertainty for water sustainability and urban climate change adaptation. *Sustainability* **2015**, *7*, 14761–14784. [CrossRef]
16. Burgman, M. *Risks and Decisions for Conservation and Environmental Management*; Cambridge University Press: Cambridge, UK, 2005; ISBN 0521543010.
17. Power, M.; McCarty, L.S. Environmental risk management decision-making in a societal context. *Hum. Ecol. Risk Assess.* **2006**, *12*, 18–27. [CrossRef]
18. Uusitalo, L. Advantages and challenges of Bayesian networks in environmental modelling. *Ecol. Model.* **2007**, *203*, 312–318. [CrossRef]
19. Wallach, D.; Mearns, L.O.; Ruane, A.C.; Rötter, R.P.; Asseng, S. Lessons from climate modeling on the design and use of ensembles for crop modeling. *Clim. Chang.* **2016**, *139*, 551–564. [CrossRef]
20. IPCC. *Climate Change 2007: Impacts, Adaptation and Vulnerability: Contribution of Working Group II to the Fourth Assessment Report of the Intergovernmental Panel on Climate Change (Parry M.L., Canziani O.F., Palutikof J.P., van der Linden P.J. e Hanson C.E.)*; Cambridge University Press: Cambridge, UK, 2007; ISBN 0521880106.
21. Tebaldi, C.; Knutti, R. The use of the multi-model ensemble in probabilistic climate projections. *Philos. Trans. R. Soc. Lond. A Math. Phys. Eng. Sci.* **2007**, *365*, 2053–2075. [CrossRef] [PubMed]
22. Martre, P.; Wallach, D.; Asseng, S.; Ewert, F.; Jones, J.W.; Rötter, R.P.; Boote, K.J.; Ruane, A.C.; Thorburn, P.J.; Cammarano, D. Multimodel ensembles of wheat growth: many models are better than one. *Glob. Chang. Biol.* **2015**, *21*, 911–925. [CrossRef]
23. Krishnamurti, T.N.; Kishtawal, C.M.; Zhang, Z.; LaRow, T.; Bachiochi, D.; Williford, E.; Gadgil, S.; Surendran, S. Multimodel ensemble forecasts for weather and seasonal climate. *J. Clim.* **2000**, *13*, 4196–4216. [CrossRef]
24. Luo, M.; Meng, F.; Liu, T.; Duan, Y.; Frankl, A.; Kurban, A.; De Maeyer, P. Multi–Model Ensemble Approaches to Assessment of Effects of Local Climate Change on Water Resources of the Hotan River Basin in Xinjiang, China. *Water* **2017**, *9*, 584. [CrossRef]
25. Schellekens, J.; Dutra, E.; Martínez-de la Torre, A.; Balsamo, G.; van Dijk, A.; Weiland, F.S.; Minvielle, M.; Calvet, J.-C.; Decharme, B.; Eisner, S. A global water resources ensemble of hydrological models: The eartH2Observe Tier-1 dataset. *Earth Syst. Sci. Data* **2017**, *9*, 389. [CrossRef]
26. Xu, H.; Brown, D.G.; Steiner, A.L. Sensitivity to climate change of land use and management patterns optimized for efficient mitigation of nutrient pollution. *Clim. Chang.* **2018**, *147*, 647–662. [CrossRef]
27. Zuliani, A.; Zaggia, L.; Collavini, F.; Zonta, R. Freshwater discharge from the drainage basin to the Venice Lagoon (Italy). *Environ. Int.* **2005**, *31*, 929–938. [CrossRef] [PubMed]
28. Osservatorio naturalistico della Laguna del Comune di Venezia; Guerzoni, S. *Atlante Della Laguna: Venezia tra Terra e Mare*; Marsilio: Venice, Italy, 2006; ISBN 8831787640.
29. Facca, C.; Ceoldo, S.; Pellegrino, N.; Sfriso, A. Natural recovery and planned intervention in coastal wetlands: Venice Lagoon (Northern Adriatic Sea, Italy) as a case study. *Sci. World J.* **2014**, *2014*. [CrossRef] [PubMed]
30. Pesce, M.; Critto, A.; Torresan, S.; Giubilato, E.; Santini, M.; Zirino, A.; Ouyang, W.; Marcomini, A. Modelling climate change impacts on nutrients and primary production in coastal waters. *Sci. Total Environ.* **2018**, *628*, 919–937. [CrossRef] [PubMed]
31. Jacob, D.; Petersen, J.; Eggert, B.; Alias, A.; Christensen, O.B.; Bouwer, L.M.; Braun, A.; Colette, A.; Déqué, M.; Georgievski, G. EURO-CORDEX: new high-resolution climate change projections for European impact research. *Reg. Environ. Chang.* **2014**, *14*, 563–578. [CrossRef]
32. Scoccimarro, E.; Gualdi, S.; Bellucci, A.; Sanna, A.; Fogli, P.G.; Manzini, E.; Vichi, M.; Oddo, P.; Navarra, A. Effects of Tropical Cyclones on Ocean Heat Transport in a High-Resolution Coupled General Circulation Model. *J. Clim.* **2011**, *24*, 4368–4384. [CrossRef]
33. Cattaneo, L.; Zollo, A.L.; Bucchignani, E.; Montesarchio, M.; Manzi, M.P.; Mercogliano, P. Assessment of Cosmo-Clm Performances over Mediterranean Area. *SSRN Electron. J.* **2012**. [CrossRef]
34. Thomson, A.M.; Calvin, K.V.; Smith, S.J.; Kyle, G.P.; Volke, A.; Patel, P.; Delgado-Arias, S.; Bond-Lamberty, B.; Wise, M.A.; Clarke, L.E. RCP 4.5: A pathway for stabilization of radiative forcing by 2100. *Clim. Chang.* **2011**, *109*, 77. [CrossRef]
35. Riahi, K.; Rao, S.; Krey, V.; Cho, C.; Chirkov, V.; Fischer, G. RCP 8.5—A scenario of comparatively high greenhouse gas emissions. *Clim. Chang.* **2011**, *109*, 33–57. [CrossRef]

36. Sperotto, A.; Molina, J.L.; Torresan, S.; Critto, A.; Pulido-Velazquez, M.; Marcomini, A. A Bayesian Networks approach for the assessment of climate change impacts on nutrients loading. *Environ. Sci. Policy* **2019**, *100*, 21–36. [CrossRef]
37. Madsen, A.L.; Jensen, F.; Kjaerulff, U.B.; Lang, M. The Hugin tool for probabilistic graphical models. *Int. J. Artif. Intell. Tools* **2005**, *14*, 507–543. [CrossRef]
38. Bromley, J.; Jackson, N.A.; Clymer, O.J.; Giacomello, A.M.; Jensen, F.V. The use of Hugin®to develop Bayesian networks as an aid to integrated water resource planning. *Environ. Model. Softw.* **2005**, *20*, 231–242. [CrossRef]
39. Arnold, J.G.; Moriasi, D.N.; Gassman, P.W.; Abbaspour, K.C.; White, M.J.; Srinivasan, R.; Santhi, C.; Harmel, R.D.; Van Griensven, A.; Van Liew, M.W. SWAT: Model use, calibration, and validation. *Trans. ASABE* **2012**, *55*, 1491–1508. [CrossRef]
40. Marcot, B.G. Metrics for evaluating performance and uncertainty of Bayesian network models. *Ecol. Model.* **2012**, *230*, 50–62. [CrossRef]
41. Kragt, M.E. *A Beginners Guide to Bayesian Network Modelling for Integrated Catchment Management*. Available online: http://www.landscapelogic.org.au/publications/Technical_Reports/No_9_BNs_for_Integrated_Catchment_Management.pdf (accessed on 30 August 2019).
42. Molina, J.-L.; Zazo, S.; Rodríguez-González, P.; González-Aguilera, D. Innovative Analysis of Runoff Temporal Behavior through Bayesian Networks. *Water* **2016**, *8*, 484. [CrossRef]
43. Pearl, J. *Probabilistic Reasoning in Intelligent Systems: Networks of Plausible Inference*; Morgan Kaufmann: Burlington, MA, USA, 1988; ISBN 0080514898.
44. Pollino, C.A.; Woodberry, O.; Nicholson, A.; Korb, K.; Hart, B.T. Parameterisation and evaluation of a Bayesian network for use in an ecological risk assessment. *Environ. Model. Softw.* **2007**, *22*, 1140–1152. [CrossRef]
45. Pesce, M.; Critto, A.; Torresan, S.; Giubilato, E.; Pizzol, L.; Marcomini, A. Assessing uncertainty of hydrological and ecological parameters originating from the application of an ensemble of ten global-regional climate model projections in a coastal ecosystem of the lagoon of Venice, Italy. *Ecol. Eng.* **2019**, *133*, 121–136. [CrossRef]
46. Bouraoui, F.; Galbiati, L.; Bidoglio, G. Climate change impacts on nutrient loads in the Yorkshire Ouse catchment (UK). *Hydrol. Earth Syst. Sci. Discuss.* **2002**, *6*, 197–209. [CrossRef]
47. Panagopoulos, Y.; Makropoulos, C.; Mimikou, M. Diffuse surface water pollution: Driving factors for different geoclimatic regions. *Water Resour. Manag.* **2011**, *25*, 3635. [CrossRef]
48. Molina, J.-L.; Pulido-Velázquez, D.; García-Aróstegui, J.L.; Pulido-Velázquez, M. Dynamic Bayesian networks as a decision support tool for assessing climate change impacts on highly stressed groundwater systems. *J. Hydrol.* **2013**, *479*, 113–129. [CrossRef]

sustainability

MDPI

Article

Challenges for Sustainable Water Use in the Urban Industry of Korea Based on the Global Non-Radial Directional Distance Function Model

Na Wang [1,*] and Yongrok Choi [2,*]

1 Institute of Blue and Green Development, Shandong University, Weihai 264209, China
2 Global E-governance Program, Inha University, Inharo100, Nam-gu, Incheon 402-751, Korea
* Correspondence: 5sunnywang@live.cn (N.W.); yrchoi@inha.ac.kr (Y.C.)

Received: 17 June 2019; Accepted: 9 July 2019; Published: 17 July 2019

Abstract: Since water stress and industrial water pollution pose a huge threat to South Korea's sustainable water use, it is an urgent task to assess industrial water green use efficiency (GUEIW). Based on the global non-radial directional distance function (GNDDF) model, this paper calculated GUEIW in 16 Korean local governments from 2006 to 2015 using two decomposition indicators: Economic efficiency of industrial water use (ECEIW) and environmental efficiency of industrial water use (ENEIW). The growth of GUEIW is mainly driven by ECEIW, and subsequent environmental problems are obstacles to achieving green use of Korean industrial water. The regional heterogeneity of GUEIW is so important that the downstream region outperformed the upstream region in all three indicators. The government's efforts to ensure water quality inhibits industrial development in upstream areas, where incomes are much lower than in downstream areas, and downstream industrial areas have to pay upstream industrial areas extra for water. However, regarding upstream industrial areas, low prices easily promote water waste. Because of relatively high water use costs, downstream producers are encouraged to save water. To improve the economic efficiency of industrial water use in upstream areas, advanced water technology should be developed or introduced to make full use of water resources in industrial production.

Keywords: water resource; South Korean urban industry; green use efficiency of industrial water (GUEIW); global non-radial directional distance function model (GNDDF); economic efficiency of industrial water use (ECEIW); environmental efficiency of industrial water use (ENEIW)

1. Introduction

1.1. Background of the Research

In the 20th century, oil was referred to as black gold, the "core of the world market economy," but in the 21st century, "blue gold," or water, has become an increasingly valuable resource due to population explosion, climate change, and the current world economic model [1]. Water systems are vital to human well-being and offer diverse benefits to society. Water policies have long aimed at optimal water use due to the ever-increasing demand for limited water resources in many parts of the world [2,3].

However, many countries around the world still face significant challenges in managing their scarce water resources due to industrialization, urbanization, and climate change. Industrialization and urbanization, driven by population pressure, are the main factors of economic growth; meanwhile, intensive agricultural spread-out use of high pesticides has also become dominant, resulting in water shortage and degradation in many parts of the world. In addition, climate change has increased characteristics of severe spatial and temporal variations in water resources [4].

South Korea ("Korea") is no exception. Over the past few decades, the Korean economic growth rate has been astonishingly high [4,5]. Unfortunately, economic growth has come at the expense of the environment [6,7], indicated by water shortage and deteriorating water quality [8] which became serious problems in the 1990s [9]. Rapid economic development has led to urban, industrial, and agricultural pollution that has severely disrupted water supplies and ecosystems. Population and industrial growth have increased pressure on limited available water resources, resulting in water conflicts among stakeholders [10]. Sewage contains a large number of organic pollutants which affect the ecosystem in the form of toxicity, reduce dissolved oxygen in water, and endanger human health. Specific organic pollutants refer to organic compounds with high toxicity, strong accumulation, and delayed degradation that are listed as priority pollutants [11]. Organic pollutants have different characteristics. The hazardous substances are a minor fraction of wastewater, often measured in µg/L, while the less harmful organic substances may be given in hundreds or even thousands of mg/L. They are not toxic or pose a strong risk to human health but are responsible for oxygen consumption in receiving water bodies and for this reason should be reduced by treatment. The water resources management information system of the Korean government advises that the concentration of organic pollutants (BOD) needs to be kept below 10 mg/L [12]. Regarding water regulation policies, an intermediate law concerning the flow of river water was introduced in 1999, which mandates lowland water users to pay upland residents in an effort to reduce agricultural intensification (Agricultural intensification refers to a kind of agricultural operation mode in which more labor, capital, and technology are invested in a certain area of land in order to obtain more output per unit area and simultaneously reduce the labor cost per unit of product.) in the upland regions and decrease the need for more water pollution treatment facilities. However, water pollution has still occurred [13,14].

More than direct regulatory policies, therefore, water prices may play an important role in determining the efficiency of resource utilization, as suggested by Kumbhakar and Bhattacharyya [15]. Specifically, relatively low prices can easily promote water waste, and relatively high prices may lead to higher production costs, thus encouraging water consumers to save water [16]. Therefore, the price should be set within a reasonable range for sustainable use of water. Uncontrolled water wastage and numerous water pollution incidents have made the problems of water shortage and degradation more serious, which has led to a huge negative impact on social and economic development [17–19]. Therefore, improving water efficiency and reducing water pollution are crucial for sustainable water use [20–23].

As for the estimation of resource efficiency, many recent studies prefer the distance function method, which simultaneously takes multiple input and output factors into account [24–29]. There are two methods for estimating the distance function: The nonparametric data envelope analysis (DEA) method and the parametric method. The DEA method was proposed by Charnes et al. [30] and has been widely applied in recent environmental and energy-related studies [31]. A major advantage of the DEA method over the parametric method is that it does not require the underlying technology to have a specific functional form [32–35].

As for the evaluation of water use efficiency in China, Hu et al. [36] developed the pioneer empirical analytic framework DEA to evaluate water use efficiency at the national level. The concept adjustment quantity they proposed was used to determine the optimal scale of water use. Liao and Dong [37] adopted a similar method to evaluate the efficiency of provincial water resources utilization. However, these studies only took economic benefits into account and may be considered only a part of a comprehensive analysis because they ignored pollutants (such as water waste and water pollution) from industrial production. Therefore, the expected output should also be considered along with the undesirable output. By utilizing this undesirable output in our model, "green use efficiency" will be achieved [38–41]. In addition, most previous studies tended to use the radial DEA method, which aims to increase the good output and reduce the bad output at the same rate. This is inconsistent with actual production activities and often leads to the same efficiency value of 1 for many evaluated observations, making it very difficult to evaluate the observed results in an appropriate manner [42]. In addition, as pointed out by Zhang et al. [32], many relevant studies have used time series data only, but studies on cross-sectional data can be regarded as yielding more efficient and effective evaluations over the same period. Obviously, production technology varies from year to year, and results based on production technology at the same time may not be reasonable. To overcome these spatial and dynamic issues, Zhang et al. [32] and Zhang et al. [33] proposed a global non-radial directional distance function (GNDDF) method, which increased the good output and reduced the bad output at different rates and covered all contemporary technologies during the research period.

Although there are many papers on resource utilization efficiency, to our knowledge, there is no research on the analysis of the green use efficiency of industrial water (GUEIW) in Korea. Therefore, we applied a relatively advanced GNDDF model, which included undesirable output, to calculate the GUEIW in Korea from 2006 to 2015. We further analyzed water use efficiency using the economic efficiency of industrial water use (ECEIW) and the environmental efficiency of industrial water use (ENEIW) components to illuminate the major contributing factors of GUEIW. Moreover, we calculated local government-level ECEIW and ENEIW to find out which local government is more effectively promoting the efficiency of Korean industrial water use.

1.2. Geographic Features

Korea is located in East Asia, south of the Korean peninsula, and is surrounded by the East Sea and Yellow (West) Sea. The country lies between 124° and 132° longitude and between 33° and 42° latitude. This geographical location has a major influence on the country's climate, which is divided into four distinct seasons and is characterized by continental and temperate monsoon climates, depending on the region [43].

The country's land area is 99,596 square kilometers, with mountainous topography covering 70 percent of the country. Most of the mountains are in the eastern part, characterized by sharp declines, while the height of the western and southern regions drops slowly. This is why four major rivers flow from the eastern mountain areas into the West Sea (see Figure 1) [43]. The Han river is 481.7 km long, with a catchment area of 26,018 sq km, and is the largest river in South Korea; it flows through the most populous Seoul metro area, including Incheon (the third largest city), and into the West Sea. The Naktong river (506.17 km in length), with a catchment area of 23,384 sq km, is the longest river flowing into the southern sea, passing through the two metropolises of Busan (second) and Daegu (fourth), as well as several industrial cities. The Geum river, which is 394.79 km in length and covers 9912.15 sq km, begins in the central part of the country and ends in the West Sea. The cities of Daejeon and Sejong (fifth) reside on its banks. The Yeongsan, which is 115.5 km in length and 3371 sq km in basin area, is a river in southwestern Korea. It passes through Gwangju (sixth) and flows into the West Sea [43].

Despite heavy rainfall, water supplies in South Korea are limited. The Organization for Economic Co-operation and Development (OECD) classifies Korea as a water-stressed country (Figure 2), for, although the concept of water stress is relatively new, access to freshwater resources has become much more difficult over time and could lead to further depletion and deterioration of available water resources. Water shortage can be caused by climate change, droughts or floods, increased pollution, increased human demand, and excessive use of water resources. A water crisis can occur when the amount of potable, uncontaminated water in an area is less than what the area needs. Shortage of water resources is caused by two converging phenomena: The ever-increasing fresh water use and the exhaustion of available fresh water resources. Water stress can be the result of two mechanisms: Material (absolute) water stress and economic water stress. Material water stress is the result of water resources being naturally inadequate in meeting regional needs, while economic water stress is the result of mismanagement of adequate available water resources (OECD Environmental Outlook to 2050).

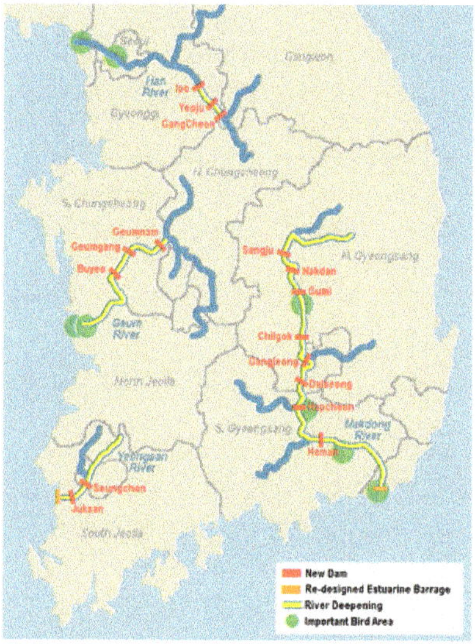

Figure 1. The four major river basins in South Korea.

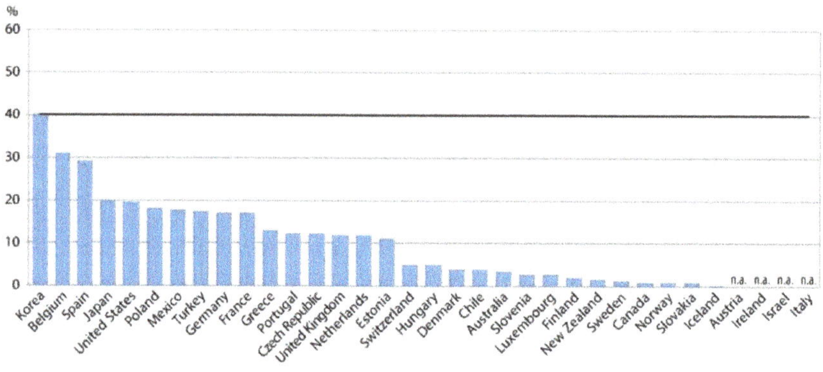

Figure 2. Water stress in Organization for Economic Co-operation and Development (OECD) countries. Note: Water stress levels: Below 10% = no stress; 10%–20% = low stress; 20%–40% = medium stress; above 40%: Severe stress.

As shown in Figure 2, compared with other OECD countries, Korea is coming close to facing a water crisis. As of 2015, Korea had used up to 33 percent of its total available water, putting the country's water balance at risk. Korea's total water resources amounted to 132.3 billion m^3 in 2014, as shown in Figure 3. However, available water resources are estimated to be slightly more than half (76 billion m^3, or 57%). During the rainy season between June and September, 42.5 percent of water (56.3 billion m^3) was discharged. In particular, heavy rain brought by summer monsoons and typhoons resulted in flooding in the downriver areas of the four major river basins (see Figure 3). Due to steep mountain slopes and covered soils, the surface runoff is fast and half of it goes directly into the Sea. Furthermore, almost half of the precipitation is lost due to evaporation and transpiration [44].

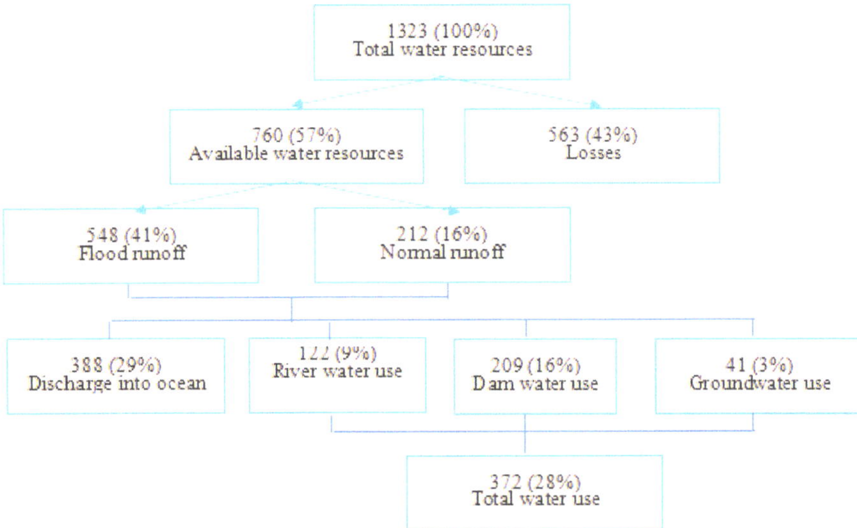

Figure 3. Distribution of water resources in South Korea, 100 million m^3/year. Note: Flood runoff is measured between June and September. Source: MoLIT (2017), The 4th Long-term Comprehensive Plan of Water Resources (2001–2020), 3rd revision.

Moreover, the proportion of water utility bills that covers production costs is far lower than total costs, with a 0.7 per cent decrease per year on average, from 84.4 percent in 2007 to 77.5 percent in 2015 (see Table 1). This continuously increasing disparity between the production costs of drinkable water and the price for which it is supplied has magnified the mismatch between water demand and supply. According to Kim [13], Korea's per capita daily water consumption of 395 liters is the highest among the 29 member nations of the OECD. The amount stands at 197.5 liters in France, 197.5 liters in Germany, and 158 liters in Denmark. Korea's low-priced water bills are likely to lead to this overuse. Therefore, in order to delineate the sustainable performance of this water policy, we will evaluate the quantitative and qualitative efficiency of water in the following section.

Table 1. Water tariff over 2007–2015.

	2007	2008	2009	2010	2011	2012	2013	2014	2015
Public water price (won/m^3)	603.9	609.3	609.9	610.2	619.3	649.1	660.4	666.9	683.4
Production costs (won/m^3)	715.4	730.7	761.6	777.2	813.4	814.7	849.3	876.5	881.7
Cost coverage ratio (%)	84.4	83.4	80.1	78.5	76.1	79.7	77.8	76.1	77.5

Source: Ministry of the Interior and Safety, https://www.mois.go.kr/.

2. Methodology

In this section, the global non-radial directional distance function (GNDDF) is presented to evaluate water stress in terms of its efficiency. This model shows how to estimate the green use efficiency of industrial water (GUEIW), the "economic" efficiency of industrial water (ECEIW), and the "environmental" efficiency of industrial water (ENEIW).

2.1. GNDDF (Global Non-Radial Directional Distance Function)

Chambers et al. [30] first introduced the traditional directional distance function (DDF). After Chung et al. [45] extended this approach to evaluate environmental efficiency, DDF became a widely accepted approach to resource and environmental efficiency assessment. As shown in Equation (1), the conventional radial directional distance function approach always assumes that all inputs and outputs are introduced at the same proportionate rate, which does not conform to real production activities. Moreover, this radial DDF approach may overestimate efficiency due to non-zero slacks existing in inputs or outputs [46,47]. To overcome these problems, this paper utilized a non-radial DDF approach, which is widely used in research on resource efficiency evaluations [48], as shown in Equation (2)

$$\vec{D}(x,y,b;g) = \sup\{\beta:((x,y,b) + g \times \beta) \in T\} \tag{1}$$

$$\vec{D}(x,y,b;g) = \sup\{W^T\beta:((x,y,b) + g \times diag(\beta)) \in T\}, \tag{2}$$

where $W^T = (x,y,b)^T$ denotes a normalized weight vector with respect to inputs and outputs, $g = \left(-g_x, g_y, -g_b\right)$ is an explicit directional vector, and $\beta = (\beta_x, \beta_y, \beta_b)^T \geq 0$ represents the vector of scaling factors. Thus, input and output may have different adjustment ratios, conforming to real production activities. The symbol *diag* denotes diagonal matrices and the environmental technology possibility set.

In this paper, let us assume that there are N local governments being evaluated. Each local government has inputs (x) to produce desirable outputs (y) and undesirable outputs (b). The regulated environmental technology T_1 for N decision making units (DMUs) can be expressed as follows:

$$T_1(x) = \left\{(x,y,b) \middle| \; x \quad can \quad produce \quad (y,b):\right\}$$
$$s.t.\begin{cases} \sum\limits_{n=1}^{N} z_n x_n \leq x, \\ \sum\limits_{n=1}^{N} \lambda_n y_n \geq y, \\ \sum\limits_{n=1}^{N} \lambda_n b_n = b, \\ \sum\limits_{n=1}^{N} \lambda_n = 1, \end{cases} \tag{3}$$
$$\lambda_n > 0, n = 1, \ldots, N,$$

where $T_1(x)$ satisfies the production function theory, implying that limited inputs can produce limited outputs only. Inactivity is assumed as well, implying that there is no production for no undesirable outputs [49]. In addition, we imposed a weak disposability hypothesis on $T_1(x)$, implying that the undesirable output cannot be reduced freely. Considering the production technical change during the experimental period, we applied a function that imposed the constraints $\sum\limits_{n=1}^{N} \lambda_n = 1$ into variable returns to scale (VRS). Because the value is estimated based on the benchmark technology of the same period, the technology is obviously different over the years. The values over different years cannot be compared with each other; therefore, we utilized a global technology benchmark to include all of the contemporary technologies over the research period. Global technology integration, as proposed by

Pastor [50], is denoted by $T_G = T_1 \cup T_2 \cup \ldots \cup T_N$, and the global benchmark technology possibility set $T_2(x)$ can be defined as follows:

$$GT_{2(x)} = \{(x, y, b)\ x \quad can \quad produce(y, b)\ \}$$
$$s.t.\begin{cases} \sum\limits_{t=1}^{T}\sum\limits_{n=1}^{N} x_n^t \lambda_n^t \le x, \\ \sum\limits_{t=1}^{T}\sum\limits_{n=1}^{N} y_n^t \lambda_n^t = y, \\ \sum\limits_{t=1}^{T}\sum\limits_{n=1}^{N} b_n^t \lambda_n^t = b, \end{cases} \tag{4}$$
$$\lambda_n > 0, n = 1, \ldots, N, t = 1, 2, \ldots, T$$

In this research, in accordance with previous research and data availability, input variables mainly included three indicators: Industrial water use (WA), industrial labor force (L), and industrial capital (K), referring to the water amount used for industrial production, the strength of the labor force in the industrial sector, and annual net industrial fixed assets, respectively. The desirable output was industrial GDP (Y), while the undesirable output was the two major water stress factors in industries: Water waste and organic pollution (OP). In order to eliminate the dilution effect of industrial labor and industrial capital, we removed it from the objective function and constrained it with the "ceteris paribus" condition. Therefore, we can calculate the GUEIW through the GNDDF model as follows.

$$\vec{D} = \max w_{IW}\beta_{IW} + w_Y\beta_Y + w_{WW}\beta_{WW} + w_{OP}\beta_{OP}$$
$$s.t.\begin{cases} \sum\limits_{t=1}^{T}\sum\limits_{n_h=1}^{N} \lambda_n^t IW_n \le (1-\beta_{IW})IW_o, \sum\limits_{t=1}^{T}\sum\limits_{n=1}^{N} \lambda_n^t L_n \le L_o, \sum\limits_{t=1}^{T}\sum\limits_{n=1}^{N} \lambda_n^t K_n \le K_o, \\ \sum\limits_{t=1}^{T}\sum\limits_{n=1}^{N} \lambda_n^t Y_n \ge (1+\beta_Y)Y_o, \\ \sum\limits_{t=1}^{T}\sum\limits_{n=1}^{N} \lambda_n^t WW_n = (1-\beta_{WW})WW_o, \sum\limits_{t=1}^{T}\sum\limits_{n=1}^{N} \lambda_n^t OP_n = (1+\beta_{OP})OP_o, \\ \beta_{IW} \ge 0, \beta_Y \ge 0, \beta_{WW} \ge 0, \beta_{OP} \ge 0, \end{cases} \tag{5}$$
$$n = 1, \ldots, N; t = 1, 2, \ldots, T; \lambda_n^t \ge 0, \sum_{n=1}^{N} \lambda_n^t = 1$$

where subscript 0 refers to the evaluation of the local government. β_{IW} is the input adjustment ratio of industrial water, β_Y is the adjustment radio of desirable output of industrial GDP's, and β_{WW} and β_{OP} are the adjustment ratios of undesirable outputs of industrial water waste and organic pollutants. Subscript n refers to the number of local governments in the sample; superscript t is the year. When β is zero, it means the local government is located on the production frontier.

2.2. GUEIW (Green Use Efficiency of Industrial Water)

In accordance with Wang et al. [51], we set the weight vector as (1/3, 1/3, 1/6, 1/6) since we included two undesirable outputs. By solving the following equations, we can estimate the GUEIW and its two decompositions ECEIW and ENEIW [51]. The estimated DMU is valid in industrial water when the value of the GUEIW is equal to 1 and is ineffective when the GUEIW is less than 1. ECEIW and ENEIW are the same case.

$$GUEIW = \frac{(1-\beta_{IW}) + (1-\beta_Y)\frac{1}{2}[(1-\beta_{WW}) + (1-\beta_{OP})]}{3} = 1 - \frac{\beta_{IW} + \beta_Y + \frac{1}{2}(\beta_{WW} + \beta_{OP})}{3}$$
$$ECEIW = \frac{(1-\beta_{IW}) + (1-\beta_Y)}{2} = 1 - \frac{\beta_{IW} + \beta_Y}{2} \tag{6}$$

$$ENEIW = \frac{(1-\beta_{IW}) + \frac{1}{2}[(1-\beta_{WW}) + (1-\beta_{OP})]}{2} = 1 - \frac{\beta_{IW} + \frac{1}{2}(\beta_{WW} + \beta_{OP})}{2}$$

3. Empirical Results

In this section, we present the data including the inputs and outputs of the GNDDF framework. We collected annual industry data for all 16 Korean local government levels from 2006–2015. In this chapter, we calculated GUEIW based on the non-radial directional distance function (NDDF) and GNDDF models for comparison. In order to examine the main catalyst for GUEIW growth in Korea, we calculated the two forms of decomposition of the GUEIW: ECEIW and ENEIW.

3.1. Data Collection

In the DEA model, production inputs include two basic types: Industrial labor (L) and industrial capital (K); we added industrial water (IW) as the third input. We selected industrial GDP as the desirable output and selected water waste (WW) and organic pollutants (OP) {Organic pollutant (kg/year) = BOD (mg/l) × water waste (m^3/year) × 10^3} as two undesirable outputs. Table 2 shows the descriptive statistics for each variable. The data were derived from the Korean Statistical Information Service (KOSIS). The industrial GDP and capital were converted based on 2010 prices.

Table 2. Descriptive statistics for variables.

	Variable	Unit	Mean	Std. Dev.	Min	Max	Obs.
Input	Labor	10^3 person	166.43	183.20	3.14	885.85	160
	Industrial water	10^7 m^3	11.16	10.91	1.68	43.90	160
	Capital	10^{12} KRW	6.11	14.82	0.02	82.88	160
Desirable output	GDP	10^{12} KRW	172.09	157.39	11.02	753.95	160
Undesirable output	water waste	10^6 m^3	11.90	10.54	0.29	46.87	160
	Organic pollutant	10^3 ton	5.63	4.81	0.19	20.98	160

Sources: Korean Statistical Information Service (KOSIS) (http://kosis.kr/).

3.2. Results and Discussion

As demonstrated in Figure 4, the GUEIW calculated from the GNDDF model is always lower than the value based on the NDDF model during the study period, with average values of 0.929 and 0.864. This indicates that the GUEIW based on the NDDF model might be over-estimated, as it is based on contemporary environmental production technology. The NDDF measures relative efficiency and its evaluation criteria are determined by the evaluated DMUs. The efficiency scores obtained under different evaluation criteria are not comparable. To compare efficiency intertemporally, they must be carried out on the same benchmark (frontier). For example, when evaluating the efficiency of 16 local governments in 2006, relevant data constituted a frontier surface A, and the efficiency of each DMU was relative to the evaluation benchmark A. When evaluating the efficiency of 16 local governments in 2007, relevant data constituted a new frontier surface B, and the efficiency of each DMU was relative to the evaluation benchmark B. Generally, A and B are not the same, so the efficiency values in 2006 and 2007 were not comparable. If intertemporal comparisons are to be made, the efficiency of the two years must be based on the same frontier. The GNDDF in this paper refers to the data of the same region sampled in the ten years from 2006 to 2015 as different DMUs. The 160-sample data are then combined to create a common frontier on which the efficiency may be measured. However, the GNDDF model encompasses all concurrent technologies during the sample period. The calculated value is more reasonable because the most efficient node 1 is based on global environmental production technology. Therefore, the following analysis was based on the GNDDF model. In Figure 4, the GUEIW trend on GNDDF exhibits an m-shaped curve, which can be divided into three phases. The first phase is 2006–2007, the second 2008–2012, and the third 2013–2015. The first phase shows a rising trend from 2006–2007, indicating that industrial water use efficiency had achieved obvious improvement during the two years, but decreased in 2008 because of the global economic crisis. A rising trend can be seen again after 2010, following strong support for President Lee's "green growth policies." However, this upward trend only lasted two years; in 2013, a sharp decline occurred again following less emphasis on the environment by President Park's administration. It is clear that the GUEIW for Korea experienced two notable increase–decrease phases over the sample period. Measurements during 2007 and 2012 show that water use was more efficient; this may stem from the impacts of ENEIW and ECEIW. Therefore, in order to further understand the dynamic changes in GUEIW, it is necessary to analyze the trends of its contributing factors.

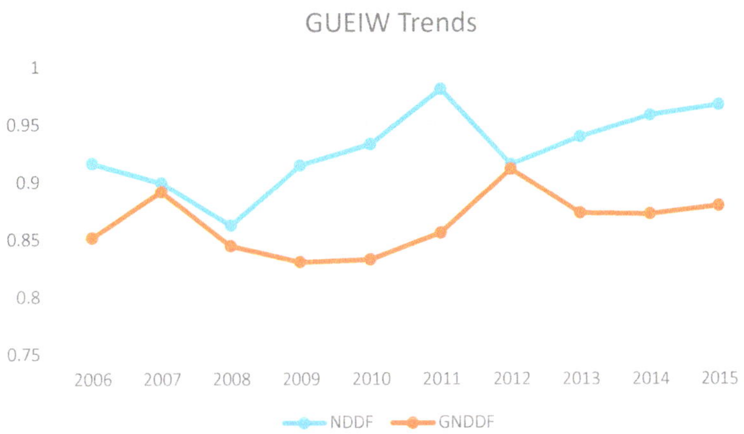

Figure 4. The green use efficiency of industrial water (GUEIW) for Korea based on the non-radial directional distance function (NDDF) and global non-radial directional distance function (GNDDF) models, 2006–2015.

Based on Equation (6) and as shown in Figure 5, we were able to outline the trends in ECEIW and ENEIW for Korea. It is clear that they exhibited similar trends as the GUEIW. There was no large gap during the study period, but the value of ECEIW was higher than that of ENEIW, with an average value of 0.870, while that of ENEIW was 0.864. This implies that environmental efforts may present some opportunity costs that are expressed in the gap between ECEIW and ENEIW. The year 2010 exhibited the biggest gap between ECEIW and ENEIW, with average values of 0.845 and 0.826. This implies that the growth of GUEIW is mainly driven by ECEIW and that environmental issues function as constraints in the achievement of green use efficiency of industrial water in Korea. This is consistent with the argument of Wang et al. [51], who stated that environmental protection of industrial sectors needs to be improved in China too. In order to improve the economic efficiency of industrial water use, advanced water technology should be developed or introduced to make full use of water resources in industrial production. For example, in Denmark, through technology integration, "wastewater treatment plants" could be turned into "energy factories", biomass power could be produced by extraction of organic matter in sewage, and enough electricity could be supplied not only to run the motor, but also to pump water and have a surplus. In addition, waste water treatment plants use the money earned from waste power generation and heating to support the operation of the sewage treatment plant and even to achieve a break-even point [52]. In addition, it is noteworthy that ENEIW showed a significant upward trend during 2006–2007 and 2010–2012, indicating a significant environmental improvement in Korea.

This can be attributed to a series of laws and regulations. As the River Basin Law was amended in 2007 under the Long-term Comprehensive Water Resources Plan, the Korean government worked to ensure a stable supply of water in the event of climate change. The Korean government implemented the river-oriented national land improvement project (four-river restoration project) and the new national land development paradigm (low-carbon green growth) [53]. Nevertheless, the unanticipated international financial crisis in 2008 had a significant negative impact on Korea's industrial exports, which caused many factories to close down and workers to lose their jobs, resulting in sharp decreases in both ECEIW and ENEIW. This forced the central government to introduce a series of economic stimulus plans, the components of which included expanding investment in small- and medium-sized industries to ensure industrial growth, thus boosting the recovery of the ECEIW [54]. Under the stimulus plan, some small- and medium-sized industries in Korea have overemphasized temporary doubts about the monitoring and implementation of environmental policies; they take chances on not being regulated, causing serious industrial pollution to many rivers. This is consistent with our findings. Figure 5 demonstrates the rapid upward trend of the ECEIW after 2008, as well as a large gap in the ENEIW in 2010. As Water Vision 2020 was revised for the second time in 2011 [55], both the ECEIW and ENEIW exhibit upward trends. The period between 2008–2012 exhibits these stricter environmental regulations that can stimulate technological innovation and improve production efficiency, leading to resource savings and environmentally friendly industrial production. This provides evidence for the Porter hypothesis [56]. Unfortunately, since 2013, President Park Geun-hye changed the paradigm from "green growth" to "creative economy." Environmental policies became passive and loose compared with those during Lee Myung-bak's administration. These changes led to the decline of the GUEIW.

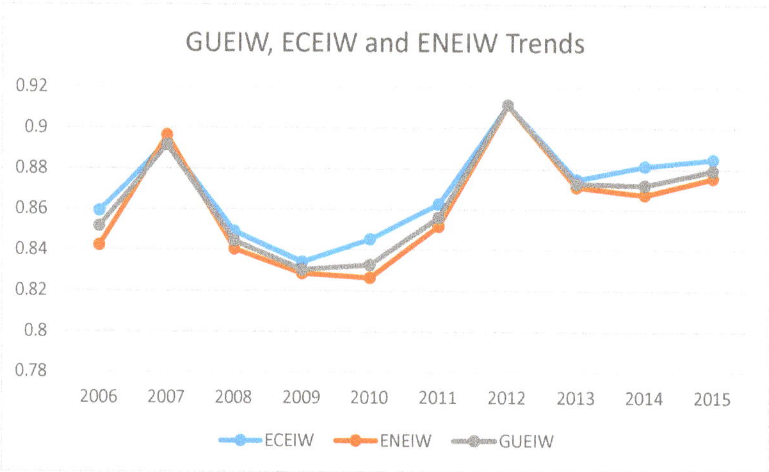

Figure 5. The GUEIW, economic efficiency of industrial water use (ECEIW), and environmental efficiency of industrial water use (ENEIW) for Korea, 2006–2015.

Figure 6 shows the average values of the GUEIW, ECEIW, and ENEIW at the local government levels. All three values are close to 1 in many governments; this is because Korea's water management system is centralized. In other words, all of these local governments share the same management experience and are under the same laws and regulations, so their efficiency values are extremely close. Therefore, there is no competition among local governments to enhance water management, resulting in similar performances. During the study period, Ulsan showed the highest value of 1 and Gangwon showed the lowest value of 0.351. It is worth noting that 10 local governments with a GUEIW value of over 0.9 are located on the south and west coasts of South Korea (Chungnam, Busan, Jeonnam, Gyeonggi, Gyeongnam, Seoul, Gyeongbuk, Jeju, Gwangju, Ulsan). These local governments are situated downstream of rivers. Moreover, Gangwon, Daejeon, Daegu, Chungbuk, and Jeonnbuk are situated upstream, in the water source protection area. The results are not surprising, as the Han River Law was established in 1999 [57] to improve water quality, manage drinking water sources, and support the compensation of local governments in the upper reaches of the river basin as an economic incentive to reduce chemical use in factories and farms. In the same year, the River Basin Law envisaged the comprehensive improvement of national development plans [56]. This has led downstream industries to pay a kind of water environmental tax to upstream industries. Water users in downstream areas, such as Seoul, Gyeonggi, Chungnam, Busan, Jeonnam, Gwangju, Ulsan, and Gyeongnam pay for upstream water protection compensation in conservation areas that supply them with more environmentally friendly water. Gangwon, Chungbuk, Daejeon, Daegu, and Jeonnbuk provinces are the areas that protect upstream areas and their people by regulating economic activity (such as environmentally friendly housing and industries). It is worth mentioning that Incheon, although in the downstream.

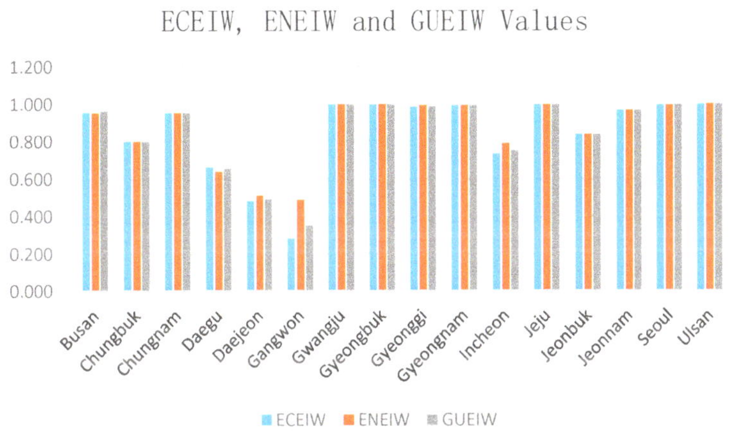

Figure 6. Average values of GUEIW, ECEIW, and ENEIW for Korea local governments.

m region, has a GUEIW value of 0.752, lower than other downstream regions. This result is not surprising. Most of Incheon's cities have been developed as industrial zones for Seoul; thus, many of these zones, such as Namdong or Gajwa, consist of small- and medium-sized enterprises in the country. Compared with large enterprises in other cities such as Ulsan, industries in Incheon do not have sufficient technologies and facilities for environmentally friendly production processes, resulting in low ECEIW and ENEIW values. This situation also indicates that small- and medium-sized enterprises in the area can learn more from big companies by a spill-over effect.

Except for Incheon, all downstream regions outperformed the upstream region in all three indicators, particularly ECEIW. The government's efforts to ensure water quality inhibits industrial development in upstream areas, which negatively impacts the incomes of upstream areas; the relatively high water use costs in downstream areas encourage producers to save water. While the government focuses on compensating the upstream areas for water quality management with the water tax from downstream areas, such supplies cannot be sustainable because the government is only responsible for the initial stage of installation and does not provide any incentives for industries to reduce emissions. Regulations should be strengthened and effective incentives should be implemented to allow upstream industries to provide their own environmental protection activities. Therefore, the upstream region should focus on environmental protection as a valuable activity for sustainable performance and develop suitable industries with lower pollution and higher output. To develop such an industry, the water emission trading system (ETS) could be a good option for sustainable performance because the industry may reduce water use (quantitative efficiency) as well as produce less pollution (qualitative efficiency) for enhanced GUEIW. Here, ETS is the market-oriented carbon emission regulatory mechanism for the participating unit to reduce its carbon emission for better compensation in the ETS market. A water ETS could be constructed in a similar direction with caps and a trading system based on water management of industries.

4. Conclusions

Since water stress—which mainly comes from industrial water waste and water pollution—poses a significant threat to sustainable water management in Korea, it is urgent to assess GUEIW and make useful policy recommendations. Based on the GNDDF model, this paper calculated GUEIW in 16 Korean local governments from 2006 to 2015 based on two decomposition indicators: ECEIW and ENEIW.

Empirical results showed that GUEIW exhibited fluctuations during the study and that the growth of GUEIW was mainly driven by ECEIW. In order to improve the economic efficiency of industrial water use, advanced water technology should be developed or introduced to make full use of water resources in industrial production. Local governments should employ advanced technology and help less developed local governments achieve enhanced industrial water use.

For the ENEIW, environmental issues might be constraints for achieving green use of industrial water. Stricter environment regulations can stimulate technological innovation and improve water management efficiency, leading to water resource savings and environmentally friendly industrial production. This result supports the arguments related to the Porter hypothesis. It also demonstrates that environmental deregulation leads to a decline in GUEIW.

Regarding local government heterogeneity, the downstream region outperformed the upstream region in all three indicators, particularly in ECEIW. The government's efforts to ensure water quality inhibits industrial development in upstream areas. Upstream area incomes are much lower than in downstream areas, as downstream industries have to pay upstream industries for upstream water quality management. With relatively high water use costs, producers in downstream regions promote conservation of water. Upstream industries, however, may easily create water waste due to relatively low prices of water use. In order to improve the economic efficiency of industrial water use in upstream areas, an advanced water technology should be developed and introduced to make full use of water resources in industrial production. The local government in upstream regions should help local industries acquire advanced technology. In addition, the government can provide incentives for producers to save water to encourage industrial water saving. Water ETS could be one of the best alternatives for these performance-oriented policies.

Author Contributions: The authors are contributed each part of a paper by Conceptualization, Y.C.; Methodology, N.W.; Software, N.W.; Validation, Y.C.; Formal Analysis, N.W.; Investigation, Y.C.; Resources & Data Curation, N.W.; Writing—Original Draft Preparation, N.W.; Writing—Review & Editing, Y.C.; Visualization, N.W.; Supervision, Y.C.; Project Administration, Y.C.; Funding Acquisition, Y.C.

Funding: This research received no external funding.

Conflicts of Interest: The authors declare no conflict of interest.

References

1. Kim, B.W.; Kim, G.B.; Kim, D.G.; Koh, Y.K.; Choi, D.H. Current states of the Global water market and considerations for the Groundwater industry in South Korea. *J. Eng. Geol.* **2014**, *24*, 431–440. [CrossRef]
2. Baron, J.S.; Poff, N.L.; Angermeier, P.L.; Dahm, C.N.; Gleick, P.H.; Hairston, N.G.; Jackson, R.B.; Johnston, C.A.; Richter, B.D.; Steinman, A.D. Meeting ecological and socieal needs for freshwater. *Ecol. Appl.* **2002**, *12*, 1247–1260. [CrossRef]
3. Vugteveen, P.; Lenders, H.J. The duality of integrated water management: Science, policy or both? *J. Integr. Environ. Sci.* **2009**, *6*, 51–67. [CrossRef]

4. Choi, I.-C.; Shin, H.-J.; Nguyen, T.T.; Tenhunen, J. Water Policy Reforms in South Korea: A Historical Review and Ongoing Challenges for Sustainable Water Governance and Management. *Water* **2017**, *9*, 717. [CrossRef]
5. Han, S.-Y.; Kwak, S.-J.; Yoo, S.-H. Valuing environmental impacts of large dam construction in Korea: An application of choice experiments. *Environ. Impact Assess. Rev.* **2008**, *28*, 256–266. [CrossRef]
6. Hoang, V.-N.; Seo, B.; Nguyen, T.T.; Hoang, V. Cost and environmental efficiency of rice farms in South Korea. *Agric. Econ.* **2012**, *43*, 369–378.
7. Organization for Economic Co-Operation and Development (OECD). *Evaluation of Agricultural Policy Reforms in Korea*; OECD Publishing: Paris, France, 2008.
8. Cho, C.J. The Korean growth-management programs: Issues, problems and possible reforms. *Land Use Policy* **2002**, *19*, 13–27. [CrossRef]
9. Kwak, S.J.; Russell, C.S. Contingent valuation in Korean environmental planning: A pilot application to the protection of drinking water quality in Seoul. *Environ. Resour. Econ.* **1994**, *4*, 511–526. [CrossRef]
10. Labadie, J.W.; Fontane, D.G.; Lee, J.H.; Ko, I.W. Decision support system for adaptive river basin management: Application to the Geum River basin, Korea. *Water Int.* **2007**, *32*, 397–415. [CrossRef]
11. Xi, D.L.; Sun, Y.S. *Environmental Monitoring*, 4th ed.; Higher Education Press: Beijing, China, 2010; Volume 7, pp. 127–135.
12. Water Resources Management Information System. Available online: http://www.wamis.go.kr/wke/wke_wqbase_lst.aspx#menu02 (accessed on 1 July 2019).
13. Kim, K.M. *Improvement of the Han River Watershed Management Fund Policies*; National Assembly Research Service (NARS) Issue Report 160; NARS: Seoul, Korea, 2012.
14. Shin, H.J.; Jeon, C.H.; Choi, I.C.; Yeon, I.C. Estimation of beneficiary's willingness to pay in mid and down-stream area to the water quality improvements in upper Bukhan River Basin. *Seoul Stud.* **2009**, *10*, 91–106. (In Korean)
15. Kumbhakar, S.C.; Bhattacharyya, A. Price distortions and resource-use efficiency in indian agriculture: A restricted profit function approach: A restricted profit function approach. *Rev. Econ. Stat.* **1992**, *74*, 231–239. [CrossRef]
16. Li, J.; Ma, X. The utilization efficiency of industrial water under the dual constraints of resource and environment—An empirical study based on sbm-undesirable and meta-frontier model. *J. Nat. Resour.* **2014**, *29*, 920–933. (In Chinese)
17. Cheng, H.; Hu, Y.; Zhao, J. Meeting China's water shortage crisis: Current practices and challenges. *Environ. Sci. Technol.* **2009**, *43*, 240–244. [CrossRef] [PubMed]
18. Xie, H.; Wang, W. Spatiotemporal differences and convergence of urban industrial land use efficiency for China's major economic zones. *J. Geogr. Sci.* **2015**, *25*, 1183–1198. [CrossRef]
19. Cai, Y.; Yue, W.; Xu, L.; Yang, Z.; Rong, Q. Sustainable urban water resources management considering life-cycle environmental impacts of water utilization under uncertainty. *Resour. Conserv. Recycl.* **2016**, *108*, 21–40. [CrossRef]
20. Gao, H.; Wei, T.; Lou, I.; Yang, Z.; Shen, Z.; Li, Y. Water saving effect on integrated water resource management. *Resour. Conserv. Recycl.* **2014**, *93*, 50–58. [CrossRef]
21. Liu, J.; Li, Y.P.; Huang, G.H.; Zeng, X.T. A dual-interval fixed-mix stochastic programming method for water resources management under uncertainty. *Resour. Conserv. Recycl.* **2014**, *88*, 50–66. [CrossRef]
22. Hu, Z.; Chen, Y.; Yao, L.; Wei, C.; Li, C. Optimal allocation of regional water resources: From a perspective of equity–efficiency. *Resour. Conserv. Recycl.* **2016**, *109*, 102–113. [CrossRef]
23. Tu, Y.; Zhou, X.; Gang, J.; Liechty, M.; Xu, J.; Lev, B. Administrative and market-based allocation mechanism for regional water resources. *Resour. Conserv. Recycl.* **2015**, *95*, 156–173. [CrossRef]
24. Zhang, N.; Xie, H. Toward green IT: Modeling sustainable production characteristics for Chinese electronic information industry, 1980–2012. *Technol. Forecast. Soc. Chang.* **2015**, *96*, 62–70. [CrossRef]
25. Shao, Y. Analysis of energy savings potential of China's nonferrous metals industry. *Resour. Conserv. Recycl.* **2016**. [CrossRef]
26. Wu, H.; Shi, Y.; Xia, Q.; Zhu, W. Effectiveness of the policy of circular economy in China: A DEA-based analysis for the period of 11th five-year-plan. *Resour. Conserv. Recycl.* **2014**, *83*, 163–175. [CrossRef]
27. Bian, Y.; Yan, S.; Xu, H. Efficiency evaluation for regional urban water use and wastewater decontamination systems in China: A DEA approach. *Resour. Conserv. Recycl.* **2014**, *83*, 15–23. [CrossRef]

28. Geissler, B.; Mew, M.C.; Weber, O.; Steiner, G. Efficiency performance of the world's leading corporations in phosphate rock mining. *Resour. Conserv. Recycl.* **2015**, *105*, 246–258. [CrossRef]
29. Jaeger, S.D.; Rogge, N. Cost-efficiency in packaging waste management: The case of Belgium. *Resour. Conserv. Recycl.* **2014**, *85*, 106–115. [CrossRef]
30. Charnes, A.; Cooper, W.W.; Rhodes, E. Measuring the efficiency of decision making units. *Eur. J. Oper. Res.* **1978**, *2*, 429–444. [CrossRef]
31. Ren, J.; Manzardo, A.; Mazzi, A.; Fedele, A.; Scipioni, A. Emergy analysis and sustainability efficiency analysis of different crop-based biodiesel in life cycle perspective. *Sci. World J.* **2013**, *1*, 1–12. [CrossRef] [PubMed]
32. Zhang, N.; Choi, Y. A note on the evolution of directional distance function and its development in energy and environmental studies 1997–2013. *Renew. Sustain. Energy Rev.* **2014**, *33*, 50–59. [CrossRef]
33. Zhang, N.; Kong, F.; Kung, C.C. On modeling environmental production characteristics: A slacks-based measure for China's Poyang lake ecological economics zone. *Comput. Econ.* **2014**, *46*, 1–16. [CrossRef]
34. Zhang, N.; Kong, F.; Choi, Y. Measuring sustainability performance for China: A sequential generalized directional distance function approach. *Econ. Model.* **2014**, *41*, 392–397. [CrossRef]
35. Ren, J.; Tan, S.; Dong, L.; Mazzi, A.; Scipioni, A.; Sovacool, B.K. Determining the life cycle energy efficiency of six biofuel systems in China: A data envelopment analysis. *Bioresour. Technol.* **2014**, *162*, 1–7. [CrossRef] [PubMed]
36. Hu, J.; Wang, S.; Yeh, F. Total-factor water efficiency of regions in China. *Resour. Policy* **2006**, *31*, 217–230. [CrossRef]
37. Liao, H.; Dong, Y. Utilization efficiency of water resources in 12 western provinces of china based on the DEA and Malmquist TFP index. *Resour. Sci.* **2011**, *33*, 273–279.
38. Zhao, X.; Chen, B.; Yang, Z.F. National water footprint in an input–output framework—A case study of China 2002. *Ecol. Model.* **2009**, *220*, 245–253. [CrossRef]
39. Manzardo, A.; Ren, J.; Piantella, A.; Mazzi, A.; Fedele, A.; Scipioni, A. Integration of water footprint accounting and costs for optimal chemical pulp supply mix in paper industry. *J. Clean. Prod.* **2014**, *72*, 167–173. [CrossRef]
40. Wang, W.; Xie, H.; Jiang, T.; Zhang, D.; Xie, X. Measuring the total-factor carbon emission performance of industrial land use in China based on the global directional distance function and non-radial luenberger productivity index. *Sustainability* **2016**, *8*, 336. [CrossRef]
41. Xie, H.; Wang, W.; Yang, Z.; Choi, Y. Measuring the sustainable performance of industrial land utilization in major industrial zones of China. *Technol. Forecast. Soc. Chang.* **2016**. [CrossRef]
42. Zhou, P.; Poh, K.L.; Ang, B.W. A non-radial DEA approach to measuring environmental performance. *Eur. J. Oper. Res.* **2007**, *178*, 1–9. [CrossRef]
43. Min, K.J. The Role of the State and the Market in the Korean Water Sector: Strategic Decision Marking Approach for Good Governance. Ph.D. Thesis, University of Bath School of Management, Bath, UK, January 2011.
44. Ministry of Land, Infrastructure and Transportation of Korea (MOLIT); Korea Water Resources Corporation (K-Water). *Water for the Future: Water and Sustainable Development*; K-Water: Daejeon, Korea, 2015.
45. Chung, Y.H.; Färe, R.; Grosskopf, S. Productivity and undesirable outputs: A directional distance function approach. *J. Environ. Manag.* **1997**, *51*, 229–240. [CrossRef]
46. Zhang, N.; Wang, B.; Chen, Z. Carbon emission reductions and technology gaps in the world's factory, 1990–2012. *Energy Policy* **2016**, *91*, 28–37. [CrossRef]
47. Zhang, N.; Wang, B.; Liu, Z. Carbon emissions dynamics, efficiency gains, and technological innovation in China's industrial sectors. *Energy* **2016**, *99*, 10–19. [CrossRef]
48. Chang, T.P.; Hu, J.L. Total-factor energy productivity growth, technical progress, and efficiency change: An empirical study of China. *Appl. Energy* **2010**, *87*, 3262–3270. [CrossRef]
49. Färe, R.; Grosskopf, S. *New Directions: Efficiency and Productivity*; Springer: New York, NY, USA, 2005.
50. Pastor, J.T.; Lovell, C.A.K. A Global Malmquist Productivity index. *Econ. Lett.* **2005**, *88*, 266–271. [CrossRef]
51. Wang, W.; Xie, H.; Dong, X.; Zhang, N. Sustainable water use and water shadow price in China's urban industry. *Resour. Conserv. Recycl.* **2018**, *128*, 489–498. [CrossRef]
52. Ministry of Land, Infrastructure and Transportation of Korea (MOLIT). (2011–2020): Water Vision 2020. In *The Long-Term Comprehensive Water Resource Plan (2011–2020): The Second Revised Water Vision 2020*; MOLIT: Sejong, Korea, 2011. (In Korean)

53. Ministry of Environment. Available online: http://eng.me.go.kr/ndg/web/main.do (accessed on 15 July 2019).
54. Song, P.; Zhu, M. Korean economy in two economic crises and its countermeasures. *Shangdong Soc. Sci.* **2011**, *191*, 7. (In Chinese)
55. Oh, B.H. *2012 Modularization of Korea's Development Experience: Korea's River Basin Management Policy*; Korea Development Institute School of Public Policy and Management: Seoul, Korea, 2013.
56. Porter, M.E.; van der Linde, C. Toward a new conception of the environment: Competitiveness relationship. *J. Econ. Perspect.* **1995**, *9*, 97–118. [CrossRef]
57. Choi, J.; Hearne, R.; Lee, K.; Roberts, D. The relation between water pollution and economic growth using the environmental Kuznets curve: A case study in South Korea. *Water Int.* **2015**, *40*, 499–512. [CrossRef]

sustainability

MDPI

Article

Evaluation of Regression Analysis and Neural Networks to Predict Total Suspended Solids in Water Bodies from Unmanned Aerial Vehicle Images

Tainá T. Guimarães [1], Maurício R. Veronez [2,3,4,*], Emilie C. Koste [2], Eniuce M. Souza [2,3], Diego Brum [2,3], Luiz Gonzaga Jr. [2,3] and Frederico F. Mauad [1]

1 Graduate Programme in Environmental Engineering Sciences, São Carlos Engineering School, University of São Paulo, São Carlos 13566-590, Brazil; tainathomg@usp.br (T.T.G.); mauadffm@sc.usp.br (F.F.M.)

2 Advanced Visualization & Geoinformatics Lab—VizLab, Unisinos University, São Leopoldo 93022-750, Brazil; emiliek@edu.unisinos.br (E.C.K.); eniuce@edu.unisinos.br (E.M.S.); diebrum@unisinos.br (D.B.); lgonzaga@unisinos.br (L.G.J.)

3 Graduate Programme in Applied Computing, Unisinos University, São Leopoldo 93022-750, Brazil

4 Graduate Programme in Biology, Unisinos University, São Leopoldo 93022-750, Brazil

* Correspondence: veronez@unisinos.br; Tel.: +55-51-3591-1100 (ext. 1619)

Received: 14 March 2019; Accepted: 22 April 2019; Published: 5 May 2019

Abstract: The concentration of suspended solids in water is one of the quality parameters that can be recovered using remote sensing data. This paper investigates the data obtained using a sensor coupled to an unmanned aerial vehicle (UAV) in order to estimate the concentration of suspended solids in a lake in southern Brazil based on the relation of spectral images and limnological data. The water samples underwent laboratory analysis to determine the concentration of total suspended solids (TSS). The images obtained using the UAV were orthorectified and georeferenced so that the values referring to the near, green, and blue infrared channels were collected at each sampling point to relate with the laboratory data. The prediction of the TSS concentration was performed using regression analysis and artificial neural networks. The obtained results were important for two main reasons. First, although regression methods have been used in remote sensing applications, they may not be adequate to capture the linear and/or non-linear relationships of interest. Second, results show that the integration of UAV in the mapping of water bodies together with the application of neural networks in the data analysis is a promising approach to predict TSS as well as their temporal and spatial variations.

Keywords: suspended solids; unmanned aerial vehicle; spectral imaging; artificial neural networks

1. Introduction

The typical methodology for investigating water quality involves collecting water samples directly from various locations and laboratory analyses. While this method may result in accurate assessments of water body quality with limited areas, it is time consuming and expensive, and difficult to apply in large areas. Moreover, because the results are punctual, they do not necessarily reflect the quality of the whole site [1,2].

Alternative measures for in situ monitoring of water quality in lakes, dikes, and reservoirs can be obtained by means of remote sensing techniques. Such an application is only possible due to the presence of optically active components in the water. These substances can be identified via sensor systems in that their presence in a water body results in different absorption and backscattering patterns of the incident light, which are characteristic of each component. Among the parameters of water quality, suspended inorganic sediments, organic chlorophyll-a, and dissolved organic material are the main agents of absorption and scattering of electromagnetic radiation in a water body [3,4].

It should be noted that these components are directly related with the quality of the aquatic ecosystem and its surroundings. For example, total suspended solids (TSS), which represents the total amount of inorganic or organic particles drifting or floating in water [5], may be related to water pollution since these can serve as a transporting and storage agent of various pollutants, as well as erosive processes in a river basin (resulting in silting of major rivers and reservoirs) [4]. TSS concentration is often related to total primary production, heavy-metal and micro-pollutant flows, and in many turbid regions, is directly linked to sediment transport problems and the light available for primary production [6].

An indirect measurement of TSS in water bodies via remote sensing can compensate for deficiencies in manual water quality monitoring by being fast, allowing for continuous monitoring of large areas [2,7,8]. Most of the studies published on the TSS prediction from remote sensing involve the use of spectral data retrieved from satellite images. Because of its medium spatial resolution (30 m), in the studies of remote water sensing, one the most common satellites are Landsat, such as found in Qun et al. [2], Kong et al. [8], Din et al. [9], and Amanollahi et al. [10]. Song et al. [6] tested the images for medium spatial resolution IRS-P6 (Indian Remote Sensing Satellite) as well. Besides these, Wang et al. [7], Breuning et al. [11], and Moridnejad et al. [12] used MODIS (Moderate Resolution Imaging Spectroradiometer) satellite images and Campbell et al. [3] of the MERIS (Medium Resolution Imaging Spectrometer), with these having low spatial resolutions (250 and 300 m, respectively) and limited to large areas of water.

Thus, although remote sensing serves as a powerful technique for monitoring environmental and seasonal changes, and its ability to remotely monitor water resources has increased in recent decades because of the improved quality and availability of satellite imagery data [13], the analysis of small water bodies may not be adequate due to the medium image resolution of the most usual commercial satellites [1]. In this case, the use of aerial images obtained by unmanned aerial vehicle (UAVs) for monitoring small bodies of water has presented good results and becomes promising for producing greater detail due to high spatial resolution and the possibility of constant monitoring [14,15].

Although some applications of UAVs for water quality parameters monitoring, such as chlorophyll-a [1,15–17], organic matter [18], and suspended solids [1,18–20], have been demonstrated in the literature, there are still few studies focused on this application. For suspended solids monitoring, for example, Veronez et al. [18] and Saénz et al. [19] used regression analyses between TSS values measured in the laboratory and the UAV responses in the visible and near infrared (NIR) regions to generate their prediction models. Saenz et al. [19] explored relations between individual bands and combinations between them (as NIR-red, for example), and Veronez et al. [18] chose to relate to vegetation indexes such as normalized difference vegetation index (NDVI) and normalized difference water index (NDWI). Although both mentioned studies have shown positive results, the modeling of these parameters in complex environments is not always possible through regression analysis.

Therefore, cognizant of the limitations that techniques such as regression analysis has, the need for research and improvement of inland waters monitoring techniques integrated with the facilities provided by the technologies developed and available in the market can be seen. Among these modern techniques that can provide the support for the monitoring of waters via remote sensing is the artificial intelligence with the use of neural networks.

Approaches involving neural networks are promising in the area of remote sensing and the development of water quality models because they can be more sensitive and robust than other traditional regression techniques, with the ability to capture both linear and non-linear relationships between the involved parameters [8,12,13,21]. However, results presented in the literature on artificial neural networks (ANN) approaches in water bodies use mainly satellite imagery of low [12] to medium [8,13,21] spatial resolution.

No papers were found that included the application of artificial neural networks to the analysis of high spatial resolution images obtained using UAVs, and therefore, this study intends to fill this gap. The aim of this article was to use remote sensing technologies to evaluate water quality, identifying an alternative method for monitoring and quantifying the concentration of suspended solids in water, through the correlation between UAV images and limnological data using regression analysis (RA) and artificial neural networks (ANN). Furthermore, this study aims to contribute to the development of temporal and spatial water quality monitoring techniques through modern remote sensing tools and artificial intelligence.

The manuscript is structured as follows: Section 2 contains the information about the field site, the acquisition of the data, and its subsequent analyses; in Section 3, we present and discuss the results of the research on the concentration of TSS, regression models, and ANN; and finally, Section 4 presents our conclusions regarding the study, with indications of its importance and its continuity.

2. Materials and Methods

The method that we are proposing can be structured according to the following steps: GNSS (Global Navigation Satellite System) data acquisition, water sampling and laboratory analysis, overflight with the UAV and processing of the images, extraction of values from images UAV, regression analysis, and training and testing of the ANN. The flowchart of the proposed method is depicted in Figure 1 and detailed in the following subsections.

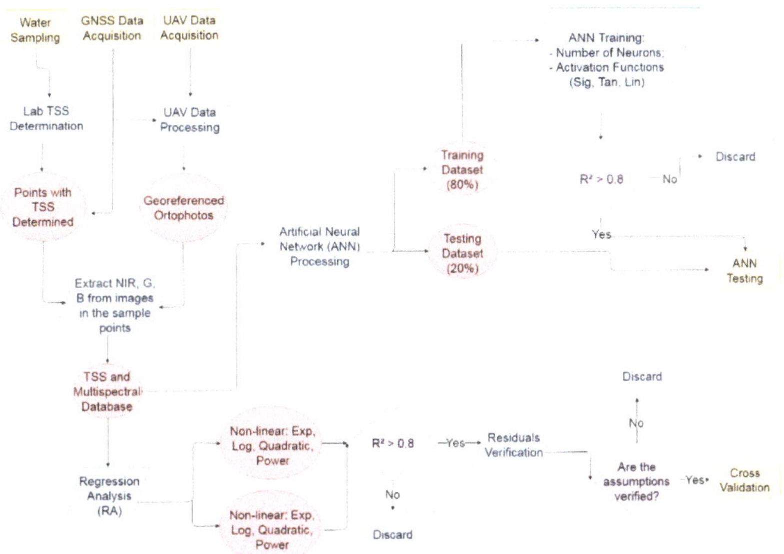

Figure 1. Flowchart of the proposed method.

2.1. Field Site

The adopted study site was the lake on the Unisinos University campus, located in the state of Rio Grande do Sul, southern Brazil (Figure 2). The lake is artificial, has an area of approximately 0.025 km^2 and maximum depth of 4 m. Although small, it is located at the lowest altitude of the campus, and because it is formed from rainwater drainage collected at the university, it contains several inorganic and organic compounds found in the form of suspended solids or organic matter from rainwater runoff [18].

The lake and its surroundings also function as an ecosystem for several species of animals, such as ducks, geese, and several other birds, as well as a great diversity of fish. Because it is a university campus, the area has several buildings, paved areas, and a large circulation of people and cars.

However, the campus also has several vegetated areas, mainly around the lake, as can be seen in Figure 2.

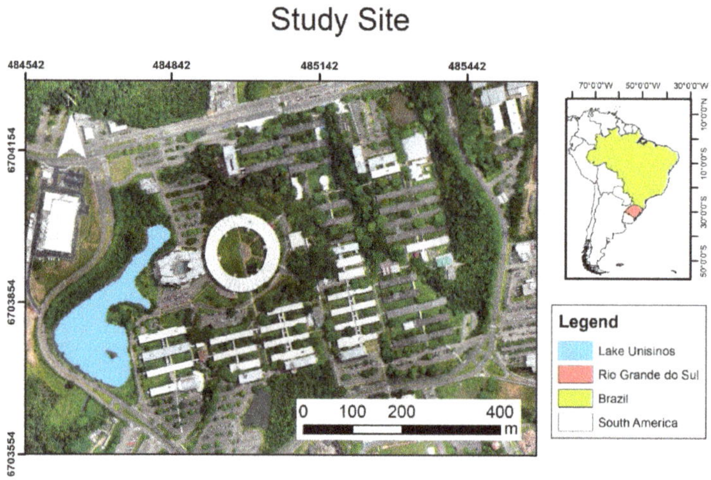

Figure 2. Location of the study site.

Studies addressing the applicability of remote sensing in the monitoring of water bodies have already been developed in this area. Guimarães et al. [17] used spectral data collected in the field and UAV images to model the chlorophyll-a concentration in the environment. Veronez et al. [18], based on UAV images, applied neural networks to estimate Landsat 8 OLI satellite bands and correlated this with data on suspended solids and dissolved organic matter.

Studies including the characterization of this lake, the behavior of the limnological variables, and their relationships with the remote sensing variables are important as they serve as pilot studies to be applied in larger water bodies.

2.2. Data Acquisition

We performed two field samplings in March 2016 and 2017 during the transition period between the seasons of summer and fall. The collections were carried out in a single day and we ensured that the climatic conditions of both days were similar. The average temperatures were between 22 and 24 °C, winds with a speed of 0.4 ms^{-1} (southeast direction), and without the occurrence of precipitation events on the days of the collections.

On the same days, the UAV overflew the area and in situ collection of water samples occurred such that the two pieces of information could be compared as being representative of the same conditions of the lake. Besides, possible temporal variations from one year to another can be evaluated for the collected data and compared to the predicted one from the analyzed RA and ANN methods.

We selected 21 sample points, as shown in Figure 3, that were spatially distributed over the lake such that surface water samples (up to 0.5 m) were collected for the laboratory determination of suspended solids using the gravimetric method described in the Standard Methods for the Examination of Water and Wastewater [22].

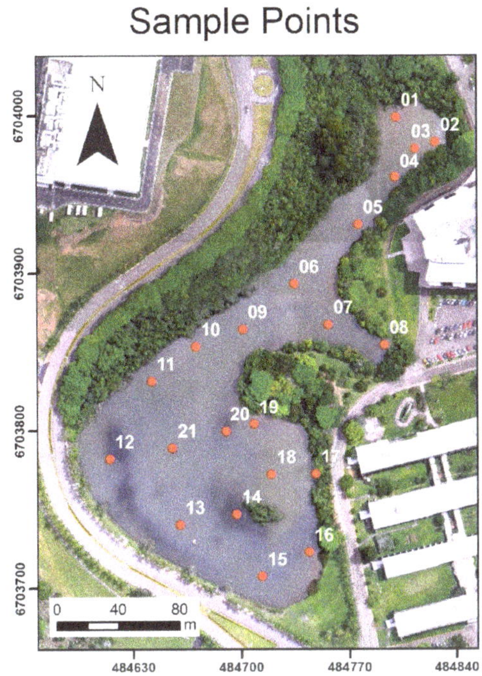

Figure 3. Sample points used in the survey.

The UAV used to take the images was the SenseFly, Swinglet CAM model (SenseFly Parrot Group, Cheseaux-sur-Lausanne, Switzerland). It was coupled to a Canon ELPH 110HS (Canon U.S.A., Inc., New York, NY, United States) camera with a 16-megapixel resolution and was factory-modified to capture the NIR band instead of the red band. Thus, mapping was in three distinct channels, namely: near infrared (NIR), green (G), and blue (B).

As well as the sampling points for water collection, in the field we also established and tracked six ground control points (GCPs), through the GNSS (Global Navigation Satellite System), based on the RTK (real time kinematic) method, located in the area of coverage of the flight such that later their positions were used in the georeferencing of the images obtained.

The images obtained using the UAV were processed using the PIX4D software, version 2.1 (Pix4D S.A., Lausanne, Switzerland), in which the images were orthorectified and georeferenced, where we adopted the SIRGAS 2000 (Geocentric Reference System for the Americas) as the reference system, in the UTM (Universal Transverse Mercator) −22S projection zone. We generated orthophotos with a pixel size of 5 cm × 5 cm.

2.3. Data Analysis

In order to perform analysis between the data collected of the water quality and those obtained via remote sensing, we plotted the sample points where samples were collected in the orthophotos generated by the overfly with the UAV and extracted the values of the pixels concerning each point for the NIR, G, and B channels. We emphasize that among the collected values of the two years ($n = 42$), four were disregarded because the points were located in a shaded area of the image. Thus, a sample of 38 points was considered for the analysis.

We used this data to predict the suspended solids concentration in Lake Unisinos using linear and non-linear regression analysis (RA) and artificial intelligence through an artificial neural network (ANN). The aim of this step was to identify a model to quantify the concentrations of suspended solids present in the water using the information obtained through remote sensing. We evaluated their performances using the following statistical metrics: coefficient of determination (R^2) and root mean square error (RMSE).

Linear and non-linear regression models were investigated. The considered non-linear functions were exponential, logarithmic, quadratic, and power (range from −1 to 1). Knowing that different concentrations of suspended solids present different responses for each wavelength, to predict TSS in RA models, we considered as independent variables each channel individually (NIR, G, and B) and the operations of bands (sum, subtraction, and ratios) to highlight the spectral characteristics of the compounds [6–8,13,21]. Thus, besides the simple regressions with the covariates included individually, multiple regressions were considered with two or more independent variables combined, taking care to avoid dependence among the covariates.

Considering the sample size of this experiment, all 38 observations were used for RA modeling because the probabilistic assumptions of this class of models. After adjustment, the usual residual verifications related to the distribution (Gaussian), independence, and homoscedasticity were checked. If the estimated model was adequate, a cross-validation step could be performed where one observation at a time was left out of the adjustment for comparison or a sample part was reserved when the sample size was large.

In the TSS prediction from the ANN, which is a distribution-free method, the neural network modeling considered two processing steps, the first being the training of the network, and the second being its subsequent testing with a data set different from the first stage. In this study, we used 80% of the data collected for ANN training and 20% for testing, which were randomly defined [23].

As the objective of the method was to create an ANN capable of recovering the concentration of suspended solids in the water from the bands of the modified Canon sensor incorporated into a UAV, we considered the normalized values of the NIR, G, and B channels as inputs to ANN, and the TSS concentration as an output at the same sampling point. We used a network of feed-forward backpropagation, with this being commonly used in remote sensing studies [8,10,21].

During the training phase, several tests were carried out in order to obtain the best ANN topology applicable to this study, choosing the network that provided the highest correlation coefficient and the lowest mean square error during training and testing. We tested different numbers of neurons (from 5 to 20) in one single hidden layer, as well as three activation functions (sigmoid, tangent, and linear), and the number of training cycles.

3. Results and Discussion

The results of the laboratory analyses were satisfactory for the research and compatible with prior knowledge of the water quality in the study area and analysis of the spatial behavior of these parameters, which would later be compared with the UAV images. Table 1 shows the descriptive statistics of the total suspended solids (TSS) analyzed in this research for March 2016 and 2017.

Table 1. Descriptive statistics of TSS.

Parameter	Value (mg/L)		
	2016	**2017**	**All**
Average	16.27	13.65	14.93
Standard deviation	2.97	3.07	3.29
Minimum value	11.67	9.33	9.33
Maximum value	23.75	20.00	23.75
Median	15.33	12.68	14.87
Variance Coefficient (%)	19.77	23.65	23.02

We observed from the analysis of the data presented in Table 1 that the characteristics of the study lake were not the same between the two collections. This was also confirmed from the Wilcoxon test at a 95% confidence level. There was a decrease in the concentration of suspended solids from 2016 to 2017, which can be seen in the averages, medians, maximum, and minimum values of Table 1.

This difference in TSS concentration, although small, can be justified because although it did not rain on the days of sampling in 2016 and 2017, there were rainfall events in the week before the collection of 2016 (85 mm according to the experimental climatological station located at Unisinos University), which did not occur in 2017. Allen et al. [24] point out that in impermeable urban areas, the flow of rainwater in the soil causes the collection of the pollutants and sediments from these surfaces, which are transported to the nearest waterways. As the lake receives the drainage of rainwater from the university campus, it is expected that in rainy periods, various compounds will be carried into it, increasing the concentration of suspended solids, for example.

As initial cartographic products, obtained via overflying with the UAV in 2016 and 2017, and by processing the images, we have the orthophotos of the area, as shown in Figure 4.

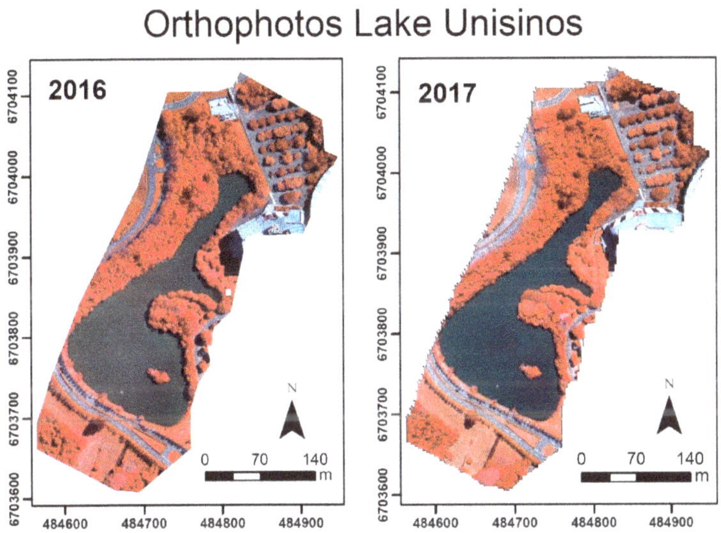

Figure 4. Orthophotos generated by the overfly with UAV in March 2016 and 2017.

The simple and multiple linear and non-linear RA described in the previous section were evaluated; however, most of the results were unsatisfactory. Table 2 shows the best results that we obtained in these analyses. Although not shown, residual analyses were performed to check the error assumptions.

Table 2. Best results of the RA Model.

Variables	Function	R^2	RMSE
NIR	Power	0.20	3.05
G/NIR	Exponential	0.19	3.07
G	Linear	0.16	3.08
B	Linear	0.13	3.14
B and G/NIR	Multiple Linear	0.20	3.00
G/NIR and NIR/B	Multiple Linear	0.18	3.05
G and B	Multiple Linear	0.16	3.08

According to Table 2, the best adjustments of the simple regression analyses were for the NIR and G/NIR variables, agreeing with Song et al. [6] and Amanollahi et al. [10], although both studies obtained better results than ours (0.7 for Amanollahi et al. [10] and above 0.9 for Song et al. [6]). Also, a combination of B and G/NIR was the best result for the multiple linear regressions.

Although the regression models in Table 2 showed statistical significance, the low R^2 values indicate that the RA models were not ideal for TSS recovery in the study area. This result can be explained by the optical complexity of the study waters such that the relations between the bands of the collected images and the concentration of TSS could not be explained by traditional regression techniques.

To improve the accuracy of TSS predictions, ANN can be effective. Kong et al. [8] emphasize that ANN models establish different weights for each input in the network and thus take full advantage of the characteristics of TSS included in the different bands.

We performed several trainings of neural networks with different topologies. In Table 3, the results include a coefficient of determination (R^2) greater than 0.5 in the training step and their respective topologies, activation functions, number of epochs, and time of training are presented.

Table 3. Results of the ANN training for $R^2 > 0.5$.

Topology [a]	Activation Function [b]	Epochs	Time [c] (s)	R^2	RMSE
3-5-1	Tangent/Linear	100	1	0.60	2.11
3-5-1	Sigmoid/Linear	400	4	0.57	2.11
3-7-1	Tangent/Tangent	300	5	0.84	1.33
3-10-1	Tangent/Tangent	300	4	0.50	2.32
3-10-1	Tangent/Linear	400	5	0.50	2.30
3-10-1	Sigmoid/Linear	100	1	0.64	2.05
3-10-1	Sigmoid/Linear	400	5	0.53	2.19
3-12-1	Tangent/Tangent	200	3	0.65	2.02
3-15-1	Tangent/Tangent	500	7	0.58	2.10
3-15-1	Tangent/Linear	400	5	0.53	2.21
3-17-1	Sigmoid/Linear	300	4	0.50	2.27
3-20-1	Tangent/Linear	100	1	0.55	2.20
3-20-1	Sigmoid/Linear	200	2	0.50	2.30

[a] with "input (3)-neurons-output (1)". [b] Hidden layer/output layer. [c] Where the computer used had the following configuration: processor—Intel® Core ™ i3-4005U CPU @ 1.70 GHz ×4, memory—4GB DDR3 1600MHz RAM.

According to Table 3, the topology in which we obtained the smallest RMSE and the highest determination coefficient was 3-7-1, with the tangent function as the activation function, and with 300 training cycles. Thus, the ANN adopted was a feed-forward backpropagation type, with three input layers (NIR, G, and B), seven neurons in a single hidden layer, and one output (TSS). As usual, the training processing time depended on the number of epochs and was not a problem in our experiment because of the reduced sample size.

The results that we found in the training and testing steps for this best ANN are presented in Table 4. The graph presented in Figure 5 demonstrates the comparison between the data measured in the laboratory and those estimated through ANN.

The ANN training stage resulted in an R^2 of 0.84 and RMSE of 1.33, while during testing, these values were 0.57 and 2.97, respectively. As shown in Table 4 and Figure 5, considering all the data collected in this study as inputs to the ANN, the R^2 was 0.75 and the RMSE was 1.81.

As expected, the results showed a significant improvement in the prediction of suspended solids data in the study area through the use of ANN in place of the simple and multiple linear and non-linear investigated RA.

Table 4. Results of the ANN.

Steps	n	R^2	RMSE
Training	30	0.84	1.33
Testing	8	0.57	2.97
All	38	0.75	1.81

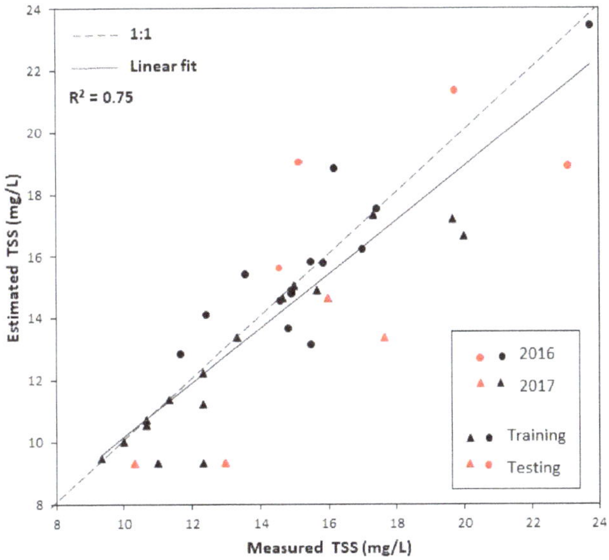

Figure 5. Comparison between TSS measurements and estimated values from the ANN model in training (black) and testing (red) set using March 2016 (circular shape) and 2017 (triangular shape) samplings.

Although several studies show good results using regression methods to predict TSS [2,7,11,15–19], others, such as Song et al. [6], Amanollahi et al. [10], Moridnejad et al. [12], and Wu et al. [25], compared the two methodologies (RA and ANN) and obtained results indicating better quality in the prediction of the data through an ANN, signaling the capacity of the neural networks to model more complex and non-linear relations between the parameters. Only Kong et al. [8] reported that an ANN did not present better results than regression methods for TSS predictions in their area of study.

Din et al. [9] used statistical correlation analysis only as a support for choosing the ideal bands of the Landsat 8 OLI satellite for an ANN input. Then, the authors decided to include also the bands of the short-wave infrared (SWIR-1 and SWIR-2) as inputs, which is not common in papers about ANN for predicting water quality parameters since only visible and near infrared regions are exploited [6,8,10,12,13,21,25,26].

Although the aforementioned approaches from the literature are similar to our paper for comparing RA and ANN for the prediction of TSS, we point out that our results differ and are highlighted by the high spatial resolution of the UAV images used in comparison to low or medium spatial resolutions of the satellite images of other studies. Thus, our method allows for giving more geographically accurate TSS predictions because of the small pixel size of the UAV images (5 cm in comparison to 30 m for Landsat, for example) and generating high quality and resolution TSS monitoring maps.

Finally, the ANN model was used to predict the TSS concentration for the whole lake using the NIR, G, and B variables for the 2016 and 2017 UAV images. Thus, the generated TSS maps for Lake Unisinos are shown in Figure 6.

While analyzing Figure 6, we noticed the highest concentrations of suspended solids in the 2016 sampling compared to the 2017 one, a situation that was already indicated in Table 1. Besides, the used data set presented a significant statistical difference between the two years, where the spatial distribution also became evident in Figure 6. The highest concentrations of TSS in 2016 were in the lower central region of the lake, whereas in 2017, they were near the lower-right margin. A large part is found in the center of the lake for 2017 with the minimum TSS values. This difference in spatial distribution, mainly showing as a large TSS concentration in 2016, is consistent with the in situ collected water samples and is also explained by the rain that occurred in the previous week of the field collection in 2016. Figure 6 also shows that TSS concentrations were in the same range (9.33 to 23.75 mg/L) for both years. In this sense, to verify if the statistical characteristics of the prediction, data remain close to the observed ones, where Figure 7 shows the box plot of TSS concentrations for both observed and predicted values in March 2016 and 2017.

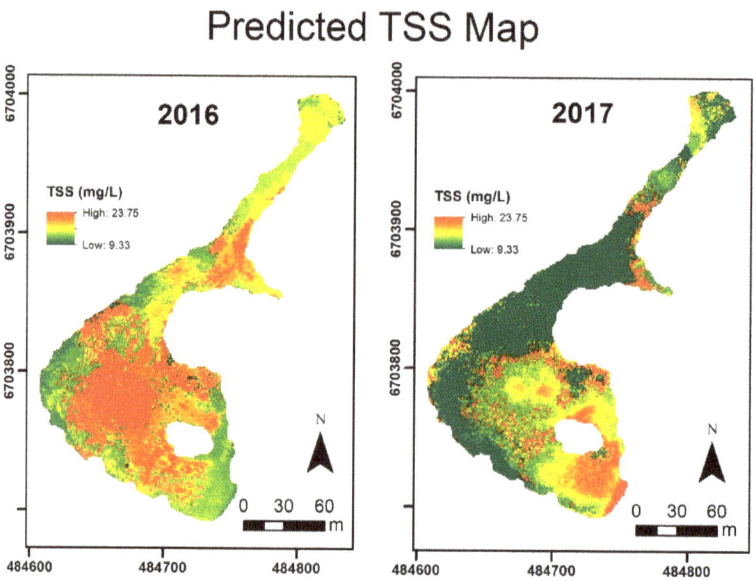

Figure 6. Maps of predicted TSS based on the ANN model for March, 2016 and 2017.

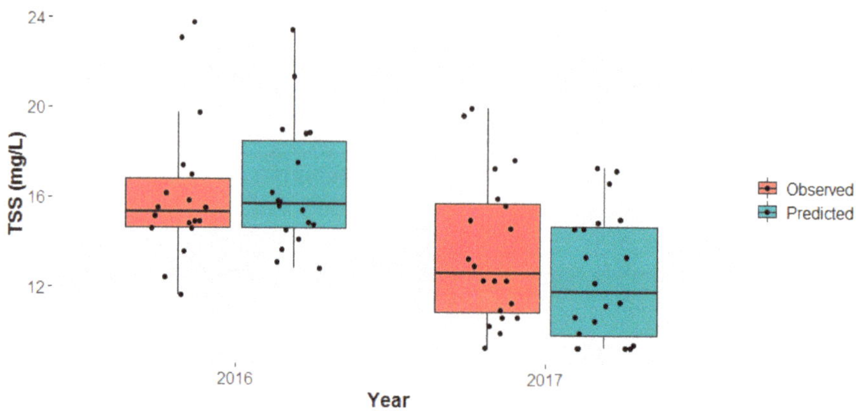

Figure 7. Box plot of TSS concentrations for observed and predicted values in 2016 and 2017.

From Figure 7, we can see the similarity between the observed and predicted distributions, even though this was not a large sample. From the Wilcoxon test, which is adequate for asymmetric distributions, the null hypothesis of equal medians was not rejected at a 5% significance level (p-value = 0.74). From the same test, the statistically significant difference between the years already seen for the observed TSS values was maintained for the predicted TSS (p-value = 0.0007). The observed average that was 16.27 and 13.65 (Table 1) became 15.72 and 12.51 for the predicted TSS in March 2016 and 2017. The variance coefficient also remained similar, 17.51% and 22.56%, which are close to 19.77% and 23.65% presented in Table 1.

Although the results of this study confirm the viability of the prediction of the concentration of TSS from remote sensing data and ANN, we emphasize that because it is a new methodology and that is still under development it has some limitations that should be considered.

For example, since each water body has its own characteristics (hydraulic, physical, chemical, and biological), which are related to its surroundings and the region's climate, the proposed model in this study was trained relative to these conditions of the study area. Thus, it is necessary to develop regional models adapted to the area of interest of the study. Other authors like Kong et al. [8] and Chen et al. [26] also point out the absence of a standard model for different regions.

In relation to temporal variation, we point out that the field samplings were carried out in March of the two years and therefore the seasonal variation of TSS (not considered in this study) may indicate that a single model trained with data from only one season is not sufficient to predict other values throughout the year. Another factor that stands out is that besides the seasonal variation of TSS, other changes can occur in the natural environments over the years. Once the environmental characteristics are modified, it is not possible to affirm the capability of the trained neural network to predict data in the long term at this time. Thus, the monitoring of TSS from remote sensing does not rule out laboratory analyses from time to time. For instance, if the predictions exhibit unexpected behavior, such as a growth trend, new TSS and spectral data may be collected to check if it is a real change in the TSS or the neural network needs to be updated for current conditions.

Finally, although studies as this serve as pilot studies to be applied in larger water bodies, we emphasize that adaptations need to be made for this to occur because when flying with a UAV in lakes and dams because large homogeneous areas makes it difficult to generate orthophotos and products generated from the Structure for Motion (SfM) technique. One of the ways to minimize this problem would be to perform high altitude flights, facilitating the identification of homologous points in the images for generating the orthophoto, but which would result in a loss in resolution of the images.

The presented limitations indicate that this research needs to be continued. Nevertheless, what we have demonstrated in this article should instigate replications of this method in other water bodies such that more involved communities benefit from our positive results. This can be done through area flyovers with UAVs with RGB and NIR cameras, correct processing of acquired images, reliable data collection of water quality, and the establishment of an ANN with the ideal parameters for the prediction of interest, which can be TSS as in this study, or for example, chlorophyll-a or organic matter.

The prediction of TSS in water bodies from images acquired using a UAV and processed via an ANN should benefit managers, professionals, and researchers linked to the management and control of water resources by presenting a method for the dynamic and spatial monitoring of water quality problems, such as the presence of suspended solids.

4. Conclusions

The use of UAVs in the mapping of water quality is shown to be a promising tool because it alleviates issues found in the usual in situ monitoring, such as the insufficiency of data, high time and money costs, and modeling via remote orbital sensing, such as the low spectral and temporal resolutions. Through analysis of the response that the sensor on board the UAV collected in the regions of visible and near infrared, it was possible to model the concentration of optically active compounds, such as suspended solids, and generate maps that allowed for their temporal monitoring and spatial analysis at the study site.

We emphasize the applicability of the use of artificial intelligence through artificial neural networks to meet the need for modeling suspended solids in complex aquatic environments, where more simplistic analyses, such as the regression models presented in this study, may not be sufficient. The use of an ANN instead of RA significantly improved the quality of the results from the generated models, where R^2 values rose from 0.20 (RA) to 0.75 (ANN).

However, although the model presented could accurately predict suspended solids concentrations compatible with the statistical features of the in situ observed values, its use was limited only to the study area where the ANN was trained and calibrated, and possible adaptations to it are required for use in other environments.

The presented results are important for two main reasons. First, although regression methods have been used in remote sensing applications, they may not be adequate for capturing the linear and/or non-linear relationships of interest. Second, they show that the use of UAVs in the mapping of water bodies together with the application of neural networks in the analysis of the results obtained is a promising approach and has the potential to assist in monitoring the quality of these environments. Thus, we intend to continue monitoring the total suspended solids concentrations in Lake Unisinos by performing new overflights with a UAV in the region and simulating the data collected with the neural network.

We also emphasize the need to continue the research in order to improve the generated model, as well as to consider the interference of other optically active compounds, such as chlorophyll and organic matter, in the spectral response of water, and consequently, in the neural network generated.

Author Contributions: T.T.G., E.C.K., and D.B. were responsible for collecting and processing the images obtained from the UAV. T.T.G. and E.C.K. were responsible for water collection and for the laboratorial analysis. T.T.G., M.R.V., and L.G.J. implemented the artificial neural network. T.T.G., E.M.S., D.B., and L.G.J. were responsible for analyzing the data. T.T.G., M.R.V., and E.M.S. wrote the paper. M.R.V. and F.F.M. reviewed the paper. All the authors have read and approved the paper final version.

Acknowledgments: M.R.V. and F.F.M. thank the Brazilian Council for Scientific and Technological Development (CNPq) for the research grant. T.T.G. thanks DS/CAPES for the financial support of the MSc scholarship. D.B. thanks CAPES for the financial support of the MSc scholarship.

Conflicts of Interest: The authors declare no conflict of interest.

References

1. Kageyama, Y.; Takahashi, J.; Nishida, M.; Kobori, B.; Nagamoto, D. Analysis of Water Quality in Miharu Dam Reservoir, Japan, using UAV Data. *IEEJ Trans.* **2016**, *11* (Suppl. 1), S183–S185. [CrossRef]
2. Qun, M.; Tan, X.; Liu, Z.; Liu, C.; Li, Q. Monitoring Chlorophyll-a and Suspended Substance in Nansi Lake, China through Remote Sensing Technology. In Proceedings of the 2008 International Workshop on Education Technology and Training 2008 International Workshop on Geoscience and Remote Sensing, Shanghai, China, 21–22 December 2008; Volume 2, pp. 348–351.
3. Campbell, G.; Phinn, S.R.; Dekker, A.G.; Brando, V.E. Remote sensing of water quality in an Australian tropical freshwater impoundment using matrix inversion and MERIS images. *Remote Sens. Environ.* **2011**, *115*, 2402–2414. [CrossRef]
4. Jensen, J.R. *Sensoriamento Remoto do Ambiente: Uma Perspectiva em Recursos Terrestres*; Parêntese: São José dos Campos, Brasil, 2011.
5. United States Environmental Protection Agency (EPA). *Modeling Total Suspended Solids (TSS) Concentrations in Narragansett Bay*; Office of Research and Development—National Health and Environmental Effects Research Laboratory: Narragansett, RI, USA, 2016.
6. Song, K.; Li, L.; Wang, Z.; Liu, D.; Zhang, B.; Xu, J.; Du, J.; Li, L.; Li, S.; Wang, Y. Retrieval of total suspended matter (TSM) and chlorophyll-a (Chl-a) concentration from remote-sensing data for drinking water resources. *Environ. Monit. Assess.* **2012**, *184*, 1449–1470. [CrossRef] [PubMed]
7. Wang, J.; Tian, Q. Estimation of total suspended solids concentration by hyperspectral remote sensing in Liaodong Bay. *Indian J. Mar. Sci.* **2015**, *44*, 1137–1144.
8. Kong, J.; Shan, Z.; Chen, Y.; Yang, J.; Hu, Y.; Wang, L. Assessment of remote-sensing retrieval models for suspended sediment concentration in the Gulf of Bohai. *Int. J. Remote Sens.* **2018**, *40*, 2324–2432. [CrossRef]
9. Din, E.S.; Zhang, Y.; Suliman, A. Mapping concentrations of surface water quality parameters using a novel remote sensing and artificial intelligence framework. *Int. J. Remote Sens.* **2017**, *38*, 1023–1042.
10. Amanollahy, J.; Kaboodvandpour, S.; Majidi, H. Evaluating the accuracy of ANN and LR models to estimate the water quality in Zarivar International Wetland, Iran. *Nat. Hazards* **2017**, *85*, 1511–1527. [CrossRef]

11. Breuning, F.M.; Pereira Filho, W.; Galvão, L.S.; Wachholz, F.; Cardoso, M.A.G. Dynamics of limnological parameters in reservoirs: A case study in South Brazil using remote sensing and meteorological data. *Sci. Total Environ.* **2017**, *574*, 253–263. [CrossRef] [PubMed]
12. Moridnejad, A.; Abdollahi, H.; Alavipanah, S.K.; Samani, J.M.V.; Moridnejad, O.; Karimi, N. Applying artificial neural networks to estimate suspended sediment concentrations along the southern coast of the Caspian Sea using MODIS images. *Arab. J. Geosci.* **2015**, *8*, 891–901. [CrossRef]
13. Peterson, K.T.; Sagan, V.; Sidike, P.; Cox, A.L.; Martinez, M. Suspended Sediment Concentration Estimation from Landsat Imagery along the Lower Missouri and Middle Mississippi Rivers Using an Extreme Learning Machine. *Remote Sens.* **2018**, *10*, 1503. [CrossRef]
14. Gago, J.; Douthe, C.; Coopman, R.E.; Gallego, P.P.; Ribas-Carbo, M.; Flexas, J.; Escalona, J.; Medrano, H. UAVs challenge to assess water stress for sustainable agriculture. *Agric. Water Manag.* **2015**, *153*, 9–19. [CrossRef]
15. Cândido, A.K.; Paranhos Filho, A.C.; Haupenthal, M.R.; da Silva, N.M.; de Sousa Correa, J.; Ribeiro, M.L. Water Quality and Chlorophyll Measurement Through Vegetation Indices Generated from Orbital and Suborbital Images. *Water. Air. Soil Pollut.* **2016**, *227*, 224. [CrossRef]
16. Su, T.-C.; Chou, H.-T. Application of Multispectral Sensors Carried on Unmanned Aerial Vehicle (UAV) to Trophic State Mapping of Small Reservoirs: A Case Study of Tain-Pu Reservoir in Kinmen, Taiwan. *Remote Sens.* **2015**, *7*, 10078–10097. [CrossRef]
17. Guimarães, T.; Veronez, M.; Koste, E.; Gonzaga, L.; Bordin, F.; Inocencio, L.; Larocca, A.; de Oliveira, M.; Vitti, D.; Mauad, F. An Alternative Method of Spatial Autocorrelation for Chlorophyll Detection in Water Bodies Using Remote Sensing. *Sustainability* **2017**, *9*, 416. [CrossRef]
18. Veronez, M.R.; Kupssinskü, L.S.; Guimarães, T.T.; Koste, E.C.; Silva, J.M.; Souza, L.V.; Oliverio, W.F.M.; Jardim, R.S.; Koch, I.É.; Souza, J.G.; et al. Proposal of a Method to Determine the Correlation between Total Suspended Solids and Dissolved Organic Matter in Water Bodies from Spectral Imaging and Artificial Neural Networks. *Sensors* **2018**, *18*, 159. [CrossRef]
19. Sáenz, N.A.; Paez, D.E.; Arango, C. Local algorithm for monitoring total suspended sediments in micro-watersheds usin drones and remote sensing applications. Case study: Teusacá River, La Calera, Colombia. *Int. Arch. Photogramm. Remote Sens. Spatial Inf. Sci.c* **2015**, *XL-1/W4*, 159–165.
20. Roig, H.L.; Ferreira, A.M.R.; Menezes, P.H.B.J.; Marotta, G.S. Uso de câmeras de baixo custo acopladas a veículos aéreos leves no estudo do aporte de sedimentos no Lago Paranoá. In *Anais XVI Simpósio Brasileiro de Sensoriamento Remoto—SBSR*; INPE: Foz do Iguaçu, Brasil, 2013; pp. 9332–9339.
21. Guo, Q.; Wu, X.; Bing, Q.; Pan, Y.; Wang, Z.; Fu, Y.; Wang, D.; Liu, J. Study on Retrieval of Chlorophyll-a Concentration Based on Landsat OLI Imagery in the Haihe River, China. *Sustainability* **2016**, *8*, 758. [CrossRef]
22. American Public Health Association (APHA). *Standard Methods for Examination of Water and Wastewater*; APHA: Washington, DC, USA, 1995.
23. Haykin, S.S. *Neural Networks and Learning Machines*, 3rd ed.; Pearson: Upper Saddle River, NJ, USA, 2009.
24. Allen, D.; Arthur, S.; Haynes, H.; Olive, V. Multiple rainfall event pollution transport by sustainable drainage systems: The fate of fine sediment pollution. *Int. J. Environ. Sci. Technol.* **2017**, *14*, 639–652. [CrossRef]
25. Wu, J.L.; Ho, C.R.; Huang, C.C.; Srivastav, A.L.; Tzeng, J.H.; Lin, Y.T. Hyperspectral Sensing for Turbid Water Quality Monitoring in Freshwater Rivers: Empirical Relationship between Reflectance and Turbidity and Total Solids. *Sensors* **2014**, *14*, 22670–22688. [CrossRef] [PubMed]
26. Chen, C.J.; Zhu, W.; Tian, Y.Q.; Yu, Q.; Zheng, Y.; Huang, L. Remote estimation of colored dissolved organic matter and chlorophyll-a in Lake Huron using Sentinel-2 measurements. *J. Appl. Remote Sens.* **2017**, *11*, 036007. [CrossRef]

sustainability

MDPI

Review

Reducing Groundwater Contamination from On-Site Sanitation in Peri-Urban Sub-Saharan Africa: Reviewing Transition Management Attributes towards Implementation of Water Safety Plans

Felix R. B. Twinomucunguzi [1,*], Philip M. Nyenje [1], Robinah N. Kulabako [1], Swaib Semiyaga [1], Jan Willem Foppen [2] and Frank Kansiime [3]

[1] Department of Civil and Environmental Engineering, Makerere University, P.O. Box 7062 Kampala, Uganda; nyenje@gmail.com (P.M.N.); rkulaba@gmail.com (R.N.K.); semiyaga@gmail.com (S.S.)

[2] Department of Water Science and Engineering, IHE Delft Institute for Water Education, P.O. Box 3015, 2601 DA Delft, The Netherlands; j.foppen@un-ihe.org

[3] Department of Environmental Management, Makerere University, P.O. Box 7062 Kampala, Uganda; fkansiime@gmail.com

* Correspondence: twinof@gmail.com; Tel.: +256-772-588-651

Received: 15 April 2020; Accepted: 19 May 2020; Published: 21 May 2020

Abstract: High urbanization in Sub-Saharan Africa (SSA) has resulted in increased peri-urban groundwater contamination by on-site sanitation. The World Health Organization introduced Water Safety Plans (WSP) towards the elimination of contamination risks to water supply systems; however, their application to peri-urban groundwater sources has been limited. Focusing on Uganda, Ghana, and Tanzania, this paper reviews limitations of the existing water regime in addressing peri-urban groundwater contamination through WSPs and normative attributes of Transition Management (TM) towards a sustainable solution. Microbial and nutrient contamination remain prevalent hazards in peri-urban SSA, arising from on-site sanitation within a water regime following Integrated Water Resources Management (IWRM) principles. Limitations to implementation of WSPs for peri-urban groundwater protection include policy diversity, with low focus on groundwater; institutional incoherence; highly techno-centric management tools; and limited regard for socio-cultural and urban-poor aspects. In contrast, TM postulates a prescriptive approach promoted by community-led frontrunners, with flexible and multi-domain actors, experimenting through socio-technical tools towards a shared vision. Thus, a unified risk-based management framework, harnessing attributes of TM and IWRM, is proposed towards improved WSP implementation. The framework could assist peri-urban communities and policymakers in formulating sustainable strategies to reduce groundwater contamination, thereby contributing to improved access to safe water.

Keywords: contamination; integrated water resources management; groundwater; pollution; Sub-Saharan Africa; transition management; water safety plan

---

1. Introduction

Groundwater contamination by human activities is a growing global concern in light of increasing population, urbanization, industrialization, and agriculture [1,2]. Over one-third of the global water use is derived from groundwater, which is under increasing stress from anthropogenic contamination and diminishing resource quantities due to over-extraction and climate change effects [2]. On-site sanitation practices for disposal of human excreta and greywater are major contributors to groundwater contamination, especially in peri-urban areas where settlement patterns are dense [3,4]. On-site sanitation facilities, mainly pit latrines and septic tanks, remain the primary form of improved sanitation in rural and peri-urban areas in most of the developing world, including Sub-Saharan Africa (SSA); South, East, and Central Asia; Southern and Middle America, and Oceania [4–6]. While such facilities have promoted access to improved sanitation to the peri-urban communities, their increased number, and usually poor construction and maintenance, results in increased groundwater contamination [4,7–13]. Groundwater contamination by on-site sanitation systems in SSA is, thus, a significant hindrance to providing access to safe water to vulnerable populations [7–9]. Indeed, SSA lags behind other regions of the world in the struggle to meeting sustainable development goal (SDG) number 6 on universal access to safe water and sanitation [5,14].

Sub-Saharan Africa continues to register unprecedented urbanization, mainly occurring in peri-urban areas, which have low access to safe water and sanitation [10–12]. Groundwater, drawn mainly from boreholes, shallow wells, and springs, is the predominant source of water for domestic consumption, small scale industry, and irrigation to the majority of peri-urban residents in SSA due to its relatively low cost of abstraction and high availability [3,7,9]. Despite this high reliance on groundwater, on-site sanitary practices have increased contamination, leading to severe human health and ecological consequences [12,15–17]. For instance, Murphy et al. [18] reported that over 60% of groundwater sources tested in Kampala (Uganda) were positive for *Escherichia coli* (*E. coli*), which is commonly attributed to fecal contamination, during a typhoid outbreak in 2015, which affected more than 10,000 people. The scale and nature of the contamination and associated risks increase in complexity with a growing population.

The failure to address groundwater resource challenges, such as the increasing stress of groundwater resources due to contamination by on-site sanitation in peri-urban areas, can be attributed to limitations in the groundwater management (governance) framework [12,19,20]. Vardy et al. [20] described groundwater governance to include elements of institutional setting; availability and access to information and science; robustness of civil society; and economic and regulatory frameworks. Generally, groundwater governance has evolved within the overall realm of water governance, dating back to the Helsinki Rules on the use of waters of international rivers in 1966 [19]. Globally, there have been several groundwater governance approaches, depending upon the geographical scale (local, regional, or transboundary). These have included approaches based on 1) international water law and political ecology perspectives, especially in transboundary aquifer governance [19,21–23]; institutionalism, including management of "common pool resources" through polycentric approaches advanced by Elinor Ostrom and associates [24–26]; economic and market regulation perspectives [20,27]; and socio-ecological approaches [19,28]. Generally, the current water governance framework in SSA is mainly influenced by Integrated Water Resources Management (IWRM) principles, adopted after the International Conference on Water and Environment (Dublin) and the United Nations Conference on Environment and Development (Rio de Janeiro), both in 1992 [29]. However, despite the high importance of groundwater resources, it has always received less profiling within the overall water governance frameworks [19,22,30–32].

Many countries in SSA embraced IWRM as a multi-stakeholder framework towards addressing water resources challenges, including contamination [33–35]. IWRM is defined as 'a process which promotes the coordinated development and management of water, land and related resources, in order to maximize the resultant economic and social welfare in an equitable manner without compromising the sustainability of vital ecosystems' [29]. IWRM introduced improved integrated policies, and improved participatory and gender-responsive approach, which have provided an enabling environment for implementing water resources improvement initiatives. However, implementation of IWRM in SSA has faced several challenges, including; (1) a high level of diversity (ambiguity), leading to complexity and limited action to address specific societal challenges [30,33–36]; (2) contested institutional framework based on catchment-based structures [32,37–40]; (3) ineffective stakeholder participation and cooperation frameworks [30,37]; (4) undefined monitoring framework for assessing results [33,39]; (5) ineffective cost recovery mechanisms [35]; and (6) limited resources to implement the ambitious targets of IWRM approaches [40]. These IWRM experiences have influenced the adoption and implementation of water resources management initiatives within the water sector in SSA [40].

In 2004, the World Health Organization (WHO) introduced Water Safety Plans (WSPs) as an instrument for identification, prevention, and management of contamination risks to water supply systems, which has been adopted in SSA as well [41–43]. The WSP approach is based on the principles and steps of the "Multiple-Barrier" concept for prevention of contamination to water sources and the "Hazard Analysis and Critical Control Points (HACCP)" concept, adopted from food safety management systems [44]. The WSPs are intended to ensure continuous provision of safe water, free from any contamination, for all levels and types of community water supply systems [41,44]. WSPs have been implemented, mainly by water utilities, in all regions of the world, voluntarily or by regulation [42,45]. Uganda was among the first countries in SSA to implement WSPs for prevention of pollution to public water supplies [45]. Ghana and Tanzania have also recently adopted implementation of WSPs for protection of public water supplies [42]. Although WSPs have mainly been implemented for conventional urban water supply systems, with complex infrastructure, there have also been efforts to apply them for improvement of the safety of basic water supply infrastructure in rural and peri-urban areas [46]. However, like other management instruments, adoption and implementation of WSPs is greatly influenced by the existing water governance regime [47–51]. The influence of the existing water regime (IWRM) towards the implementation of WSPs for addressing peri-urban groundwater contamination, in particular, has not been comprehensively documented. An analysis of the challenges to the existing framework in implementing WSPs to address the growing challenge of peri-urban groundwater contamination would identify the gaps, which can be addressed by emerging management concepts.

Transition Management (TM) is an emerging management framework, within the context of sustainability science, which has been explored in various developing and developed countries to address complex socio-technical sustainability challenges [52–54]. TM is described as a 'prescriptive and descriptive, complex-based governance framework towards long-term social change through small steps basing on searching, learning and experimenting' [52]. TM has evolved in the past two decades in the realm of sustainability science, attempting to provide solutions to societal complex and persistent problems [52,53]. Groundwater contamination by on-site sanitation in peri-urban SSA is certainly one such challenge, which could benefit from developments of the emerging field of TM. TM acknowledges that societal problems are getting more complex with increased pressures like population growth, climate change, and technological advancement, and traditional management approaches are ill-equipped for this complexity [55]. Since 2015, a project has been implemented by the research team, T-group, focusing on adapting TM approach towards improved peri-urban groundwater management in Uganda, Ghana, and Tanzania [56,57].

Through a critical review, this paper aims to highlight the existing water regime challenges towards implementing WSPs for protecting peri-urban groundwater against contamination by on-site sanitation and explore normative attributes of TM towards a sustainable solution. The paper first illustrates the complex socio-technical system influencing peri-urban groundwater contamination arising from on-site sanitation in peri-urban SSA and then reviews the challenges of IWRM framework and the normative attributes of TM framework towards improved risk management. Using the Entity-Relationship Diagramming (ERD) technique, complementary attributes of IWRM and TM are demonstrated in a proposed risk-based management framework for reducing peri-urban groundwater contamination by on-site sanitation through WSPs for small communities. This framework could be a sound tool for comprehensive assessment and formulation of strategies to improve adoption and implementation WSPs targeted at reducing peri-urban groundwater contamination in SSA.

2. Methodology

The literature survey was conducted in accordance with the Preferred Reporting Items for Systematic Reviews and Meta-Analyses (PRISMA) guidelines [58]. Relevant documentation (both published and unpublished) were searched, with no date restrictions, from Google Scholar, Google, Scopus, Water Safety Portal, and Web of Science. Titles of all retrieved documents were reviewed to remove any duplications followed by an analysis of abstracts for eligibility. Eligible documents included those that contained information on groundwater contamination/pollution, on-site sanitation, water contamination risk management, water safety plans, integrated water resources management, groundwater management/governance, aquifer management, and transition management. The full-text analysis was only undertaken for the documents with implication to SSA context, with particular focus on Ghana, Uganda, and Tanzania. Relevant subject articles and documents from WHO and other sources pertinent to WSP and groundwater contamination by on-site sanitation were also included in the review. The documents reviewed also provided additional sources, which were assessed for eligibility. The review was conducted between January 2019 to May 2020. Figure 1 shows the flow diagram for the literature review.

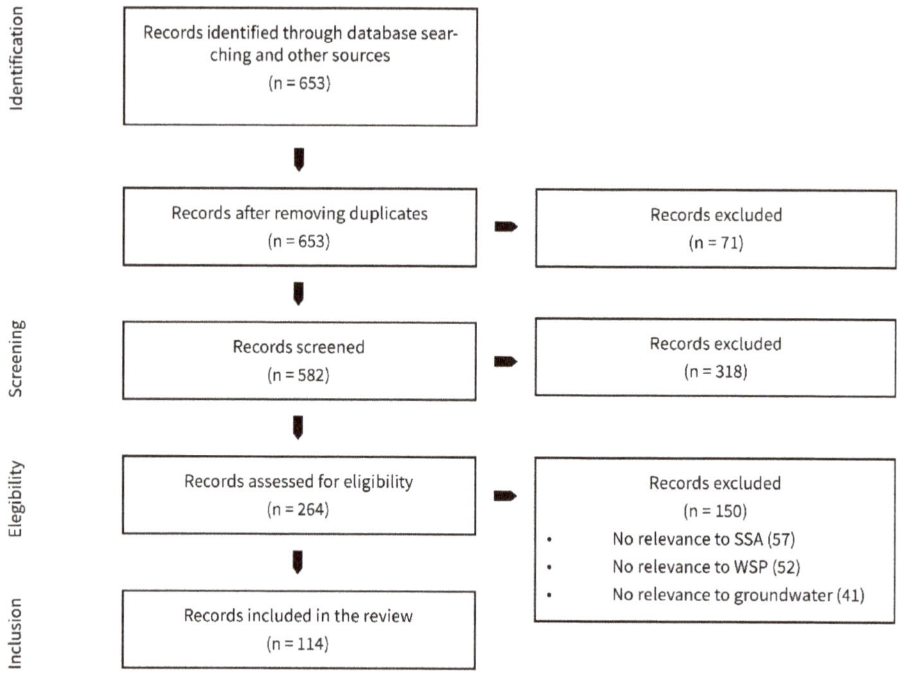

Figure 1. PRISMA (Preferred Reporting Items for Systematic Reviews and Meta-Analyses) flow diagram for the literature survey.

From the comprehensive review, the governance factors influencing implementation of WSPs for reducing peri-urban groundwater contamination in SSA were analyzed and represented using Entity-Relationship Diagramming technique (ERD). ERD can be used to visually display qualitative data on system entities, their relationships, and attributes [59]. Content validation of the proposed framework was achieved through expert opinion and stakeholder consultation at a stakeholder workshop held in Entebbe (Uganda), in February 2020. The workshop was attended by 37 participants drawn from government institutions, private sector, and research institutions. Iterative improvements were made to the final version presented.

3. Understanding the Complex Socio-Technical System Influencing Groundwater Contamination by On-Site Sanitation in Peri-Urban SSA

On-site sanitation, through pit latrines and septic tanks, is the predominant sanitation system used in peri-urban areas for disposal of human waste in most of SSA, and the developing world, particularly in Asia, Middle America, and Oceania [3–8]. In the developed world, including Europe and North America, on-site facilities, mainly in the form of decentralized treatment options, are mostly used in the rural setting [60,61]. While on-site systems have contributed to the increased access to improved sanitation in peri-urban areas, they have been reported to be a source of contamination hazards to groundwater, which is used by many peri-urban dwellers [1–10,12,15–18,60,61]. Table 1 summarizes recently documented cases of peri-urban groundwater chemical and microbial contamination from on-site sanitation in the three focus countries of Uganda, Tanzania, and Ghana.

Table 1. Recently documented cases of urban groundwater contamination by on-site sanitation in Uganda, Tanzania, and Ghana.

Country	Reported Contaminants from On-Site Sanitation in Peri-Urban Areas [1]	Reference
Uganda	Nutrient contamination (nitrate of up to 94.6 mg/L; orthophosphate up to 2.4 mg/L) in shallow groundwater in Kampala and Lukaya	[3,9,62]
	Microbiological contamination (*E. coli*, fecal coliforms, viruses, salmonella) detected from springs in Kampala	[18,63]
Tanzania	Nutrient contamination (nitrate of up to 445 mg/L in Dar es Salaam; 449 mg/L in Dodoma; 100 mg/L in Tanga and 180 mg/L in Manyara)	[64]
	Microbial contamination (Fecal coliforms, *E. coli* and fecal streptococci) in Arusha, Dar es Saalam, and Babati	[12,16,65]
Ghana	Nutrient contamination (nitrate of up to 170 mg/L in Volta Region)	[66]
	Microbial contamination (Fecal coliforms, *E. coli*, and salmonella) in Kumasi, Ashanti Region, and Accra	[10,67–69]

[1] Data covering the period of 2010–2020.

From Table 1, it can be noted that *E. coli*, fecal coliforms, and salmonella are widely studied in peri-urban groundwater matrices in SSA. Due to increased access to modern analytical methods in SSA, previously un-detected microbial groundwater contaminants like viruses are also increasingly being analyzed and reported [70]. Chemical contamination has also been well documented, arising from on-site sanitation facilities, especially with respect to nitrate contamination (Table 1). High nutrient (nitrate and phosphorous) and microbial groundwater contamination by on-site sanitation are also widely reported in countries of other regions of the world, including India [4], France [60], and Sweden [61]. Shivendra and Ramaraju [4] reported microbial contamination in the majority of wells sampled from Kanakapura town (India), with nitrate concentration of up to 45.9 mg/L, which is comparable to the Table 1 data. In the developed world setting, there is growing concern over emerging organic and inorganic contaminants, including pharmaceutical and personal care products, which could be attributed to on-site sanitation systems [17,71]. However, there is still limited information on occurrence, distribution, and ecological effects of emerging contaminants in peri-urban groundwater in the SSA context, which could also potentially be attributed to on-site sanitation practices [17,71–73].

The WHO introduced WSPs as a management tool towards identification, reduction, and prevention of such hazards (Table 1) in a water supply system, from catchment to consumer [41]. Rickert et al. [44] provided a simplified WSP process for small communities encompassing six steps from assembling a WSP team, describing the community water supply system, hazard assessment, deriving improvement plans, monitoring, and finally reviewing the WSP process (Figure 2). The process acknowledges consideration of societal and technical factors across the entire catchment, which may affect the quality of a water supply system. The process, thus, attempts to integrate comprehensive technical and socio-institutional aspects influencing risks to water supply stems. A socio-technical approach to resolving groundwater resources challenges has gained prominence due to the complexity of interactions [74,75].

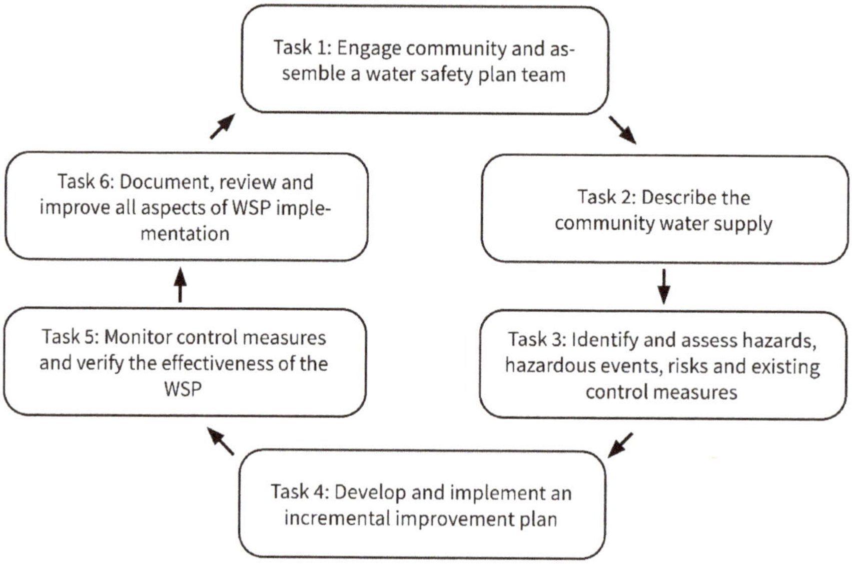

Figure 2. Tasks to develop a Water Safety Plan (WSP) for small communities (adapted from World Health Organization (WHO) [44]).

These socio-technical regime factors, thus, significantly influence the adoption and implementation of technologies, processes, and management tools, such as WSPs. Socio-technical regime factors can be summarized into five dimensions of (1) policy and regulation, (2) institutions, (3) science and technology, (4) user and market dynamics, and (5) socio-cultural considerations [76–78]. Addressing a complex system, therefore, requires careful consideration of both technical and socio-institutional factors influencing the challenge. The WSP process has been implemented in Uganda, Ghana, and Tanzania within the last decades [42,50,51], with varying levels of enabling and inhibiting regime factors, as reviewed in the next section. In assessing a given socio-technical system, it is important to understand the system boundaries or geographical scale [79]. This review focuses on small community groundwater-based supplies in peri-urban areas, which may be part of larger aquifer systems.

4. Challenges of IWRM and Attributes of TM in Implementing WSPs to Reduce Peri-Urban Groundwater Contamination by On-Site Sanitation in SSA

Like many countries in SSA, the existing water regime in Ghana, Uganda, and Tanzania generally follows the principles of IWRM [34,37,38]. IWRM promotes a multi-stakeholder approach to water resources management through a framework of creating an enabling environment, developing management instruments, and defining institutional roles for action [29]. IWRM promotes integrated management of water and land resources through four principles, namely; (1) fresh water is a finite and vulnerable resource, essential to sustain life, development, and the environment; (2) water development and management should be based on a participatory approach, involving users, planners, and policymakers, at all levels; (3) women play a central part in the provision, management, and safeguarding of water; and (4) water has an economic value in all its competing uses and should be recognized as an economic good [29]. These principles have been adopted in the policy and regulation framework, institutional set-up, and influence practices within the water sector in most of SSA. While the explicit implementation of WSPs in SSA is still growing [42,51], there is extensive documentation on implementation of the constituent tasks of the WSP process (Figure 2), influenced by the IWRM regime.

Transition Management, on the other hand, is still unexplored in the context of SSA, despite promising attributes at addressing persistent societal challenges [56,57]. TM has been explored, especially in the global north, in addressing sustainability problems in the transport, energy, water, and environment sectors, among others [56,80,81]. Rotmans and Loorbach [55] summarized eight principles of TM as; (1) creating space for niche developments; (2) empowering the niches; (3) focus on frontrunners; (4) guided variation and selection (system flexibility); (5) radical change in incremental steps; (6) learning by doing and doing by learning; (7) multi-domain approach; and (8) anticipation and adaptation. Transition Management, in practice, is usually implemented through concepts of Transition Management Cycle (TMC), Multi-Level Perspective (MLP), Multi-Phase Perspective (MPP), and Multi-Pattern Approach (MPA), among others [82–84]. The normative attributes of these TM approaches towards re-enforcing the principles and application of IWRM in implementing WSPs for reducing groundwater contamination in peri-urban areas in SSA, are further discussed.

4.1. Policy and Regulation

Ghana enacted the Water Resources Commission Act in 1996, which was the first legislation to embrace IWRM principles of managing water resources through a multidisciplinary and participatory approach [34]. Uganda developed the Water Action Plan in the period 1993–1995, basing on IWRM principles, culminating into the Water Policy of 1999 [85]. Adoption of IWRM principles in Tanzanian legislation can also be dated to as early as 1991, with the first river basin organization (Pangani River Basin Office, Moshi), and later entrenched in the Water Resources Management Act 2009 and the Water Supply and Sanitation Act 2009 [86]. The improved policy and legislation recognized holistic management of activities within a catchment and the interlinkages with the hydrological cycle, including groundwater, within overall sector objectives and documentation. This improved the appreciation of effects of human activities in the catchment on the quality of groundwater, and thus provided a holistic approach for groundwater contamination risk analysis, assessment, and management. This improved policy framework, thus, provided an enabling environment for the introduction of WSPs. Due to this holistic policy framework, Uganda was one of the first countries to adopt WSPs in SSA, undertaken in Kampala in 2005, which resulted in the national framework and guidelines for water source protection in 2013 [47,51]. In Ghana, the first WSPs were implemented in 2014, leading to the national drinking water quality management framework in 2016 [42]. In Tanzania, the first WSPs were undertaken in Dar es Salaam in 2014, and development of national guidelines is still underway [42].

However, in an attempt to include multiple stakeholders and multiple sectors towards water and land resources management, IWRM policies are conceptually diverse, without clear boundaries nor guidance to address specific societal challenges [32,33,87,88]. As a result, groundwater-specific issues have not been adequately addressed in policy and strategic sector direction [32]. Komakech and de Bont [89] pointed out that while the water management policies in Tanzania refer to groundwater protection, anthropogenic contamination of groundwater in urban areas continues unregulated, as depicted in Table 1. Within the maze of many issues affecting water resources management, it is therefore difficult to have sufficient scope analysis for the problem of peri-urban groundwater contamination, thus insufficient risk management strategies. Several pilot interventions on WSP implementation have been undertaken in Uganda, Ghana, and Tanzania, however, they are not specific to groundwater, despite the wide use of groundwater by the peri-urban majority, albeit, a low-income population [42,51,82]. Low funding for WSPs is also a manifestation of low prioritization to water safety, which is even more severe for the groundwater resources [90].

Transition Management approach is a prescriptive, stepwise, approach usually intended for a particular societal persistent (wicked) problem [91]. Loorbach [52] described the Transition Management Cycle to transition society from an undesirable state to a desired equilibrium state through; (a) establishing a transition arena, (b) establishing a transition agenda, (c) experimenting, and (d) monitoring and evaluating progress. Considering the extensive use of groundwater in peri-urban areas, the persistent hazards identified (Table 1) constitute a societal challenge that deserves particular attention. Through a TMC, the problem can be critically analyzed through its historical perspectives, identifying the critical formal and informal stakeholders and thereby addressing all socio-technical aspects. Through incremental steps and effective monitoring, the process is adapted until sustainability on this particular issue is achieved. The Multi-Phase Perspective (MPP) is a tool proposed under TM approaches for monitoring a transition process through stages of pre-development, take-off, acceleration, and finally stabilization, but aiming to avoid undesired scenarios of system lock-in, backlash, or system break down [92]. It is not clear to what extent WSPs have contributed to the elimination of contamination to public water supply systems in SSA, due to limited audits and follow-up processes [47,50]. Specific WSPs for sustainable protection of the groundwater resources against contamination by on-site sanitation in peri-urban areas could also be formulated, and implemented through small incremental steps until sustainability is attained by reduced contamination levels.

The need for prescriptive, flexible policy and regulation arrangements to groundwater management can be drawn from hydropolitics experiences in transboundary aquifer management in the Disi and Guarani aquifers, among others [22,23]. In an effort to address water scarcity in Jordan, a co-operation agreement was negotiated, with several geopolitical tradeoffs, between Jordan and Saudi Arabia for exploitation of the Disi aquifer to supply drinking water to the Jordanian capital, Aman, and other towns [23]. The co-operation agreement between Argentina, Uruguay, Paraguay, and Brazil towards improved management of the shared Guarani aquifer, reached after a decade of negotiations, also shows the benefits of policy specificity to handle a particular problem [22]. These examples from transboundary groundwater management show that specific and flexible policy and regulation instruments are required to address particular groundwater challenges. Thus, specific policies and regulations, within the local hydropilitics of the urban/peri-urban areas, could be helpful towards improved adoption and implementation of WSPs for reducing groundwater contamination.

4.2. Institutions

The IWRM principle of subsidiarity, management of the water resources at the lowest level, has been mainly constructed around catchment/basin organizations. Uganda, Tanzania, and Ghana have all made progress towards operationalizing of the catchment-based institutions, with varying degrees of success [40,85,86]. However, the adequacy of catchment/ basin organizations for groundwater management has been severely contested [37,38,40]. Foster and Ait-Kadi [32] argued that for effective groundwater management, the hydrogeological delineation would serve better than river/lake basin delineation. The effectiveness and coordination between various institutions and governance entities, including town authorities, has not been well addressed [37,93]. Effectiveness of the basin organizations in Uganda, Ghana, and Tanzania is as well in infancy stages, trying to establish legitimacy in the already existing institutional framework [37,40,85]. 'Top-down' centralized planning and control of resources has also remained prevalent in all the countries reviewed, leaving the created basin institutions deprived of technical and financial resources to make meaningful contributions to issues affecting peri-urban areas [37,89,94]. This institutional conundrum, therefore, has resulted in failure to address groundwater contamination challenges in peri-urban areas, among other challenges.

Implementation of WSPs has not been shielded from these institutional challenges. The WSPs developed in Uganda have mainly been undertaken within the National Water and Sewerage Corporation (NWSC), with minimal involvement of external stakeholders in the process [51,95–97]. In Ghana, the implementation of a WSP for a small-town water supply system (Assin Fosu), not operated by the national utility company, showed limited capacity of the local government and community beneficiaries to participate in a formal WSP process effectively [98]. Adoption and implementation of an effective WSP process for community water supplies in a peri-urban area requires a diverse section of actors, owing to the complex interlinkages required to make a sustainable societal change [99]. In Tanzania, Herslund and Mguni [100] argued that the water utility in Dar es Salaam city, the Dar es Salaam Water and Sewerage Corporation (DAWASCO), had done little to improve sanitation in low-income areas, despite high levels of contamination of the groundwater used by the peri-urban communities. This is evidenced by the microbial and nutrient contamination reviewed in Table 1, with a similar situation in urban centers in Uganda and Ghana. From all the cases, there is an institutional vacuum to supporting peri-urban communities in implementing WSPs for water supply options, which are usually off the conventional water supply network.

From TM concepts, the multi-domain principle encourages multiple pathways through both 'top-down' and 'bottom-up' approaches towards realizing an intended societal goal, as advanced by the Multi-Pattern Approach (MPA) [94,99]. The MPA advances a systematic analytic framework for review of different interrelated processes (patterns) aimed at achieving a desired societal goal by a different configuration of structures, cultures, and processes [84,99]. A specific system and its goal must first be defined, and then the relevant stakeholders and actors identified. In this regard, the societal goal is universal access to safe water (SDG 6). Through this approach, the tensions between different agencies in the water management initiatives can be minimized by exploring the most feasible and practical institutional constellation (formal and informal) for achieving the desired goal. De Haan and Rotmans [99] described that transitions could occur along three patterns of empowerment (bottom-up), re-constellation (top-down), or adaptation (internally induced change). Thus, this may include a co-existence of 'bottom-up' approaches aimed at improving community water sources in un-served low-income areas, while advances are also being explored by the existing regime (through public water utilities) for universal network coverage.

The concept of frontrunners can also be helpful to identify individuals from any pertinent organization, with commitment and capacity to contribute to the WSP process. This would ensure diversity of representatives in the process and thus improved community ownership of the processes. This approach is also supported by the attribute of guided variation and selection, which advocates for system flexibility, to devise alternative system configurations where a problem persists. De Haan and Rogers [84] documented several socio-institutional constellations that have been explored to address water management challenges in Melbourne (Australia), emphasizing the need for system flexibility to emerging challenges. Such a flexible institutional configuration, which may involve formal and informal agents towards a particular objective, is also supported by the growing literature on polycentric governance approaches [24–26]. Polycentric structures include multiple, interdependent autonomous agents (formal and informal), with a defined conflict resolution mechanism, working towards a common goal [25,26]. It is argued that polycentric governance approaches offer advantages of a context-specific institutional fit, enhanced system adaptability to emerging challenges, eliminate redundant actors, and improve local participation and accountability [26]. There is therefore sufficient theoretical grounding for an alternative institutional configuration to support community level initiatives for implementation of WSPs to protect their groundwater sources, with collaboration with the local governments and utilities providing water and sanitation services.

4.3. Science and Technology

Through IWRM, research has also been supported to develop and promote groundwater quality management instruments and processes in SSA such as groundwater quality/risk mapping, modelling, and protection/zoning [8,15,94,101]. However, extensive risk analysis for emerging organic and inorganic contamination from on-site sanitation is still limited [71,72]. Aquifer vulnerability assessments have been conducted, mainly through DRASTIC approach (representing parameters of depth, net recharge, aquifer media, soil media, topography, impact of vadose zone and hydraulic conductivity), as a step in the risk management process [15]. Risk assessment tools, mainly Quantitative Microbial Risk Assessment, have also been explored in SSA [67]. However, most of the management instruments/tools developed under the IWRM regime are still techno-centric, being water-sector specific, with limited integration with socio-institutional parameters [1,101,102]. Management instruments that integrate the socio-institutional aspects and hydrogeological assessments are limited, which complicates holistic decision making.

Water professionals have mainly been responsible for the WSP processes in Uganda, Tanzania, and Ghana, where the process tends to focus on the water treatment processes [95,97,98]. For instance, Howard et al. [95] noted that the WSPs should be based on health-based targets and independent surveillance conducted to ensure water safety by the health sector, but in practice, the WSP processes are usually implemented by the water sector. The limited capacity of local governments in participating in the WSPs process in Ghana could also be reviewed as a process left to the water sector [98]. Integration of traditional and non-traditional approaches, from various sectors and actors, could result in new, improved socio-technical approaches [103,104].

Transition Management concepts stem from complex systems theory trying to represent the complex relations between nature, science, society, and technology [55]. Complexity theory contends that such systems tend to be non-linear, continually evolving with changes in the environment and adapt to new situations. It is, thus, usually impossible to generalize or predict system behaviors [99]. Practical experimentation and learning from each experiment to adapt the system to an improved configuration is one of the prime attributes of TM. Peri-urban communities are usually served by groundwater supplies which are off the primary water supply network, thus not included in municipal physical models. Implementation of WSPs for such communities would be implemented in an experimental approach, learning lessons from incremental steps, which can be assimilated by the regime upon demonstrated successes. Such experimentation can be led by any relevant stakeholder, especially in the domain neglected by the primary grid, such as peri-urban areas. This variable approach could result in novel approaches, which could be adopted by the water regime. In analyzing requirements for a transition to sustainable urban wastewater management principles in Stellenbosch Municipality (South Africa), Malisa et al. [103] recommend for adoption of both traditional sources and alternatives such as rainwater harvesting, aquifer storage, and stormwater and wastewater reuse.

4.4. User and Market Dynamics

It is also argued that stakeholder participation under IWRM has been largely focused to institutional and high-level stakeholders, with limited active involvement of lower-level community stakeholders (usually economically vulnerable) in decision making processes [37,87,105]. Stakeholder participation has been promoted through Multi-Stakeholder Platforms (MSPs), in the form of Catchment Management Committees, which are decision making bodies (voluntary or statutory) comprising different stakeholders aimed at interdependence in solving water resources management problems [106,107]. Stakeholders in groundwater should encompass all entities (institutional, individual, association), whose actions directly or indirectly affect groundwater, with or without there consciousness [107]. However, it is noted that individual households are vital sources of peri-urban groundwater contamination, and thus their involvement in key decision-making is crucial to the resolution of the problem. Water User Associations in Uganda were an attempt to involve the communities at the lowest level, but these structures were not supported to grow and remained inactive, especially in urban areas [85]. In Tanzania, Pantaleo et al. [12] showed that despite the high level of microbial contamination in shallow wells in Babati town, the local community was not aware of the contamination, nor the potential risks. The involvement of individual households or local communities in WSP process in SSA has equally been limited [43,51].

The Multi-Level Perspective (MLP) has gained prominence in structuring transition processes at different levels of influence, structured as landscape, regime, and niche levels [82,83]. The landscape level encompasses the external environment, which puts pressure on the regime processes to create space for niche innovations to emerge towards the desired sustainability [82]. In this regard, international commitment to the attainment of SDGs can be regarded as one of the vital landscape pressures influencing national processes to develop innovations at all levels towards the attainment of SDG targets. Such innovations may include implementation of WSPs to improve community water sources in peri-urban areas, which have previously been abandoned by the urban water utilities. The national processes (socio-technical regime) need to allow for protected spaces (niches) for various innovations to incubate, which can be assimilated by the regime upon successful demonstration [83]. Creating space for niches and empowering them significantly improves community participation at the lowest level of society, as emphasized by the TM approach. Niches are created at the local level, with the participation of willing individuals. Frontrunners, as well, could be derived from the community, which offers an opportunity for individual participation in vision creation and decision making. Poustie et al. [80] shared experiences in designing a transition experiment for leapfrogging to sustainable urban water management in the city of Port Vila through niche experiments. Through a TM process, it was analyzed that unsustainability of the urban water infrastructure in Port Vila was as a result of the focus on technical and institutional capacities, with limited stakeholder networks and collaborations. Participation in a new vision creation and transition pathways analysis was achieved through a transition arena composed of community members, state actors, and private sector actors. Individual households could, therefore, through such an approach, be engaged to champion sustainability transition towards improved on-site sanitation facilities to reduce the risk of peri-urban groundwater contamination.

Implementation of the IWRM economic principle (polluter-pays-principle) also remains to be effective in SSA and almost non-existent for peri-urban groundwater context, which is affected by communities with limited economic capacity to pay. Groundwater management strategies should take into consideration the socio-economic position of the community [20,27]. Even in the conventional utility networks, raising resources for WSPs is a cross-cutting challenge [47,51,96]. Komakech and de Bont [89] attested to the inability of peri-urban communities in Tanzania to pay for such services. The principle would essentially imply that individual households owning the on-site sanitation facilities would meet the costs of implementing the WSP. While the principle may be applicable to industrial discharges, it is impractical to implement for community-level sources, such as unlined pit latrines and illicit municipal waste dumping by impoverished communities. Through the system flexibility principle of TM approach, "polluter-pays-principle" in the context of peri-urban groundwater management can be assessed depending upon the community characteristics such as willingness and ability to pay for such services; the level of existing public sanitation infrastructure in the area and settlement (housing) patterns. An alternative framework for compensating polluting activities could be assessed and recommended for the peri-urban areas depending upon the economic vulnerability of the population. An example of promoting resource recovery approaches to a circular economy in the Dutch wastewater system transition demonstrates a policy shift from "polluter-pays-principle" [108]. In this regard, a WSP process that advocates for resource recovery from the on-site products would be recommended. The recovered products could then be marketed for nutrient recovery in agricultural production and energy recovery, which benefits the society [6,61], in comparison to requesting the community to pay for the pollution. The economic principle of IWRM has found limited application to groundwater, especially in low-income communities, thus, system flexibility is required to find alternative remedies.

4.5. Socio-Cultural Considerations

Under IWRM, adequate consideration of socio-cultural aspects in the prevention of groundwater contamination in peri-urban areas in SSA has been rather limited [105,109]. Yeleliere et al. [110] highlighted the immense challenges of implementing integrated water resources policies in regulating peri-urban groundwater pollution in Ghana due to a disconnect with the customary practices in the region, which regulated equitable access to water resources and prevented community contamination of water resources. Such customary practices (like customary rights over water and gender customs in water utilization) are usually addressed by informal stakeholders such as traditional leaders, community elders, and unregistered well drillers (and diggers), who are unregulated by the regime, yet influential to a certain extent considering their high numbers in the SSA context [105,109]. Mapunda et al. [111] estimated that over 68% of the population from 20 cities in Tanzania is covered by informal providers, mainly from unregulated groundwater sources, which are not part of the city public water infrastructure. Such informal services are, thus, operated through social-cultural arrangements for water pricing, pollution control, operational schedules, and protection against vandalism. Culture (socio and institutional) issues have found limited space in the WSP processes to date [112]. Van Koppen et al. [86] argued that the implementation of IWRM in Tanzania removed the customary rights to water enshrined in previous arrangements. Consideration of socio-cultural factors in the implementation of WSPs in SSA has equally been limited, primarily focused on the technical factors [112].

Creating space for niches and empowering the niches to develop at the community level, as advocated by TM approaches, provides for an opportunity to incorporate and address socio-cultural issues, particular to the society [57]. These may include customary rights and practices to water and sanitation, and customary land ownership, which may influence groundwater protection zoning and community mobilization towards water source protection [57,105]. In assessing strategies for developing the transformative capacity of the urban water management sector in the city of Melbourne (Australia), Brodnik and Rebekha [113] emphasized the inclusion of organizational and socio-cultural factors in the transition process. Geels [82] also stressed the relevance of niche level innovations in achieving a transition process, usually emanating from the community, well grounded in the societal socio-cultural practices. Mukherji and Shah [28] also present strong arguments for active involvement of community users in groundwater management, drawing on experiences from India, Pakistan, Bangladesh, China, Spain, and Mexico, which have intensive use of groundwater.

However, the TM approach is also faced with unresolved socio-institutional challenges, including limited regard of "politics, power, and conflicts" in achieving democratic participation and ensuring equal and influential participation of weaker members of society in transition processes [56,114]. Other noted challenges include uncertainty in the framing of regimes and system boundaries, low legitimacy of frontrunners, concern for individual perspectives, and inadequate capacity for steering and monitoring societal changes [56]. Considering the complexity of the problems addressed in TM, the development of appropriate management instruments is a work-in-progress [84]. Globally, there are also still limited empirical prescriptive interventions by TM. In view of these limitations, TM alone, as well, may not be the "silver bullet" to addressing the complex dynamic challenge of peri-urban groundwater contamination in SSA. A complementary approach harnessing attributes of both IWRM and TM could offer complementarity towards improved uptake and implementation of WSPs for peri-urban groundwater management.

5. Proposal for a Risk-Based Management Framework towards Reducing Peri-Urban Groundwater Contamination by On-Site Sanitation in SSA

Remedies to complex problems need an interdisciplinary and multifaceted approach [33]. While several studies have advocated for alternative approaches to IWRM [33,39,88], in light of the achievements and gains realized in the global efforts in the implementation of IWRM, others have argued for the strengthening of the concept with attributes of emerging frameworks in order to address emerging challenges [30,38,40]. From the achievements of IWRM in Uganda, Tanzania, and Ghana, this review advocates for strengthening of IWRM towards addressing the challenge of peri-urban groundwater contamination by on-site sanitation. From the analysis of attributes of IWRM and TM, a unified risk-based management framework is proposed towards improved adoption and implementation of WSPs for reducing groundwater contamination from peri-urban areas in SSA (Figure 3).

The socio-technical regime factors that influence groundwater contamination in peri-urban areas in SSA are presented in Figure 3, described with a simplified Entity-Relationship Diagramming (ERD) approach. These factors must be taken into consideration when designing the WSP process for protecting peri-urban groundwater sources against contamination from on-site sanitation. The tenets of natural resources sustainability are hinged on the balance between economic efficiency, social equity and environmental (ecological) sustainability, as advocated by IWRM. TM approach emphasizes a prescriptive approach, which should target a particular persistent societal problem. The proposed framework is, thus, envisioned to contributed to reduce peri-urban groundwater contamination by on-site sanitation. The attributes of this vision are hinged on the sustainability tenets of ecological sustainability, social equity, and economic efficiency. Simultaneous maximization of the three attributes, as suggested by IWRM, has not been practical, leading to ambiguity in actions [33,36]. Giordano and Shah [88] advocated for context-specific approaches in consideration of the attributes. In this context, considering the social-economic marginalization of peri-urban communities, it can be argued that the proposed framework maximizes social equity, while taking into reasonable consideration the economic and ecological attributes. The focus on social equity is also in accordance with the transformative principle of the SDGs of "leave no one behind" [5].

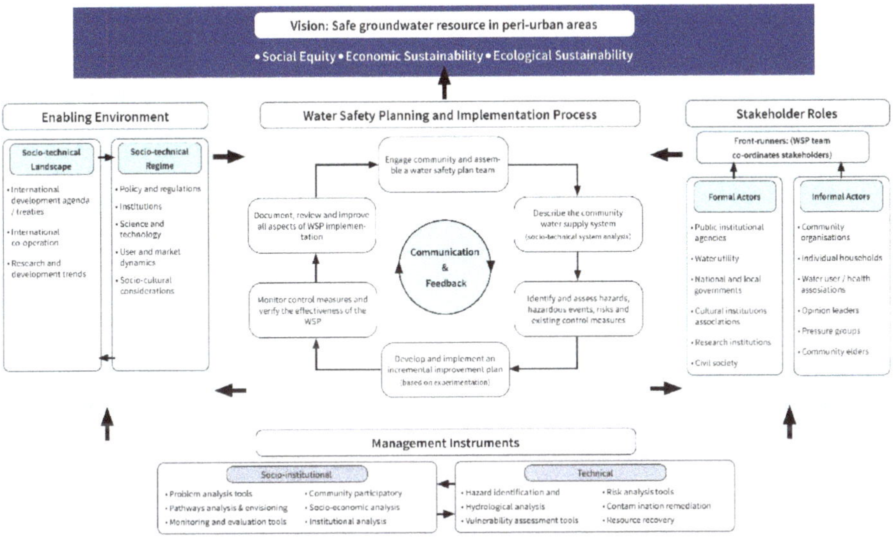

Figure 3. Proposed risk-based management framework for reducing groundwater contamination by on-site sanitation peri-urban Sub-Saharan Africa (SSA) (Adapted from [29,44,52,82]).

The WSP process is in line with the WHO process for small communities [44]. The WSP team may represent the frontrunners, according to TM approach, who follow the process through. The team should be composed of stakeholders from diverse background, different sections of the community, institutions, and leaders, with willingness and ability to participate in the process [56]. In describing the water supply system, both technical and socio-institutional factors affecting community water supply need to be critically analyzed. The Catchment Management Committees, developed under IWRM approaches, have been critiqued for the limited representation of vulnerable community members, power imbalances in decision making, and high cost of operationalizing the committees [106]. The MLP perspective offers a structured approach in analyzing the system components at landscape, regime, and niche levels, ensuring effective representation from all levels. The niche level initiatives at community level, to some extent, address representation and power imbalances. Improved socio-technical management processes and instruments should be used to comprehensively identify and assess hazards, hazardous events, risks, and existing control measures. Experimental measures should be explored, including radical steps, which may not be in the main system configuration, but demonstrated to improve water safety of the peri-urban communities. The process and activities should be monitored, along the MPP approach, with a focus on the sustainability goal [83]. All aspects of the WSP process are continually reviewed and adapted to emerging situations and the process is repeated continuously to achieve sustainable reduction of groundwater contamination from on-site sanitation. Throughout the WSP process, communication and feedback to and from all stakeholders is emphasized, while continually adapting tools, processes, and focus to emerging hazards and events. This continuous influence and adapting of processes are reflected in the bi-directional influence arrows between the various entities in Figure 3.

As postulated by IWRM, the process is influenced by a framework of 'enabling environment', 'management instruments', and 'institutional roles' [29]. Using concepts of TM approach, these elements are re-enforced for practical analysis and formulation of strategies for reducing groundwater contamination in peri-urban areas by on-site sanitation. In the enabling environment, the MLP differentiates between the socio-technical landscape and socio-technical regime to influence transitions towards sustainability. The landscape pressures, including international development agenda (SDGs), multinational development agencies, and international political/trade treaties are avenues for influencing regimes towards the sustainable achievement of groundwater quality. IWRM has been strong at mobilizing the international agenda towards holistic management of water resources [30,36]. The socio-technical regime, comprised of national level policies, structures, political dynamics, user preferences, national development, and cultures, can be rallied towards the support of groundwater protection initiatives. While the external environment influences the contamination risk management process, emerging innovations should equally impact the external environment for improved adaptation of the entire system. This provides for the possibility of upscaling promising niche level experiments to be adopted by the system regime and subsequently, the landscape level. This also ensures continuous adaptation of the landscape and regime forces towards emerging situations.

An array of stakeholders from both formal (under regulation by the regime) and informal (unregulated by the regime) settings, derived from the individual level, the national level, and the international level, are required for driving the transition towards the reduction of peri-urban groundwater contamination by on-site sanitation in SSA. This polycentric arrangement would enable a resilient and accountable institutional framework to support the particular challenges faced by peri-urban communities in managing groundwater resources. Management boundaries need to be specified between different stakeholders to avoid scenarios of inaction and conflicts. While IWRM advocates for catchment-based boundary systems, TM advocates for flexibility in setting the boundary of assessment, depending upon a given challenge. The peri-urban areas could be assessed as a specific sub-system, with particular socio-economic characteristics, within the overall urban water supply system. TM advocates for an approach that starts at the lowest community level, within the existing societal arrangements for change action. The informal arrangements with the community need to be given due consideration for any management arrangements proposed for implementation of the WSP.

Appropriate socio-technical management instruments are required to manage the entire process, stakeholders, and the environment towards realization of the sustainability objective. Prediction and modelling tools and techniques developed under IWRM could be strengthened, which could be used to study the overall system, including the technical and socio-institutional aspects and the complex interactions towards achieving the desired sustainability goal. Such management tools may include laboratory analysis techniques, problem analysis tools, pathways analysis tools such as back-casting, and vulnerability assessment tools. The process of tools development is continuous with knowledge advancement. Experimentation with various processes and adaptation of the WSP process to emerging issues from the peri-urban communities is an integral part of the process.

6. Conclusions

Groundwater is a vital resource for most of the population in low-income peri-urban areas in SSA. However, the resource is threatened by increasing contamination arising from on-site sanitation facilities, in light of a growing population. From the review of literature, it is evident that microbial and chemical contamination is prevalent and well documented in SSA. The WHO introduced WSPs to assist communities to manage such contamination by managing the technical and socio processes from catchments to the water user. However, adoption and implementation of the WSPs have been met with several successes and challenges influenced by the existing management framework (regime), which is aligned to the IWRM principles [36,37,85].

IWRM introduced improved holistic water resource management policies, improved public participation, improved management tools, and gender-sensitive approaches, which provided an enabling environment for the adoption of WSPs, which were multi-stakeholder by design [44]. However, implementation of IWRM has had a low focus on groundwater-specific problems and implemented through an incoherent institutional framework based on catchment management structures within already existing governance structures. The practicality of such structures to address groundwater-specific issues has been severely questioned [34,37]. The management instruments developed under IWRM have also, to date, remained techno-centric, with limited integration of socio-technical factors. There has also been limited regard for socio-cultural aspects and the economically vulnerable society segments, reflected in low financing of WSPs initiatives [51,98]. These limitations have affected the adoption and implementation of WSPs for groundwater interventions in peri-urban areas in SSA. The WSPs have been mainly implemented by urban water utilities for the conventional piped water supplies, with no consideration to community sources within the peri-urban areas.

TM advances normative strengths of long-term planning for societal transformation and implementation of the plan through incremental short-term measures, which has shown promise at addressing such complex adaptive systems. It provides a prescriptive approach, which targets a particular societal concern and works towards a shared vision, guided by frontrunners, which addresses the challenge of system diversity. TM processes are also multi-domain and flexible to institutional and actor dynamics, with emphasis on community-level (niche) initiatives [82]. These attributes can eliminate institutional tensions, ensure incorporation of community vulnerable persons, and integration of socio-cultural elements in a transition process towards the elimination of groundwater contamination. The process is improved through learning by doing and doing by learning, through a multi-domain approach, further adapting the system in light of increased pressures like increased urbanization and climate change. The normative strengths proposed by TM are consistent with emerging approaches in polycentric governance approaches, postulated to improve system adaptability to emerging challenges and accountability for results to the local population in attainment of set goals.

Based on the normative strengths of IWRM and TM, a unified risk-based management framework is advanced by this review towards the improved adoption and implementation of WSPs for reducing peri-urban groundwater contamination from on-site sanitation in SSA. The framework strengthens the vision of the WSP process, illuminates the enabling environment along landscape and regime aspects while refining the WSP process at the community level for incremental steps towards sustainability. Socio-technical management instruments are identified, which could be strengthened and applied in improved WSP implementation. The stakeholders, coordinated by a specific team of frontrunners, are also structured into formal and informal stakeholders towards improved WSP process implementation to emphasize community-level interventions. Experimentation of the proposed framework is recommended for empirical validation, which could then be adopted towards the improved implementation of WSPs for reducing groundwater contamination by on-site sanitation in peri-urban SSA, in the effort towards attaining sustainable development goal number 6 for universal access to safe water.

Key research gaps identified by the review include; the need for a further understanding of emerging contaminants in peri-urban groundwater, in light of increased urbanization, which have not been addressed by the WSP processes. There is also need for further assessment of the context and regional specific socio-institutional factors influencing adoption and implementation of WSPs in addressing the persistent challenge of groundwater contamination by on-site sanitation in SSA, such as cultures, power imbalances, societal practices, which differ between various socio-technical settings. While the integration of TM and IWRM approaches for improved implementation of WSPs is recommended, tensions between attributes of the approaches may also arise, which need further study through experimentation. There are also limited management instruments that incorporate analysis of socio-technical aspects of groundwater contamination, in the light of emerging challenges.

Author Contributions: Conceptualization, F.K., R.N.K., J.W.F., and F.R.B.T.; methodology, F.R.B.T.; software, F.R.B.T. and S.S.; validation, P.M.N., S.S., and F.R.B.T.; formal analysis, F.R.B.T. and F.K.; investigation, F.R.B.T.; resources, R.N.K. and J.W.F.; data curation, F.R.B.T.; writing—original draft preparation, F.R.B.T.; writing—review and editing, F.R.B.T., P.M.N., and S.S.; visualization, P.M.N.; supervision, R.N.K. and F.K.; project administration, R.N.K.; funding acquisition, J.W.F. All authors have read and agreed to the published version of the manuscript.

Funding: This research was funded by the UK Department for International Development (DFID), the Economic and Social Research Council (ESRC), and the Natural and Environment Research Council (NERC) under the UPGro Programme, grant number NE/M008045/1. The APC was funded by T-Group project under the UPGro Programme.

Acknowledgments: The authors are also grateful to Maryam Nastar for her critical input to the paper and Ismail Bukenya for his support in production of figures. We are also grateful to the two anonymous reviewers for their critical comments, which have significantly improved the quality of this paper.

Conflicts of Interest: The authors declare no conflict of interest.

References

1. Gorelick, S.M.; Zheng, C. Global change and the groundwater management challenge. *Water Resour. Res.* **2015**, *51*, 3031–3051. [CrossRef]
2. WHO. *Protecting Groundwater for Health: Managing the Quality of Drinking Water Sources*; IWA Publishing: London, UK, 2006.
3. Nayebare, J.G.; Owor, M.M.; Kulabako, R.; Campos, L.C.; Fottrell, E.; Taylor, R.G. WASH conditions in a small town in Uganda: How safe are on-site facilities? *J. Water Sanit. Hyg. Dev.* **2020**, *10*, 96–110. [CrossRef]
4. Shivendra, B.T.; Ramaraju, H.K. Impact of onsite sanitation system on groundwater in different geological settings of peri-urban areas. *Aquat. Procedia* **2015**, *4*, 1162–1172. [CrossRef]
5. WHO; UNICEF. *Progress on Drinking Water, Sanitation and Hygiene: 2017 Update and SDG Baseline*; World Health Organization: Geneva, Switzerland, 2017; Volume 66. [CrossRef]
6. Semiyaga, S.; Okure, M.A.E.; Niwagaba, C.B.; Katukiza, A.Y.; Kansiime, F. Decentralized options for faecal sludge management in urban slum areas of Sub-Saharan Africa: A review of technologies, practices and end-uses. *Resour. Conserv. Recycl.* **2015**, *104*, 109–119. [CrossRef]
7. Gaye, C.B.; Tindimugaya, C. Review: Challenges and opportunities for sustainable groundwater management in Africa. *Hydrogeol. J.* **2019**, *27*, 1099–1110. [CrossRef]
8. Back, J.O.; Rivett, M.O.; Hinz, L.B.; Mackay, N.; Wanangwa, G.J.; Phiri, O.L.; Songola, C.E.; Thomas, M.A.S.; Kumwenda, S.; Nhlema, M.; et al. Risk assessment to groundwater of pit latrine rural sanitation policy in developing country settings. *Sci. Total Environ.* **2018**, *613–614*, 592–610. [CrossRef]
9. Nyenje, P.M.; Foppen, J.W.; Kulabako, R.; Muwanga, A.; Uhlenbrook, S. Nutrient pollution in shallow aquifers underlying pit latrines and domestic solid waste dumps in urban slums. *J. Environ. Manag.* **2013**, *122*, 15–24. [CrossRef]
10. Berendes, D.M.; de Mondesert, L.; Kirby, A.E.; Yakubu, H.; Adomako, L.; Michiel, J.; Raj, S.; Robb, K.; Wang, Y.; Doed, B.; et al. Variation in *E. coli* concentrations in open drains across neighborhoods in Accra, Ghana: The influence of onsite sanitation coverage and interconnectedness of urban environments. *Int. J. Hyg. Environ. Health* **2020**, *224*, 113433. [CrossRef]
11. Grönwall, J. Self-supply and accountability: To govern or not to govern groundwater for the (peri-) urban poor in Accra, Ghana. *Environ. Earth Sci.* **2016**, *75*, 1–10. [CrossRef]
12. Pantaleo, P.A.; Komakech, H.C.; Mtei, K.M.; Njau, K.N. Contamination of groundwater sources in emerging African towns: The case of Babati town, Tanzania. *Water Pract. Technol.* **2018**, *13*, 980–990. [CrossRef]
13. Nyarko, K.B.; Oduro-Kwarteng, S.; Dwumfour-Asare, B.; Boakye, K.O. Incentives for water supply to the urban poor and the role of the regulator in Ghana. *Int. J. Water* **2016**, *10*, 267–280. [CrossRef]
14. Nhamo, G.; Nhemachena, C.; Nhamo, S. Is 2030 too soon for Africa to achieve the water and sanitation sustainable development goal? *Sci. Total Environ.* **2019**, *669*, 129–139. [CrossRef] [PubMed]
15. Anim-Gyampo, M.; Anornu, G.K.; Agodzo, S.K.; Appiah-Adjei, E.K. Groundwater Risk Assessment of Shallow Aquifers within the Atankwidi Basin of Northeastern Ghana. *Earth Syst. Environ.* **2019**, *3*, 59–72. [CrossRef]

16. Elisante, E.; Muzuka, A.N.N. Sources and seasonal variation of coliform bacteria abundance in groundwater around the slopes of Mount Meru, Arusha, Tanzania. *Environ. Monit. Assess.* **2016**, *188*, 395. [CrossRef] [PubMed]
17. Lapworth, D.J.; Nkhuwa, D.C.W.; Okotto-Okotto, J.; Pedley, S.; Stuart, M.E.; Tijani, M.N.; Wright, J. Urban groundwater quality in sub-Saharan Africa: Current status and implications for water security and public health. *Hydrogeol. J.* **2017**, *25*, 1093–1116. [CrossRef]
18. Murphy, J.L.; Kahler, A.M.; Nansubuga, I.; Nanyunja, E.M.; Kaplan, B.; Jothikumar, N.; Routh, J.; Gómez, G.A.; Mintz, E.D.; Hill, V.R.; et al. Environmental survey of drinking water sources in Kampala, Uganda, during a Typhoid Fever outbreak. *Appl. Environ. Microbiol.* **2017**, *83*, e01706-17. [CrossRef]
19. Conti, K.I.; Gupta, J. Global governance principles for the sustainable development of groundwater resources. *Int. Environ. Agreem.* **2016**, *16*, 849–871. [CrossRef]
20. Varady, R.G.; Zuniga-Teran, A.A.; Gerlak, A.K.; Megdal, S.B. Modes and approaches of groundwater governance: A survey of lessons learned from selected cases across the globe. *Water* **2016**, *8*, 417. [CrossRef]
21. Stephan, R.M. Climate change considerations under international groundwater law. *Water Int.* **2017**, *42*, 757–772. [CrossRef]
22. Hussein, H. The Guarani aquifer system, highly present but not high profile: A hydropolitical analysis of transboundary groundwater governance. *Environ. Sci. Policy* **2018**, *83*, 54–62. [CrossRef]
23. Hussein, H. Yarmouk, Jordan, and Disi basins: Examining the impact of the discourse of water scarcity in Jordan on transboundary water governance. *Mediterr. Polit.* **2018**, *24*, 269–289. [CrossRef]
24. Jones, S.D. Bridging political economy analysis and critical institutionalism: An approach to help analyse change for rural water services. *Int. J. Commons* **2015**, *9*, 65–86. [CrossRef]
25. Thiel, A.; Moser, C. Toward comparative institutional analysis of polycentric social-ecological systems governance. *Environ. Policy Gov.* **2018**, *28*, 269–283. [CrossRef]
26. Carlisle, K.; Gruby, R.L. Polycentric systems of governance: A theoretical model for the commons. *Policy Stud. J.* **2019**, *47*, 927–952. [CrossRef]
27. Megdal, S.B.; Dillon, P.; Seasholes, K. Water banks: Using managed aquifer recharge to meet water policy objectives. *Water* **2014**, *6*, 1500–1514. [CrossRef]
28. Mukherji, A.; Shah, T. Groundwater socio-ecology and governance: A review of institutions and policies in selected countries. *Hydrogeol. J.* **2005**, *13*, 328–345. [CrossRef]
29. Global Water Partnership. *Integrated Water Resources Management*; Global Water Partnership: Stockholm, Sweden, 2000.
30. Allouch, J. The birth and spread of IWRM—A case study of global policy diffusion and translation. *Water Altern.* **2016**, *9*, 412–433.
31. Mechlem, K. Groundwater governance: The role of legal frameworks at the local and national level – Established practice and emerging trends. *Water* **2016**, *8*, 347. [CrossRef]
32. Foster, S.S.D.; Ait-Kadi, M. Integrated Water Resources Management (IWRM): How does groundwater fit in? *Hydrogeol. J.* **2012**, *20*, 415–418. [CrossRef]
33. Biswas, A.K. Integrated Water Resources Management: Is It Working? *Int. J. Water Resour. Dev.* **2008**, *24*, 5–22. [CrossRef]
34. Agyenim, J.B.; Gupta, J. IWRM and developing countries: Implementation challenges in Ghana. *Phys. Chem. Earth* **2012**, *47–48*, 46–57. [CrossRef]
35. Mehta, L.; Alba, R.; Bolding, A.; Denby, K.; Derman, B.; Hove, T.; Manzungu, E.; Movik, S.; Prabhakaran, P.; van Koppen, B.; et al. The politics of IWRM in Southern Africa. *Int. J. Water Resour. Dev.* **2014**, *30*, 528–542. [CrossRef]
36. Molle, F. Nirvana concepts, Narratives and Policy models: Insights from the Water Sector. *Water Altern.* **2008**, *1*, 131–156.
37. Gudaga, J.L.; Kabote, S.J.; Tarimo, A.K.P.R.; Mosha, D.B.; Kashaigili, J.J. Effectiveness of groundwater governance structures and institutions in Tanzania. *Appl. Water Sci.* **2018**, *8*, 77. [CrossRef]
38. Mehta, L.; Bolding, A.; Derman, B.; Manzungu, E. Introduction to the Special Issue—Flows and Practices: The Politics of Integrated Water Resources Management (IWRM) in Southern Africa. *Water Altern.* **2016**, *9*, 389–411.
39. Petit, O. Paradise Lost? The difficulties in defining and monitoring Integrated Water Resources Management Indicators. *Curr. Opin. Environ. Sustain.* **2016**, *21*, 58–64. [CrossRef]

40. Duncan, E.A.; de Vries, N.; Nyarko, K.B. The effectiveness of water resources management in Pra Basin. *Water Policy* **2019**, *21*, 787–805. [CrossRef]
41. WHO. *Water Safety Plans: Managing Drinking-Water Quality from Catchment to Consumer*; World Health Organization Press: Geneva, Switzerland, 2005.
42. WHO. *Global Status on Water Safety Plans: A Review of Proactive Risk Assessment and Risk Management Practices to Ensure the Safety of Drinking-Water*; World Health Organization Press: Geneva, Switzerland, 2017.
43. Rondi, L.; Sorlini, S.; Collivignarelli, M.C. Sustainability of Water Safety Plans Developed in Sub-Saharan Africa. *Sustainability* **2015**, *7*, 11139–11159. [CrossRef]
44. Rickert, B.; Schmoll, O.; Rinehold, A.; Barrenberg, E. *Water Safety Plan: A Field Guide to Improving Drinking-Water Safety in Small Communities*; World Health Organization Press: Geneva, Switzerland, 2014.
45. Tsitsifli, S.; Tsoukalas, D. Water safety plans and HACCP implementation in water utilities around the world: Benefits, drawbacks and critical success factors. *Environ. Sci. Pollut. Res.* **2019**. [CrossRef]
46. Li, H.; Smith, D.C.; Cohen, A.; Wang, L.; Li, Z.; Zhang, X.; Zhong, G.; Zhang, R. Implementation of water safety plans in China: 2004–2018. *Int. J. Hyg. Environ. Health* **2020**, *223*, 106–115. [CrossRef]
47. Roeger, A.; Tavares, A.F. Water safety plans by utilities: A review of research on implementation. *Util. Policy* **2018**, *53*, 15–24. [CrossRef]
48. Rugemalila, R.; Gibbs, L. Urban water governance failure and local strategies for overcoming water shortages in Dar es Salaam, Tanzania. *Environ. Plan. C Gov. Policy* **2015**, *33*, 412–427. [CrossRef]
49. Kayser, G.; Loret, J.F.; Setty, K.; Blaudin De Thé, C.; Martin, J.; Puigdomenech, C.; Bartram, J. Water safety plans for water supply utilities in China, Cuba, France, Morocco and Spain: Costs, benefits, and enabling environment elements. *Urban Water J.* **2019**, *16*, 277–288. [CrossRef]
50. Summerill, C.; Smith, J.; Webster, J.; Pollard, S. An international review of the challenges associated with securing 'buy-in' for water safety plans within providers of drinking water supplies. *J. Water Health* **2010**, *8*, 387–398. [CrossRef]
51. Kanyesigye, C.; Marks, S.J.; Nakanjako, J.; Kansiime, F.; Ferrero, G. Status of water safety plan development and implementation in Uganda. *Int. J. Environ. Res. Public Health* **2019**, *16*, 4096. [CrossRef]
52. Loorbach, D. Transition Management for Sustainable Development: A Prescriptive, Complexity—Based Governance Framework. *Governance* **2010**, *23*, 161–183. [CrossRef]
53. Rauschmayer, F.; Bauler, T.; Schäpke, N. Towards a thick understanding of sustainability transitions—Linking transition management, capabilities and social practices. *Ecol. Econ.* **2015**, *109*, 211–221. [CrossRef]
54. Wen, B.; Van Der Zouwen, M.; Horlings, E. Transitions in urban water management and patterns of international, interdisciplinary and intersectoral collaboration in urban water science. *Environ. Innov. Soc. Transit.* **2015**, *15*, 123–139. [CrossRef]
55. Rotmans, J.; Loorbach, D. Complexity and transition management. *J. Ind. Ecol.* **2009**, *13*, 184–196. [CrossRef]
56. Nastar, M.; Abbas, S.; Rivero, C.A.; Jenkins, S. The emancipatory promise of participatory water governance for the urban poor: Reflections on the transition management approach in the cities of the emancipatory promise of participatory water governance for the urban poor: Reflections on the transition. *Afr. Stud.* **2018**, *77*, 504–525. [CrossRef]
57. Silvestri, G.; Wittmayer, J.M.; Schipper, K.; Kulabako, R.; Oduro-Kwarteng, S.; Nyenje, P.; Van Raak, R. Transition Management for Improving the Sustainability of WASH Services in Informal Settlements in Sub-Saharan Africa—An Exploration. *Sustainability* **2018**, *10*, 4052. [CrossRef]
58. Moher, D.; Liberati, A.; Tetzlaff, J.; Altman, D.G. Preferred Reporting Items for Systematic Reviews and Meta-Analyses: The PRISMA Statement (Reprinted from Annals of Internal Medicine). *Phys. Ther.* **2009**, *89*, 873–880. [CrossRef]
59. Brady, M.; Loonam, J. Exploring the use of entity-relationship diagramming as a technique to support grounded theory inquiry. *Qual. Res. Organ. Manag.* **2010**, *5*, 224–237. [CrossRef]
60. Olivier, L.; Dubois, Y.; LeGat, Y.; Boutin, C. Statistical analysis of the effluent of 231 on-site sanitation facilities in France monitored during a 6-year period. *Water Sci. Technol.* **2019**, *8*, 203–212. [CrossRef]
61. Vidal, B.; Hedström, A.; Barraud, S.; Kärrman, E.; Herrmann, I. Assessing the sustainability of on-site sanitation systems using multi-criteria analysis. *Environ. Sci. Water Res. Technol.* **2019**, *5*, 1599–1615. [CrossRef]

62. Bakyayita, G.K.; Norrström, A.C.; Kulabako, R.N. Assessment of levels, speciation, and toxicity of trace metal contaminants in selected shallow groundwater sources, surface runoff, wastewater, and surface water from designated streams in Lake Victoria Basin, Uganda. *J. Environ. Public Health* **2019**, *2019*, 6734017. [CrossRef]
63. Lutterodt, G.; Foppen, J.W.A.; Uhlenbrook, S. Escherichia coli strains harvested from springs in Kampala, Uganda: Cell characterization and transport in saturated porous media. *Hydrol. Process.* **2014**, *28*, 1973–1988. [CrossRef]
64. Elisante, E.; Muzuka, A.N.N. Assessment of sources and transformation of nitrate in groundwater on the slopes of Mount Meru, Tanzania. *Environ. Earth Sci.* **2016**, *75*, 277. [CrossRef]
65. Mushi, D.; Byamukama, D.; Kirschner, A.K.T.; Mach, R.L.; Brunner, K.; Farnleitner, A.H. Sanitary inspection of wells using risk-of-contamination scoring indicates a high predictive ability for bacterial faecal pollution in the peri-urban tropical lowlands of Dar es Salaam, Tanzania. *J. Water Health* **2012**, *10*, 236–243. [CrossRef]
66. Akoto, O.; Agbeshie, J.; Diane, T. Chemical characteristics and health hazards of heavy metals in shallow groundwater: Case study Anloga community, Volta Region. *Appl. Water Sci.* **2019**, *9*, 36. [CrossRef]
67. Takal, J.K.; Quaye-ballard, J.A. Bacteriological contamination of groundwater in relation to septic tanks location in Ashanti. *Cogent Environ. Sci.* **2018**, *4*, 1556197. [CrossRef]
68. Machdar, E.; van der Steen, N.P.; Raschid-Sally, L.; Lens, P.N.L. Application of Quantitative Microbial Risk Assessment to analyze the public health risk from poor drinking water quality in a low income area in Accra, Ghana. *Sci. Total Environ.* **2013**, *449*, 134–142. [CrossRef] [PubMed]
69. Aboagye, D.; Zume, J.T. Assessing groundwater quality in peri-urban localities of Kumasi, Ghana. *Afr. Geogr. Rev.* **2019**, *38*, 390–405. [CrossRef]
70. Katukiza, A.Y.; Temanu, H.; Chung, J.W.; Foppen, J.W.A.; Lens, P.N.L. Genomic copy concentrations of selected waterborne viruses in a slum environment in Kampala, Uganda. *J. Water Health* **2013**, *11*, 358–370. [CrossRef]
71. Sorensen, J.P.R.; Lapworth, D.J.; Nkhuwa, D.C.W.; Stuart, M.E.; Gooddy, D.C.; Bell, R.A.; Chirwa, M.; Kabika, J.; Liemisa, M.; Chibesa, M.; et al. Emerging contaminants in urban groundwater sources in Africa. *Water Res.* **2015**, *72*, 51–63. [CrossRef]
72. K'oreje, K.O.; Okoth, M.; van Langenhove, H.; Demeestere, K. Occurrence and treatment of contaminants of emerging concern in the African aquatic environment: Literature review and a look ahead. *J. Environ. Manag.* **2020**, *254*, 109752. [CrossRef]
73. Buekes, L.S.; King, T.L.B.; Schmidt, S. Assessment of pit latrines in a peri-urban community in KwaZulu-Natal (South Africa) as a source of antibiotic resistant *E. coli* strains. *Int. J. Hyg. Environ. Health* **2017**, *220*, 1279–1284. [CrossRef]
74. Troy, T.J.; Konar, M.; Srinivasan, V.; Thompson, S. Moving sociohydrology forward: A synthesis across studies. *Hydrol. Earth Syst. Sci.* **2015**, *19*, 3667–3679. [CrossRef]
75. Hynds, P.; Regan, S.; Andrade, L.; Mooney, S.; O'Malley, K.; DiPelino, S.; O'Dwyer, J. Muddy waters: Refining theway forward for the 'sustainability science' of socio-hydrogeology. *Water* **2018**, *10*, 1111. [CrossRef]
76. Geels, F.W. From sectoral systems of innovation to socio-technical systems: Insights about dynamics and change from sociology and institutional theory. *Res. Policy* **2004**, *33*, 897–920. [CrossRef]
77. Fuenfschilling, L.; Binz, C. Global socio-technical regimes. *Res. Policy* **2018**, *47*, 735–749. [CrossRef]
78. Baum, R.; Bartram, J. A systematic literature review of the enabling environment elements to improve implementation of water safety plans in high-income countries. *J. Water Health* **2018**, *16*, 14–24. [CrossRef] [PubMed]
79. Da Silva, B.P.L.; Hussein, H. Production of scale in regional hydropolitics: An analysis of La Plata River Basin and the Guarani aquifer system in South America. *Geoforum* **2019**, *99*, 42–53. [CrossRef]
80. Poustie, M.S.; Frantzeskaki, N.; Brown, R.R. A transition scenario for leapfrogging to a sustainable urban water future in Port Vila, Vanuatu. *Technol. Forecast. Soc. Chang.* **2016**, *105*, 129–139. [CrossRef]
81. Loorbach, D.; Shiroyama, H. The challenges of sustainable urban development and transforming cities. In *Governance of Urban Sustainability Transitions*; Loorbach, D., Shiroyama, H., Fujino, J., Eds.; Springer: Tokyo, Japan, 2016; pp. 3–12.
82. Geels, F.W. The multi-level perspective on sustainability transitions: Responses to seven criticisms. *Environ. Innov. Soc. Transit.* **2011**, *1*, 24–40. [CrossRef]
83. Geels, F.W.; Schot, J. Typology of sociotechnical transition pathways. *Res. Policy* **2007**, *36*, 399–417. [CrossRef]

84. De Haan, F.J.; Rogers, B.C. The Multi-Pattern Approach for Systematic Analysis of Transition Pathways. *Sustainability* **2019**, *11*, 318. [CrossRef]
85. Nicol, A.; Odinga, W. IWRM in Uganda—Progress after Decades of Implementation. *Water Altern.* **2016**, *9*, 627–643.
86. Van Koppen, B.; Tarimo, A.K.P.R.; van Eeden, A.; Manzungu, E.; Sumuni, P.M. Winners and losers of IWRM in Tanzania. *Water Altern.* **2016**, *9*, 588–607.
87. Anokye, N.A.; Gupta, J. Reconciling IWRM and water delivery in Ghana—The potential and the challenges. *Phys. Chem. Earth* **2012**, *47–48*, 33–45. [CrossRef]
88. Giordano, M.; Shah, T. From IWRM back to integrated water resources management. *Int. J. Water Resour. Dev.* **2014**, *30*, 364–376. [CrossRef]
89. Komakech, H.C.; de Bont, C. Differentiated Access: Challenges of Equitable and Sustainable Groundwater Exploitation in Tanzania. *Water Altern.* **2018**, *11*, 623–637.
90. Peletz, R.; Kumpel, E.; Bonham, M.; Rahman, Z.; Khush, R. To what extent is drinking water tested in sub-Saharan Africa? A comparative analysis of regulated water quality monitoring. *Int. J. Environ. Res. Public Health* **2016**, *13*, 275. [CrossRef]
91. Roorda, C.; Wittmayer, J.; Henneman, P.; van Steenbergen, F.; Frantzeskaki, N.; Loorbach, D. *Transition Management in the Urban Context: Guidance Manual*; Dutch Research Institute for Transitions, Erasmus University: Rotterdam, The Netherlands, 2014.
92. Van der Brugge, R.; Rotmans, J. Towards transition management of European water resources. *Water Resour. Manag.* **2007**, *21*, 249–267. [CrossRef]
93. Msuya, T.S.; Lalika, M.C.S. Linking Ecohydrology and Integrated Water Resources Management: Institutional challenges for water management in the Pangani Basin, Tanzania. *Ecohydrol. Hydrobiol.* **2018**, *18*, 174–191. [CrossRef]
94. Nsubuga, F.W.N.; Namutebi, E.N.; Nsubuga-Ssenfuma, M. Water Resources of Uganda: An Assessment and Review. *J. Water Resour. Prot.* **2014**, *6*, 1297–1315. [CrossRef]
95. Howard, G.; Godfrey, S.; Tibatemwa, S.; Niwagaba, C. Water safety plans for piped urban supplies in developing countries: A case study from Kampala, Uganda. *Urban Water J.* **2005**, *2*, 161–170. [CrossRef]
96. Parker, A.; Summerill, C. Water safety plan implementation in East Africa: Motivations and barriers. *Waterlines* **2013**, *32*, 113–124. [CrossRef]
97. Tibatemwa, S.; Godfrey, S.; Niwagaba, C.; Kizito, F. Implementing water-safety plans in Urban piped-water supplies in Uganda. *Waterlines* **2005**, *23*, 8–10. [CrossRef]
98. Obeng, P.A.; Obeng, P.A.; Awere, E. Water Safety Planning and Implementation in a Ghanaian Small-scale Water Supply System. *Int. J. Environ. Clim. Chang.* **2020**, *10*, 1–18. [CrossRef]
99. De Haan, F.J.; Rotmans, J. Pattern in transitions: Understanding complex chains of change. *Technol. Forecast. Soc. Chang.* **2011**, *78*, 90–102. [CrossRef]
100. Herslund, L.; Mguni, P. Examining urban water management practices—Challenges and possibilities for transitions to sustainable urban water management in Sub-Saharan cities. *Sustain. Cities Soc.* **2019**, *48*, 101573. [CrossRef]
101. Ouedraogo, I.; Defourny, P.; Vanclooster, M. Mapping the groundwater vulnerability for pollution at the pan African scale. *Sci. Total Environ.* **2016**, *544*, 939–953. [CrossRef] [PubMed]
102. Oke, S.A.; Fourie, F. Guidelines to groundwater vulnerability mapping for Sub-Saharan Africa. *Groundw. Sustain. Dev.* **2017**, *5*, 168–177. [CrossRef]
103. Malisa, R.; Schwella, E.; Kidd, M. From 'government' to 'governance': A quantitative transition analysis of urban wastewater management principles in Stellenbosch Municipality. *Sci. Total Environ.* **2019**, *674*, 494–511. [CrossRef] [PubMed]
104. Re, V.; Sacchi, E.; Kammoun, S.; Tringali, C.; Trabelsi, R.; Zouari, K.; Daniele, S. Integrated socio-hydrogeological approach to tackle nitrate contamination in groundwater resources. The case of Grombalia Basin (Tunisia). *Sci. Total Environ.* **2017**, *593–594*, 664–676. [CrossRef] [PubMed]
105. Osumanu, A.; Yelfaanibe, I.K.; Galaa, Z.S. How is Integrated Water Resources Management Working at the Local Level? Perspectives from the Black Volta Basin of the Lawra District, Ghana. *J. Environ. Earth Sci.* **2014**, *4*, 27–39.
106. Daré, W.; Venot, J.; Le Page, C.; Aduna, A. Problemshed or Watershed? Participatory modeling towards IWRM in North Ghana. *Water* **2018**, *10*, 721. [CrossRef]

107. Faysse, N. Troubles on the way: An analysis of the challenges faced by multi-stakeholder platforms. *Nat. Res. For.* **2006**, *30*, 219–229. [CrossRef]
108. Ampe, K.; Paredis, E.; Asveld, L.; Osseweijer, P.; Block, T. A transition in the Dutch wastewater system? The struggle between discourses and with lock-ins discourses and with lock-ins. *J. Environ. Policy Plan.* **2019**, *22*, 155–169. [CrossRef]
109. Shemsanga, C.; Martz, A.N.N.M.L.; Mcharo, H.K.E. Indigenous knowledge on development and management of shallow dug wells of Dodoma Municipality in Tanzania. *Appl. Water Sci.* **2018**, *8*, 59. [CrossRef]
110. Yeleliere, E.; Cobbina, S.J.; Duwiejuah, A.B. Review of Ghana's water resources: The quality and management with particular focus on freshwater resources. *Appl. Water Sci.* **2018**, *8*, 93. [CrossRef]
111. Mapunda, D.W.; Chen, S.S.; Yu, C. The role of informal small-scale water supply system in resolving drinking water shortages in peri-urban Dar es Salaam, Tanzania. *Appl. Geogr.* **2018**, *92*, 112–122. [CrossRef]
112. Omar, Y.Y.; Parker, A.; Smith, J.A.; Pollard, S.J.T. Risk management for drinking water safety in low and middle income countries-cultural influences on water safety plan (WSP) implementation in urban water utilities. *Sci. Total Environ.* **2017**, *576*, 895–906. [CrossRef]
113. Brodnik, C.; Brown, R. Strategies for developing transformative capacity in urban water management sectors: The case of Melbourne, Australia. *Technol. Forecast. Soc. Chang.* **2018**, *137*, 147–159. [CrossRef]
114. Avelino, F. Empowerment and the challenge of applying transition management to ongoing projects. *Policy Sci.* **2009**, *42*, 369–390. [CrossRef]

sustainability

MDPI

Review

Rivers' Temporal Sustainability through the Evaluation of Predictive Runoff Methods

José-Luis Molina [1,*], Santiago Zazo [1], Ana-María Martín-Casado [2] and María-Carmen Patino-Alonso [2]

1 IGA Research Group. High Polytechnic School of Engineering, University of Salamanca, Av. de los Hornos Caleros, 50, 05003 Ávila, Spain; szazo@usal.es
2 IGA Research Group. Department of Statistics, University of Salamanca, Campus Miguel de Unamuno, C/Alfonso X El Sabio s/n, 37007 Salamanca, Spain; ammc@usal.es (A.-M.M.-C.); carpatino@usal.es (M.-C.P.-A.)
* Correspondence: jlmolina@usal.es

Received: 31 January 2020; Accepted: 22 February 2020; Published: 25 February 2020

Abstract: The concept of sustainability is assumed for this research from a temporal perspective. Rivers represent natural systems with an inherent internal memory on their runoff and, by extension, to their hydrological behavior, that should be identified, characterized and quantified. This memory is formally called temporal dependence and allows quantifying it for each river system. The ability to capture that temporal signature has been analyzed through different methods and techniques. However, there is a high heterogeneity on those methods' analytical capacities. It is found in this research that the most advanced ones are those whose output provides a dynamic and quantitative assessment of the temporal dependence for each river system runoff. Since the runoff can be split into temporal conditioned runoff fractions, advanced methods provide an important improvement over classic or alternative ones. Being able to characterize the basin by calculating those fractions is a very important progress for water managers that need predictive tools for orienting their water policies to a certain manner. For instance, rivers with large temporal dependence will need to be controlled and gauged by larger hydraulic infrastructures. The application of this approach may produce huge investment savings on hydraulic infrastructures and an environmental impact minimization due to the achieved optimization of the binomial cost-benefit.

Keywords: runoff; temporal dependence; rivers' sustainability; predictive methods; causal reasoning; runoff fractions; water management

1. Introduction

In order to implement a real integrated water resource management (IWRM), it is required the proper use and organization of a great range and amount of data sources and methods. In this context, climatic and hydrological variables are essential [1,2].

Hydrological processes' variability is increasing on a global and local scale [3,4]. Consequently, those events, with values and trends far from the historical average behavior, are more recurrent [5–7]. In order to address this new hydrological reality and aimed to anticipate and forecast it, new approaches, methodologies and techniques have emerged, incorporating this changing behavior. However, in order to build tools for the present-future, not all the reasons that explain this increasing variability are brand new (climate change and projections to the future), but also, the historical behavior should be more deeply understood [8].

Hydrological temporal behavior from a stochastic view has been usually studied, and many of those concepts, assumptions and applications are still valid today. Those concepts comprise temporal-spatial correlation and statistical hydrological parameters, including basic, droughts and storage ones, as well as trends, shifts and seasonality testing, among others [8,9].

Traditionally, hydrological models and all their mathematical development are built based on a unique observation runoff record collected from a gauge station. Still, today, most of the hydrological tools officially implemented in water administrations are based on the previous and rudimentary approach [10]. However, the boom of very powerful analytical techniques and methods such as data mining (DM), artificial intelligence (AI) and machine learning (ML), among others, have produced a drastic change in several disciplines, including hydrology [8]. Consequently, river basins' hydrological behaviors can be currently characterized through those tools by means of huge amount of information coming from statistical or deterministic operations [11]. This massive amount of information is used for populating sophisticated expert systems that provide quantitative and dynamic results, incorporating and quantifying the inherent hydrological uncertainty [11]. However, those techniques have a high variety of performance, usage, utility, analytical power and/or accuracy, among other parameters. This heterogeneity is not properly revised or synthetized in academic studies or research articles, and, consequently, it is worthy to analyze and characterize it.

The most powerful and useful methods are the stochastic ones, which are able to incorporate and deal with the uncertainty of hydrological variables. In this sense, the discipline of stochastic hydrology is where those methods are included. Furthermore, ideas and features such as temporariness or persistence strongly related to the measure of the time series long-term memory (Hurst coefficient) [12], dependence-independence of internal data or structure and strength of causal relationships, are particularly significant.

This study is aimed to review the most popular, powerful and innovative approaches and methods for analyzing and predicting rivers' temporal behaviors. Furthermore, this paper is also aimed to provide the reader with an evaluation of the strengths, weaknesses, opportunities and threats (SWOT) of those methods and techniques set. Finally, this paper is also aimed to provide the reader with a rigorous suitability assessment focused on achieving the most suitable method for developing robust sustainable rivers' assessments. For that, the paper is structured as follows. Section 2 covers a revision of the main material and methods included in literature for developing a rigorous and updated analysis of rivers' temporal behavior and, by extension, their applications. Section 3 contains the methodology and the main results drawn from this research. This section includes the identification and assessment of parameters, a SWOT analysis of those techniques and methods and a suitability assessment for capturing the best methods for temporal rivers' sustainability. Finally, Sections 4 and 5 are dedicated for the discussion and conclusions, respectively.

2. Overview of Research Approaches

Traditionally, the use of hydrological models has been extensive worldwide. A model can be in the form of a physical, analog or mathematical model [13]. Currently, mathematical models are more preferred due to the rapid development of computer technology and because of including a chronological set of relations. However, as it was aforementioned before, most of the models uniquely rely on their large set of theoretical mathematical algorithms, using for their input only the historical runoff record rather than feeding them with massive high-quality information. Predictive and descriptive methods for studying and evaluating the temporal behavior of rivers may be classified according to different criteria. In this sense, main criteria can be the following: First, considering its incapacity/capacity for explicitly incorporating and quantifying the uncertainty in their whole functioning, methods can be categorized into deterministic and stochastic [14]. Then, taking into account its capacity for dealing with massive or low amounts of data, methods are grouped into DM, big data (BD) and ML or, on the other hand, scarce-data methods. Furthermore, methods can be classified into a black or white box, in view of its transparency and manageability in their processing [15]. Additionally, stochastic methods are differentiative due to its way for dealing with the uncertainty. In this sense, there are methods like causal reasoning (CR) that uses probability [11];

other stochastic methods, such as HJ-Biplot, use multivariate analysis and techniques [16], and they are able to evaluate correlation, similarity, distance/dissimilarity and covariance [16] in the datasets. Application of different multivariate statistical techniques has increased in the recent past. The most used methods are the following ones. First, HJ-Biplot, which simultaneously interprets the position of the variables, the sets and the relationships between them [16]. Furthermore, discriminant analysis (MDA) [17] and cluster analysis (CA) allow classifying the units according to similarities [18]; factor analysis (FA) explores relationships in the few principle components [19]; principal component analysis (PCA) [17] and canonical correlation analysis (CCA) [20] select variables and identify factors influencing hydrological changes.

Given the high variability, randomness and uncertainty of hydrological processes, methods that fall on the classification of stochastic and deterministic are described in detail as follows. In this sense, according to Koutsoyiannis [21], an appropriate modeling approach for any uncertain hydrological system should necessarily include quantification of its uncertainty within a stochastic framework. Consequently, for the same previous reasons, the most adequate methods are the stochastic ones [14]. However, some new deterministic approaches can be successfully applied to some river's systems, especially for very controlled and gauged rivers' basins poorly affected by climate change [14] that are explained as follows.

2.1. Deterministic Methods

2.1.1. Process-Based on Hydrological Models

This type of modeling has complex physical theory and usually needs to have a large amount of data and computational time. They have an initial given condition which is defined and parameterized and apply nonlinear partial differential equations which describe the hydrologic processes. Furthermore, they are usually accurate; however, they are often complex and need a lot of information as model inputs, such as river bathymetry; a complete set of meteorological information (air temperature, solar radiation, wind, etc.); inflow and outflow conditions; etc. As a result, application of such models for regions with limited data is impractical [22]. These models can be classified into black, grey, or white box, in view of its transparency and manageability in their processing [15]. Firstly, the lumped model (black model), which evaluates the response of the basin simply at the output and does not characterize the physical characteristics of the hydrological processes. Secondly, the semi-distributed model (grey model), which is partly permitted to change in space with division of catchment into an amount of sub-basins. It requires lesser amounts of input data in contrast with the fully distributed model. The final type is the distributed model (white model), which requires large amounts of data for parameterization; however, it presents some problems, such as the nonlinearity, scale and uncertainty [15]. One of the important advantages of the deterministic models is that they present the inside view of a process which enables better understanding of the hydrological system. On the other hand, their capacity for dealing (identifying, characterizing, evaluating or quantifying) with temporal behavior of rivers (dependence/memory) is very limited [14,23]. In this sense, traditional hydrological modeling codes such as HEC-HMS [24] or SIMPA [10], among others, have used the mathematical equations given by classic hydrology. They have implemented simulation models for being applied to hydrographic basins and then are compared to observational data records to get a good calibration. Those models follow the traditional hydrological scheme of procedure of routing for getting the final product of hydrographs [25–28].

2.1.2. Wavelets Transformation (WT)

Wavelet transform (WT) is a method that has been widely used to reveal information (signal) both over time and on a domain scale (frequency). It splits the main time series into sub-fractions, which improves the data decomposition in forecasts. Consequently, this allows capturing useful information at different levels of data resolution. It has been extensively applied in hydrology, such as rainfall–runoff relations for karstic springs [29,30], scale-dependent synthetic streamflow generation [31], temporal patterns of precipitation [32,33], variabilities of hydrological processes [34,35] and hydrological forecasting and regionalization [36,37].

The discrete wavelet decomposition is known to offer a local representation of time series data using wavelet and scaling coefficients at different resolutions obtained via Mallat's pyramidal algorithm, among others [38]. There are several researches that have employed the wavelet technique for various hydro-climatological applications [39]. For instance, Adamowski and Chan [40] developed a wavelet neural network conjunction model to forecast monthly groundwater levels using wavelet decomposed data of multiple hydrological and meteorological variables. Moosavi et al. [41] compare the forecast performance of Wavelet-ANFIS and Wavelet-ANN hybrid models to the ANFIS and ANN benchmark models. Wavelet-ANFIS model provided better groundwater level forecasts at different time horizons. Likewise, numerous groundwater level estimation studies based on hybrid Wavelet-AI models can be found in the literature [42,43]. Furthermore, such hybrid WT-ANN methods have been shown to provide good performance in hydrological studies, such as rainfall-runoff modeling [44,45] and streamflow forecasting [46,47], as well as river and groundwater level forecasting [48], among others. Several studies have also shown that the WT-ANN hybrid models perform better than some other widely used models. For example, Peng et al. [47] applied empirical wavelet transform and artificial neural networks for streamflow forecasting. They concluded that the hybrid model can capture the nonlinear features of the streamflow time series and, consequently, provide more accurate forecasts than the traditional ANN method. Adamowski and Chan [40] applied the WT-ANN method to predict groundwater level and stated that it provided better forecasting accuracy than the conventional ANN method. Similar results have also been reported by Rajaee et al. [49] and Seo et al. [50]. These studies indicate that the WT-ANN hybrid models allow users to achieve forecasts with higher accuracies.

2.2. *Deterministic Machine Learning (ML) methods*

Some of the most widely used methods have been selected to be included in this section given its high operational capacity. Most of them belong to the ML group of methods. However, there are other methods that exceed ML influence and are also worthy to include.

There are two general categories of machine learning: supervised and unsupervised. Supervised ML techniques refer to when there is a piece of data to predict or explain. This is done by using previous data of inputs and outputs to predict an output based on a new input. By contrast, unsupervised ML is aimed to search ways to relate and group data points without the use of a target variable to predict. In other words, it evaluates data in terms of traits and uses the traits to form clusters of items that are similar to one another. To keep a robust and coherent structure, ML methods have been classified in this paper into deterministic and stochastic ones.

When there is inadequate information of influential parameters, the ML methods demonstrate higher potential and efficiency, and they are more appropriate than the conventional numerical or finite element methods [38]. The ML techniques make more reliable predictions/forecasts based on their ability to identify and reveal the nonlinear, non-stationary and discriminant features from historical time-series records and also effectively handle chaos and uncertainties associated with the input data. Furthermore, currently, the development of hybrid ML techniques for modeling hydrologic time series is also a very active area of research. For instance, the ML models developed from wavelet decomposed data have demonstrated better performance than the models provided with raw data. The description of selected methods, shown as follows, is focused on their capacity for predicting increasing hydrological condition processes, such as rivers' runoff.

2.2.1. Artificial Neural Networks Methods (ANN)

During the last 20 years, ANN have been increasingly used for wider applications in forecasting time series, including geophysical, meteorological and hydrological time series. Most ANN applications are used for forecasting in situations where there is an unknown relationship between the set of input factors and the outputs. Despite there is a strong criticism about the capacity of ANN for dealing with hydrological uncertainty and its reliability for semiarid regions is seriously questioned, the ANN model has become an appreciated tool for modeling nonlinear hydrometeorological phenomena, such as precipitation forecasting [51–53], streamflow forecasting [54,55] and ground water-level simulation [56,57], as well as river water temperature forecasting [58,59]. Furthermore, in regards to more specific utilities, using ANN as a tool, the NERD (neural error regression diagnosis) model of Abramowitz et al. [60] is available for data assimilation and the SOLO (self-organizing linear output) model of Hsu et al. [61], first for classifying input features into various categories, and then for performing function mapping for each category separately. Since the hydrological conditions are increasingly changing, applications of the linear and nonlinear regression models, as well as the traditional ANN models for river water temperature modeling, frequently have limitations, especially in processing of nonstationary data [62]. In this regard, wavelet transform, as a good preprocessing method for nonstationary data, can be a potential complement for the traditional methods.

2.2.2. System Dynamics Methods (SDs)

SDs is basically a methodology for studying complex feedback systems [63]. Consequently, this method was not originally intended to be applied for hydrological forecasting [64–67]. However, due to the inherent uncertainty and complex causal and dependency relationships within the hydrological records, it is also recommended for that. Specifically, it is particularly useful when studying complex systems with interacting elements, the behavior of which cannot be easily predicted. It allows one to examine system behavior modes and response as different variables are altered. The most well-known SDM packages include: SIMILE [68], VENSIM (www.vensim.com), STELLA (www.iseesystems.com), POWERSIM (www.powersim.com) and SIMULINK—an add-on to MATLAB (www.mathworks.com) [63].

2.3. *Stochastic Methods*

2.3.1. Traditional Techniques

The traditional statistical/time series methods, such as linear and nonlinear regression models, are usually simple to develop; however, they produce, in general, large modeling errors [69]. Consequently, the way they cope with uncertainty is not recommended. It is well-known that a correlogram provides an idea of temporal dependence intensity through the correlation coefficient [70]. Furthermore, it is known that a correlogram comprises a static picture and average result of the temporal behavior of hydrological series. This fact provokes the existence of indeterminacy points for defining the temporal dependence/independence of time series [70,71]. Traditional techniques were designed exclusively for the short term through Pearson's correlation coefficient and for the long term through the Hurst coefficient, which provides an idea of the degree of dependency of hydrological events throughout time series. Within these methods, dependence has been analyzed for hydrologic studies from many perspectives and through several approaches [72]. The most common traditional measure of dependence is the Pearson's correlation coefficient, which is computed based on the assumption of normal distribution to measure the linear dependence [8]. This is commonly assumed in statistical approaches, such as autoregressive (AR) models or the autoregressive moving average (ARMA) model, to characterize the linear dependence among multivariate random variables in hydrology, such as precipitation, streamflow or runoff [73,74]. Nevertheless, the normal assumption does not hold, and the linear dependence is too basic to characterize far more complicated dependence structures [72]. Another nonparametric correlation estimator as the Spearman and Kendall has been widely used. They have also been commonly used as alternative dependence measures to characterize the nonlinear dependence of hydrological variables, as demonstrated recently with copula models [75] described later.

2.3.2. Multivariate Adaptive Regression Splines (MARS)

This procedure, proposed by Friedman [76], is a multivariate nonparametric model which searches over all possible univariate knot locations and across interactions among all variables by means of the use of local so-called basis functions. In this sense, MARS algorithm is based on three sequential steps [76,77]. In the first one, named "constructive phase", it is created through adding basis functions step by step. Then, in order to improve the model, the best pairs of spline functions are selected. At each step, the split that minimized some lack of fit criterion from all the possible splits is chosen on each basis function. The iterative procedure of construction finishes when the model reaches a maximum number of basic functions. Then, in the second step, "pruning phase", a backward elimination process to redefine the model fitting, is carried out. Here, basis functions are eliminated one at a time until the lack of fit criterion based on the generalized cross validation (GCV) criterion [78] is a minimum. Finally, in the third phase, the optimal model selection is done based on an evaluation of the properties of the different models.

On the other hand, recently, MARS procedure has begun to be applied in the field of hydrological forecasting due to its flexibility in modeling high-dimensional data. It is recommended and applied over a wide range of hydrological issues, like runoff prediction [79,80], drought prediction [81], water pollution prediction [82], pan evaporation modeling [83] and prediction of scour depth below free overfall spillways [84]. In most of these previous researches, the raw time-series data are fed as input. The performance of MARS was better than that of the ANN, ANFIS, support vector machine (SVM) and M5 Tree models [38].

2.3.3. ARIMA/ARMA Methods

ARIMA/ARMA methods are usually used for generating synthetic series from the historical record [8]. Before tackling this generation, historical time series need to be normalized, which is a drawback of this technique [85]. Once the normalization is done, by means of the previous model, several synthetic series are generated with the same occurrence probability of the historical one [70]. Frequently, these synthetic series are used to populate a further model/method, such as CR [23]. Consequently, this method is usually hybridized with other techniques of artificial intelligence, which provide more accurate results in the modeling of complex natural processes [73,74,86,87]. This approach is applied to the hybridization with Bayesian methods for developing CR [77].

2.3.4. Causal Reasoning (CR) Methods

CR is based on the propagation of probability along a Bayesian implementation through a causal model. This model is based on decision variables who represent probability distributions for temporal runoff. Most of the CR approaches are based on Bayesian networks (BNs) and dynamic Bayesian networks (DBNs). Those methods are based on probabilistic graphical models [88,89]. They easily and automatically allow defining relationships between parts into complex models, through conditional probabilities, due to both models being based on Bayes' theorem [90]. BNs are probabilistic graphical models that offer compact representations of the joint probability distribution over sets of random variables [91,92]. This property has been taken to use BNs as a decision support system (DSS), for example in [93,94], to address problems within the IWRM paradigm [95]. Dynamic Bayesian networks (DBN) are a generalization of hidden Markov models (HMM) [8,96]. One of the main advantages of BNs is that they compute inference omni-directionally. Given an observation with any type of evidence on any of the networks' nodes (or a subset of nodes), BNs can compute the posterior probabilities of all other nodes in the network, regardless of arc direction, through observational inference [97].

The inherent ability of BNs to explicitly model uncertainty makes them suitable for temporal hydrological series analysis. Another feature of BNs is the unsupervised structural learning [98], which means probabilistic relationships between a large number of variables without having to specify input or output nodes. This can be seen as a quintessential form of knowledge discovery, as no assumptions are required to perform these algorithms on unknown datasets. Furthermore, this is strongly related to machine learning that has been applied successfully in the hydrological context in papers such as [99,100]. Consequently, the resulting product has many similarities with a neuro-fuzzy system or adaptive neuro-fuzzy inference system (ANFIS) that has been applied in works such as [101]. Finally, an important contribution of this method comprises the identification, characterization and quantification of temporal conditioned runoff fractions [8]. This represents an outstanding advance over classic or alternative methods.

2.3.5. Copulas Methods

In hydrology, copula applications started after the work of De Michele and Salvadori [102], who tested Frank copulas for a joint study of the negatively associated storm intensity and duration. Other research, such as that of Zhang and Singh [103], incorporated copulas for an extreme analysis of rainfall and drought events. The study of the dependence's modeling between extreme events is widely studied nowadays [104]. Dependence in extreme events needs to be evaluated through techniques focused on an extreme distribution, such as the tail-dependence coefficient. This has been commonly used in investigating hydroclimatic extremes, such as precipitation and temperature [105].

Another important and recent area of research is the construction of a multivariate distribution in modeling different dependence structures [106]. This is applied to hydrology in the form of frequency analysis, downscaling, streamflow or rainfall simulation, geostatistical interpolation, bias correction and so on. Other methods, such as multivariate parametric distribution [107], entropy [108] and copula [75], have been developed to model various dependence structures of multivariate variables through the construction of joint or conditioned distribution.

On the other hand, it should be highlighted that Copula method is based on the description and modeling of the dependence structure between random variables independently of the marginal laws involved. In this sense, there are high similarities with CR. Differences with CR lies in the fact that a Copula is a bivariate function, while CR is based on conditioned probabilistic distributions.

2.3.6. Kalman and Particle Filter Methods

The Kalman filter (KF) is a method that allows estimating unobservable state variables from observable variables that may contain some measurement error. It is a recursive nature algorithm that requires two types of equations: those that relate the state variables (main equations) and those that determine the temporal structure of the state variables (state equations). The estimates of the state variables are made based on the dynamics of these variables (time dimension), as well as the measurements of the observable variables that are obtained at each moment of time (cross-sectional dimension). That is, the dynamics are summarized in two steps: estimate the state variables using its own dynamics (prediction stage). Then, improve that first estimate using the information of the observable variables (correction stage). Once the algorithm predicts the new state at time, it adds a correction term, and the new "corrected" state serves as the initial condition in the next stage, $t + 1$. In this way, the estimation of the state variables uses all the available information until that moment and not only the information until the stage before the moment in which the estimation is made [109].

Recently, some novel applications, based on KF, have emerged in the field of hydrologic predictions. This is due to its features of real-time adjustment and easy implementation within a framework dynamic state [110]; one of them is an ensemble Kalman filter (EnKF). This method approximates the distributions of the system states by random samples, named ensembles and replaces the covariance matrix by the sample covariance computed from the ensembles, which is used for state updating in the KF [111]. In this sense, it is worth noting interesting approaches in the hydrological predictions field. For example, Moradkhani et al. [112] showed a dual-state estimation for parameters and state variables in a hydrologic model. For its part, Weerts and Serafy [113] compared the capability of EnKF and particle filter (PF) methods by reducing uncertainty in the rainfall-runoff update and internal model state estimation for flooding forecasting purposes. For their part, Pathiraja et al. [114] proposed an approach to detect nonstationary hydrologic model parameters in a paired catchment system.

On the other hand, a particle filter (PF) technique has emerged as an attractive alternative to remove the unrealistic Gaussian assumption of the EnKF [115], improving the reliability of hydrologic predictions [116]. PF is able to fully represent the posterior distributions of model parameters and state variables through a number of independent random samples called particles, and the particles are weighted and propagated sequentially by assimilating available observations. Over the past few years, the PF and its variants have been receiving increasing attention from the hydrologic community due to its ability to properly estimate the state of nonlinear and non-Gaussian systems [117,118].

2.4. *Stochastic Machine Learning (ML) Methods*

2.4.1. HJ-Biplot

The HJ-Biplot [119] is an extension of the classical Biplot methods [120]. This technique allows a joint representation of a data matrix in a reduced subspace dimension, usually a plane, of the rows and columns of a matrix of multivariate data X_{nxp}. It is a symmetric simultaneous representation technique that is in some ways similar to correspondence analysis but is not restricted to frequency data.

HJ-Biplot uses markers (points/vectors), denoted as $g_1, g_2, \ldots, g_n$ for each row, and $h_1, h_2, \ldots, h_p$ for each column. The markers are obtained from the usual singular value decomposition (SVD) of the data matrix $X = UDV^T$, where U are the eigenvectors of XX^T, V is the eigenvectors of X^TX and D is the matrix diagonal of singular values, taking as row markers rows of $J = UD$ and, as column markers, rows of $H = VD$. HJ-Biplot is widely used, because it provides a higher goodness-of-fit for rows and columns of the matrix [16].

Recently, this technique has been applied in the field of hydrology. Carrasco et al. [16] examined the relationship between the physicochemical and biological variables, as well as the sampling points through different months. Although this method has a descriptive character contributing to deepening the hydro-chemical knowledge and, consequently, improving the hydrological and hydrodynamic understanding and interpretation of water systems, which allowed generating new interpretations, knowledge and evaluations about the water quality of Gatun Lake. It also offers the advantage of providing high representation quality, both for the sampling points and for the physicochemical and biological variables.

2.4.2. Principal Component Analysis (PCA)

PCA analysis reduces the dimension of the original dataset. This technique uses an orthogonal transformation to convert highly correlated variables into a set of values of linearly uncorrelated, which are called principal components. Each principal component is a linear combination of the original variables. The mathematical approach for processing can be called adaptive data analysis (Table 1) that comprises mathematical processes such as orthogonal linear transformation, rank (dimensionality) reduction and visualization of latent data structures.

On several occasions, hydrological datasets contain not only useful important information but also confusing noise. Sometimes, datasets are not normally distributed or may contain outliers. Even the hydrological parameters are usually many times correlated. The correlation indicates that some of the information contained in one variable is also contained in some of the other remaining variables. In order to reveal the logical structures of the data, there are several hydrological procedures, such as PCA, which reduces the data dimensionality and gain more effective features [121]. Therefore, in several studies, the reason for selecting PCA as a statistical method was the low noise sensitivity [121].

PCA analysis has been used in numerous hydrological studies to test the water quality assessment [122], understand the role of land-atmosphere interactions in driving climate variability and the spatiotemporal variability [123] or select the redundant input variables for a prediction model of hybrid runoff models [124].

Table 1. Parameters assessment. Measure of the predictability and reliability. Low (*), medium (**) and high (***).

Methods	Predictability	Reliability	Mathematical Approach for Uncertainty	Mathematical Approach for Processing	Amount of Processed Information	Manageability	Transparency
Process-based hydrological models	*	*	Null/Scarce	Theoretical algorithms	**	**	**
Artificial Neural Networks (ANN)	***	**	Null	Dynamic and Adaptive Mathematical expressions	**	*	*
Wavelets Transformation (WT)	***	**	Scaling coefficients at different resolutions	Dynamic and Adaptive Mathematical expressions	***	*	**
System Dynamics (SDs)	**	**	Null	Dynamic and Adaptive Mathematical expressions	***	*	*
Linear and nonlinear regression models	*	**	Regression	Constant Mathematical expressions	**	***	***
Pearson's Correlogram Coefficient (Correlogram)	*	*	Correlogram Coefficient	Constant Mathematical expressions	**	***	***
Hurst coefficient	**	**	Null	Constant Mathematical expressions	**	***	***
Multivariate Adaptive Regression Splines (MARS)	*	**	Optimization	Dynamic and Adaptive Mathematical expressions	**	**	**
ARMA and ARIMA	**	**	Mathematical Residuals coefficients	Constant Mathematical expressions	**	**	**
Causal Reasoning (CR)	***	***	Probability	Dynamic and Adaptive Mathematical expressions	***	**	*
Copulas	***	**	Bivariate function	Dynamic and Adaptive Mathematical expressions	***	**	*
Kalman, Particle Filter and Ensemble Kalman Filter	**	**	Random Samples (Ensembles)	Dynamic and Adaptive Mathematical expressions	***	**	*
HJ-Biplot	*	**	Null/Scarce	Matrixes Visualization and Cluster analysis	***	**	**
Principal Component Analysis (PCA)	*	**	Null/Scarce	Adaptive data analysis	***	**	**
Factorial Analysis (FAV)	**	***	Different Statistical Expressions	Dynamic and Adaptive Mathematical expressions	***	**	**

2.4.3. Factorial Analysis of Variance (FAV)

Factorial analysis is usually used to help in the study of the individual and interaction effects of the parameters [125]. More specifically, the factorial analysis of variance method is used to diagnose the curve relationship between the parameters and the response [126,127]. In other words, FAV technique is used for measuring the specific variations of hydrological responses in terms of posterior distributions to investigate the individual and interactive effects of parameters on model outputs [128]. To complete this description, FAV comprises a powerful statistical tool to facilitate the exploration of the main effects. This is done by measuring the variation ranges of hydrological response under the impact of individual parameters and parameter interactions by revealing the specific variations of each parameter's effects under the impact of another parameter [129]. The FAV technique can also quantify the sensitivity of model outputs to individual parameters, as well as their interactions, through addressing the curvilinear characteristic of the hydrological response when parameters vary across their multiple levels [130,131].

Consequently, the hybridization of this method with Bayesian methods is quite appropriate, direct and powerful. This is called the Bayesian-based multilevel factorial analysis (BMFAV), and it is used to assess parameter uncertainties and their effects on hydrological model outputs. In this sense, there are several other applications aimed to approximate the posterior distributions of model parameters with Bayesian inference [128,132]. Another important application of FAV was to develop a BMFAV method to address the dynamic influence of hydrological model parameters on runoff simulation [133].

3. Methodology and Results

This research comprises a methodological framework that includes the development of several consecutive phases listed as follows. First, an identification of parameters for a posterior SWOT analysis took place; then, those parameters were assessed based on a rigorous analysis of the global scientific literature; after that, a SWOT analysis was developed; finally, a suitability assessment for rivers' sustainability was developed. Results drawn from this research are largely established in terms of a comparative analysis of the aforementioned techniques aimed to reach a ranking of methods from a multidimensional way. This will allow selecting the most suitable technique for each component of rivers' sustainability included in this research.

3.1. Identification of Parameters

The parameters identified and established in this research for tackling the SWOT analysis are listed as follows. **Predictability**: There is a high variability on the prediction capacity provided by the analyzed methods and techniques. A qualitative score (high, medium and low) to rank and measure that skill is established. **Reliability**: This parameter is aimed to qualitatively score the degree of confidence and/or accuracy that the generated prediction provides from a certain method. **Mathematical approach for dealing with uncertainty**: This descriptive parameter informs about which way is the method that deals with uncertainty (probability, multivariate, etc.). **Mathematical approach for processing**: This provides insights of the algorithms type and theoretical development the method relies on. **Amount of processed information**: This parameter is aimed to qualitatively score the amount of information the method requires for generating its output. A qualitative score (high, medium and low) for ranking and measuring is established. **Manageability**: This provides an idea of the ease of handling. Consequently, it measures the complexity and level of expertise required for implementing the method. A qualitative score (high, medium and low) is established. **Transparency**: This measures the ease for an external observer to look into the internal process. The spectrum is quite wide, going all the way through total transparency to complete a black box. A qualitative score (high, medium and low) for ranking and measuring is also established.

3.2. Assessment of Parameters

Table 1 shows the scores and values for each method, based on a rigorous analysis of the global scientific literature and on our own experience. The reader may be referred to the previous section, where the insights of each method are described, and also to the discussion and conclusions sections, where a summary of the essential information drawn from this research is shown.

The interpretation of Table 1 is not trivial. The reader may conclude that there are several similarities among methods, and that is right. Consequently, it seems necessary to tackle a further assessment where the main strengths, weaknesses, opportunities and threats (SWOT) for each method are shown and summarized (Table 2). After that, a final assessment that comprises a suitability evaluation is carried out (Table 3). There, the degree of method suitability for assuring the highest level of rivers' sustainability depends on the final use that the method is designed to and the type of service that it is able to deliver.

Table 2. Strengths, weaknesses, opportunities and threats (SWOTs) analysis for methods.

Method	Strength	Weakness	Opportunity	Threat
Process-based hydrological models	Physical Knowledge Capture	Low capacity on temporally runoff understanding	Hybridization with Stochastic methods	Inaccurate manifold and not-extrapolatable usage
Artificial Neural Networks (ANN)	Computation power	Not-dealing with hydrological variability	Hybridization with Stochastic methods	Opacity
Wavelets transformation (WT)	Computation power	Hard output interpretation	Hybridization with Stochastic methods	High usage complexity
System Dynamics (SDS)	Complex systems interlinks handling	Hard output interpretation	Hybridization with Stochastic methods	Inaccurate usage for hydrological prediction
Linear and nonlinear regression models	Global known	High and rigid inputs requirements	Generation of advanced numerical methods	Inaccurate manifold and not-extrapolatable usage
Pearson's Correlogram Coefficient (Correlogram)	Global known	High and rigid inputs requirements	Generation of advanced numerical methods	Inaccurate manifold and not-extrapolatable usage
Hurst coefficient	Long-term runoff memory identification	Inaccurate prediction	Validation with alternative methods	Inaccurate usage
Multivariate Adaptive Regression Splines (MARS)	High predictive performance	High usage complexity	Flexible high dimension modeling	High usage complexity
ARMA and ARIMA	High synthetic generation performance	High and rigid inputs requirements	High operability	High usage complexity
Causal Reasoning (CR)	Computation power	Opacity	Hybridization ease	Opacity
Copulas	Dealing with hydrological uncertainty	Mathematical Complexity	High predictive performance	High usage complexity
Kalman, Particle Filter and Ensemble Kalman Filter	Dealing with hydrological uncertainty	Mathematical Complexity	Flexible modeling	High usage complexity
HJ-Biplot	Complex systems interlinks capture	Null predictive capacity	High descriptive capacity for Hydrology	Overrated usage
Principal Component Analysis (PCA)	Ability for information simplification	Null predictive capacity	Hybridization with other methods	Overrated usage
Factorial Analysis (FAV)	Computation power	High and rigid inputs requirements	Hybridization with other methods	Hybridization complexity

1 feature pear method is included.

Table 3. Suitability assessment for method implementation addressing a service.

Method/service	Average Prediction	C.C. Simulations	Temporal Dependence Evaluation	Spatio-Temporal Dependence	Extreme Prediction	Protection Actions	Global Suitability Score (GSS)
Process-based hydrological models	**	**			*	**	*
Artificial Neural Networks (ANN)	**	*	*	**	**	*	**
Wavelets transformation (WT)	**	**	**	**	***	*	**
System Dynamics (SDS)	*	*			*	**	*
Linear and nonlinear regression models	*	*	*	*	*	*	*
Pearson's Correlogram Coefficient (Correlogram)	*	*	**	*	*	*	*
Hurst coefficient			**		*	*	
Multivariate Adaptive Regression Splines (MARS)	*	*	*	*	*	*	*
ARMA and ARIMA	*	*	*	*			*
Causal Reasoning (CR)	***	***	***	***	***	*	***
Copulas	*	**	*	*	***	*	**
Kalman, Particle Filter and Ensemble Kalman Filter	*	**	*	*	**	*	**
HJ-Biplot						**	
Principal Component Analysis (PCA)						**	
Factorial Analysis (FAV)	*	*	**	**	*	*	**

3.3. *SWOT analysis*

This evaluation comprises the identification and inclusion of the main feature for each of the 4 characteristics of the SWOT analysis (Table 2). This deep analysis is aimed to provide a guide for the final users to get a general knowledge of each method applicability, so the usage for the goals to reach is optimum. There is a high variety in the traits across methods. However, it is worthy to mention some of the main patterns identified here. For the strength, several methods show great computation power, so make them able to cope with great amounts of data at very fast processing. Other methods are known worldwide that make their usage easier for non-specialists' users. Process-based hydrological models show their ability for capturing and reproducing physical knowledge. Other methods, such as MARS, reveals its high predictive nature and others; furthermore, KF, EnKF and PF, as well as Copulas, show their capacity for dealing with hydrological uncertainty. Regarding weaknesses, methods show also a great heterogeneity. Some of them, such as process-based hydrological models or ANN, show scarce capacity for dealing with hydrological variability and temporally runoff understanding, respectively. Then, WT and SDS reveal a hard output interpretation; traditional techniques show very high inputs requirements; others methods like MARS, Copulas, KF, EnKF and PF are very mathematically complex to deal with. CR show its great opacity in its functioning, and others show their low ability for predicting. In regards to the opportunity, the most frequent and important trait is hybridization with other methods. Finally, threats analysis also shows methods' deficiencies on opacity, as well as on inaccurate, manifold, overrated and complex usage, among others.

3.4. *Suitability Assessment for Rivers' Sustainability*

The measure of a method suitability is a quite complex matter, and the procedure designed and followed in this research is explained as follows. In this sense, the level of suitability depends on the final use that the method is designed to. For that reason, this analysis is divided on the different services that the method may address. These services are: average prediction (AP), current conditions simulation (CCS), temporal dependence evaluation (TDE), spatio-temporal dependence evaluation (SPTDE), extreme prediction (EP) and protection actions (PA). At the end, a global suitability score (GSS) for the sustainability assessment is provided.

The quantitative score for each service/method is explained as follows (Table 3): The highest level of suitability is symbolized with (***), the second level (**), the third level (*) and, finally, the absence of suitability leaves the cell in blank.

The analysis and interpretation of this assessment shows that the best method for predicting runoff temporal behavior is the CR. On the other hand, the worst methods for this purpose are HJ-Biplot and Hurst coefficient.

4. Discussion

Hydrological processes' variability is increasing on all levels, and those extreme events are more recurrent over time. In order to address this new hydrological reality and aimed to anticipate and forecast it, new approaches, methodologies and techniques have been analyzed in this research. This analysis is articulated in three phases: identification and assessment of parameters, SWOT analysis and suitability assessment.

Seven parameters have been identified and evaluated whose behavior assessment is discussed as follows. For predictability, the methods that behaves the best are ANN, WT, CR and copulas. On the other hand, considering these methods were not initially or specifically conceived for developing predictions, the worst ones for developing a prediction are process-based hydrological models, linear and nonlinear regression models, Pearson's correlogram coefficient (correlogram)–Spearman and Kendall estimator, multivariate adaptive regression splines (MARS) and HJ-Biplot. For reliability, the best methods are CR and FAV. On the contrary, the worst methods are process-based hydrological models and Pearson's correlogram coefficient (correlogram). Regarding the mathematical approach for dealing with uncertainty, 11 different values have been identified: null; null/scarce, scaling coefficients at different resolutions, regression, correlogram coefficient, optimization, mathematical residuals coefficients, probability, bivariate function, random samples (ensembles) and different statistical expressions. The methods that do not include any uncertainty treatment are the artificial neural networks (ANN), system dynamics (SDs) and Hurst coefficient. In regards to the mathematical approach for processing, four values are identified: theoretical algorithms, dynamic and adaptive mathematical expressions, constant mathematical expressions and matrixes visualization/cluster analysis. Then, regarding the amount of processed information, none of the studied methods requires low levels of data amounts, which is reasonable, because, usually, the more information for populating the method, the best performance level is obtained. For manageability, the easiest methods to work with are the Pearson's correlogram coefficient (correlogram) and Hurst coefficient; on the contrary, the most complicated and sophisticated methods to deal with are the ANN, WT and SDs. Finally, in regards to transparency, the clearest methods are the Pearson's correlogram coefficient (correlogram) and Hurst coefficient, while the darkest are the ANN, SDs, CR, Copulas, Kalman and particle filter.

SWOT analysis reveals a multidimensional analysis largely focused on the advantages and drawbacks for each method. In this sense, one feature has been identified and included for each of the four components of the SWOT analysis, which are strength, weakness, opportunity and threat (Table 2). Some of the most important traits are the following. On the strengths, several methods show great computation power, others are globally known, others are aimed to capture and reproduce physical knowledge, others have a clear predictive nature and others, such as Kalman and the particle filter, show their capacity for dealing with hydrological uncertainty. Regarding weaknesses, some of them show scarce capacity for dealing with hydrological variability and temporally runoff understanding. Then, hard output interpretations and very high input requirements are also common traits; others methods are, mathematically, very complex to deal with; opacity in its functioning and low predictive skills are also important weaknesses to mention. In regards to the opportunity, the most frequent and important trait is hybridization with other methods. Finally, a threat analysis also shows methods' deficiencies on opacity, as well as on inaccurate, manifold, overrated and complex usage, among others.

Suitability assessment for rivers' sustainability comprises an analysis of seven services, which are: average prediction (AP), current conditions simulation (CCS), temporal dependence evaluation (TDE), spatio-temporal dependence evaluation (SPTDE), extreme prediction (EP) and protection actions (PA). At the end, a global suitability score (GSS) for the sustainability assessment is provided. This evaluation reveals that the method that averagely behaves the best for achieving temporally river sustainability is causal reasoning. It is recommended to inform that this final score is an average, and, in this case, CR has an important drawback on the great opacity across the whole implementation.

It is worthy to highlight that none of the analyzed methods are too appropriate by themselves for implementing protection actions. In this service, the best methods are the traditional physical methods (process-based hydrological models) that allow changing the physical features of the hydrological model.

5. Conclusions

This review paper shows an extensive analysis of the most important methods for the hydrological understanding and prediction of rivers' runoff behaviors. Of course, there is a high variability of dimensions involved in this research. Consequently, there are methods more appropriate to one dimension than to others. However, in general terms, the most recommended method is causal reasoning, despite its dark (black box) nature, need of large data amount and manageability difficulty. The reasons why CR is the best-ranked method largely comprises its flexibility, articulated by the chance for incorporating dynamic and adaptive mathematical expressions, reliability, predictability, ease to hybridize with other methods and the uncertainty treatment method, which is probability; that, in itself, is a very powerful uncertainty tool.

This research may be useful for helping to reach a much larger sustainability of rivers, since their temporal hydrological behavior is far from being well understood.

Author Contributions: Conceptualization, J.-L.M.; methodology, J.-L.M., S.Z., A.-M.M.-C. and M.-C.P.-A.; software, J.-L.M., S.Z., A.-M.M.-C. and M.-C.P.-A.; validation, J.-L.M. and A.-M.M.-C.; formal analysis, J.-L.M., S.Z., A.-M.M.-C. and M.-C.P.-A.; investigation, J.-L.M., S.Z., A.-M.M.-C. and M.-C.P.-A.; resources, J.-L.M., S.Z., A.-M.M.-C. and M.-C.P.-A.; data curation, J.-L.M., S.Z., A.-M.M.-C. and M.-C.P.-A.; writing—original draft preparation, J.-L.M.; writing—review and editing, J.-L.M., S.Z., A.-M.M.-C. and M.-C.P.-A. and visualization, J.-L.M., S.Z., A.-M.M.-C. and M.-C.P.-A. All authors have read and agreed to the published version of the manuscript.

Funding: This research received no external funding.

Acknowledgments: Authors acknowledge the great support given by the High Polytechnic School of Avila, (Avila, Spain) and University of Salamanca to elaborate this paper.

Conflicts of Interest: The authors declare no conflict of interest.

References

1. Pulido-Velazquez, D.; Luis Garcia-Arostegui, J.L.; Molina, J.; Pulido-Velazquez, M. Assessment of Future Groundwater Recharge in Semi-Arid Regions under Climate Change Scenarios (Serral-Salinas Aquifer, SE Spain). Could Increased Rainfall Variability Increase the Recharge Rate? *Hydrol. Process.* **2015**, *29*, 828–844. [CrossRef]
2. Wang, Y.; Traore, S.; Kerh, T.; Leu, J. Modelling Reference Evapotranspiration using Feed Forward Backpropagation Algorithm in Arid Regions of Africa. *Irrig. Drain.* **2011**, *60*, 404–417. [CrossRef]
3. O'Gorman, P.A. Precipitation Extremes under Climate Change. *Curr. Clim. Chang. Rep.* **2015**, *1*, 49–59. [CrossRef] [PubMed]
4. Pfahl, S.; O'Gorman, P.A.; Fischer, E.M. Understanding the Regional Pattern of Projected Future Changes in Extreme Precipitation. *Nat. Clim. Chang.* **2017**, *7*, 423–427. [CrossRef]
5. Chang, B.; Guan, J.; Aral, M.M. Scientific Discourse: Climate Change and Sea-Level Rise. *J. Hydrol. Eng.* **2015**, *20*, A4014003. [CrossRef]
6. Kalra, A.; Li, L.; Li, X.; Ahmad, S. Improving Streamflow Forecast Lead Time using Oceanic-Atmospheric Oscillations for Kaidu River Basin, Xinjiang, China. *J. Hydrol. Eng.* **2013**, *18*, 1031–1040. [CrossRef]
7. Trenberth, K.E. Changes in Precipitation with Climate Change. *Clim. Res.* **2011**, *47*, 123–138. [CrossRef]
8. Molina, J.L.; Zazo, S. Assessment of Temporally Conditioned Runoff Fractions in Unregulated Rivers. *J. Hydrol. Eng.* **2018**, *23*, 04018015. [CrossRef]
9. Salas, J.; Delleur, J.; Yevjevich, V.; Lane, W.L. *Applied Modeling of Hydrologic Time Series*, 1st ed.; Water Resources Publications: Littleton, CO, USA, 1980; p. 484.
10. MITECO. Modelo SIMPA 2019. Periodo de Simulación: 1940/41 a 2017/18. Available online: https://www.miteco.gob.es/es/agua/temas/evaluacion-de-los-recursos-hidricos/evaluacion-recursos-hidricos-regimen-natural/ (accessed on 3 January 2020).
11. Molina, J.L.; Zazo, S. Causal Reasoning for the Analysis of Rivers Runoff Temporal Behavior. *Water Resour. Manag.* **2017**, *31*, 4669–4681. [CrossRef]
12. Hurst, H.E. Long-Term Storage Capacity of Reservoirs. *Trans. Am. Soc. Civil Eng. (ASCE)* **1951**, *116*, 770–808.
13. Molina, J.L.; Luis Garcia-Arostegui, J.; Bromley, J.; Benavente, J. Integrated Assessment of the European WFD Implementation in Extremely Overexploited Aquifers through Participatory Modelling. *Water Resour. Manag.* **2011**, *25*, 3343–3370. [CrossRef]
14. Farmer, W.H.; Vogel, R.M. On the Deterministic and Stochastic use of Hydrologic Models. *Water Resour. Res.* **2016**, *52*, 5619–5633. [CrossRef]
15. Jakeman, A.J.; Letcher, R.A.; Norton, J.P. Ten Iterative Steps in Development and Evaluation of Environmental Models. *Environ. Model. Softw.* **2006**, *21*, 602–614. [CrossRef]
16. Carrasco, G.; Molina, J.L.; Patino-Alons, M.; Castillo, M.D.C.; Vicente-Galindo, M.; Galindo-Villardon, M. Water Quality Evaluation through a Multivariate Statistical HJ-Biplot Approach. *J. Hydrol.* **2019**, *577*, 123993. [CrossRef]
17. Guo, X.; Zhang, X.; Yue, H. Evaluation of Hierarchically Weighted Principal Component Analysis for Water Quality Management at Jiaozuo Mine. *Int. Biodeterior. Biodegrad.* **2018**, *128*, 182–185. [CrossRef]

18. Forgy, E.W. Cluster Analysis of Multivariate Data—Efficiency vs. Interpretability of Classifications. *Biometrics* **1965**, *21*, 768–769.
19. Kim, S.E.; Seo, I.W.; Choi, S.Y. Assessment of Water Quality Variation of a Monitoring Network using Exploratory Factor Analysis and Empirical Orthogonal Function. *Environ. Model. Softw.* **2017**, *94*, 21–35. [CrossRef]
20. Chan, M.C.; Lou, I.C.; Ung, W.K.; Mok, K.M. Integrating Principle Component Analysis and Canonical Correlation Analysis for Monitoring Water Quality in Storage Reservoir. *Appl. Mech. Mater.* **2013**, *284*, 1458–1462. [CrossRef]
21. Koutsoyiannis, D. Hydrology, Society, Change and Uncertainty. *Geophys. Res.* **2014**, *16*. [CrossRef]
22. Graf, R.; Zhu, S.; Sivakumar, B. Forecasting River Water Temperature Time Series using a Wavelet-Neural Network Hybrid Modelling Approach. *J. Hydrol.* **2019**, *578*, 124115. [CrossRef]
23. Molina, J.L.; Zazo, S.; Martín-Casado, A.M. Causal Reasoning: Towards Dynamic Predictive Models for Runoff Temporal Behavior of High Dependence Rivers. *Water* **2019**, *11*, 877. [CrossRef]
24. USACE. US Army Corps of Engineers. Hydrologic Engineering Center. Available online: https://www.hec.usace.army.mil/software/hec-hms/ (accessed on 3 January 2020).
25. Liu, J.J.; Zhou, Z.H.; Yan, Z.Q.; Gong, J.G.; Jia, Y.W.; Xu, C.Y.; Wang, H. A new approach to separating the impacts of climate change and multiple human activities on water cycle processes based on a distributed hydrological model. *J. Hydrol.* **2019**, *578*, 124096. [CrossRef]
26. Pilz, T.; Delgado, J.M.; Voss, S.; Vormoor, K.; Francke, T.; Costa, A.C.; Martins, E.; Bronstert, A. Seasonal drought prediction for semiarid northeast Brazil: What is the added value of a process-based hydrological model? *Hydrol. Earth Syst. Sci.* **2019**, *23*, 1951–1971. [CrossRef]
27. Ren, Y.Q.; Liu, S.X. A simple regional snow hydrological process-based snow depth model and its application in the Upper Yangtze River Basin. *Hydrol. Res.* **2019**, *50*, 672–690. [CrossRef]
28. Zhang, Y.; Li, W.H.; Sun, G.; Miao, G.F.; Noormets, A.; Emanuel, R.; King, J.S. Understanding coastal wetland hydrology with a new regional-scale, process-based hydrological model. *Hydrol. Process.* **2018**, *32*, 3158–3173. [CrossRef]
29. Labat, D.; Ababou, R.; Mangin, A. Rainfall-Runoff Relations for Karstic Springs. Part II: Continuous Wavelet and Discrete Orthogonal Multiresolution. *J. Hydrol.* **2000**, *238*, 149–178. [CrossRef]
30. Labat, D.; Ababou, R.; Mangin, A. Introduction of Wavelet Analyses to Rainfall/Runoffs Relationship for a Karstic Basin: The Case of Licq-Atherey Karstic System (France). *Ground Water* **2001**, *39*, 605–615. [CrossRef]
31. Niu, J.; Sivakumar, B. Scale-Dependent Synthetic Streamflow Generation using a Continuous Wavelet Transform. *J. Hydrol.* **2013**, *496*, 71–78. [CrossRef]
32. Niu, J. Precipitation in the Pearl River Basin, South China: Scaling, Regional Patterns, and Influence of Large-Scale Climate Anomalies. *Stoch. Environ. Res. Risk Assess.* **2013**, *27*, 1253–1268. [CrossRef]
33. Roushangar, K.; Nourani, V.; Alizadeh, F. A Multiscale Time-Space Approach to Analyze and Categorize the Precipitation Fluctuation Based on the Wavelet Transform and Information Theory Concept. *Hydrol. Res.* **2018**, *49*, 724–743. [CrossRef]
34. Niu, J.; Chen, J. A Wavelet Perspective on Variabilities of Hydrological Processes in Conjunction with Geomorphic Analysis over the Pearl River Basin in South China. *J. Hydrol.* **2016**, *542*, 392–409. [CrossRef]
35. Sang, Y.; Sun, F.; Singh, V.P.; Xie, P.; Sun, J. A Discrete Wavelet Spectrum Approach for Identifying Non-Monotonic Trends in Hydroclimate Data. *Hydrol. Earth Syst. Sci.* **2018**, *22*, 757–766. [CrossRef]
36. Agarwal, A.; Maheswaran, R.; Sehgal, V.; Khosa, R.; Sivakumar, B.; Bernhofer, C. Hydrologic Regionalization using Wavelet-Based Multiscale Entropy Method. *J. Hydrol.* **2016**, *538*, 22–32. [CrossRef]
37. Maheswaran, R.; Khosa, R. Comparative Study of Different Wavelets for Hydrologic Forecasting. *Comput. Geosci.* **2012**, *46*, 284–295. [CrossRef]
38. Rezaie-balf, M.; Naganna, S.R.; Ghaemi, A.; Deka, P.C. Wavelet Coupled MARS and M5 Model Tree Approaches for Groundwater Level Forecasting. *J. Hydrol.* **2017**, *553*, 356–373. [CrossRef]
39. Nourani, V.; Baghanam, A.H.; Adamowski, J.; Kisi, O. Applications of Hybrid Wavelet-Artificial Intelligence Models in Hydrology: A Review. *J. Hydrol.* **2014**, *514*, 358–377. [CrossRef]
40. Adamowski, J.; Chan, H.F. A Wavelet Neural Network Conjunction Model for Groundwater Level Forecasting. *J. Hydrol.* **2011**, *407*, 28–40. [CrossRef]
41. Moosavi, V.; Vafakhah, M.; Shirmohammadi, B.; Behnia, N. A Wavelet-ANFIS Hybrid Model for Groundwater Level Forecasting for Different Prediction Periods. *Water Resour. Manag.* **2013**, *27*, 1301–1321. [CrossRef]

42. Moosavi, V.; Vafakhah, M.; Shirmohammadi, B.; Ranjbar, M. Optimization of Wavelet-ANFIS and Wavelet-ANN Hybrid Models by Taguchi Method for Groundwater Level Forecasting. *Arab. J. Sci. Eng.* **2014**, *39*, 1785–1796. [CrossRef]
43. Raghavendra, S.N.; Deka, P.C. Forecasting Monthly Groundwater Level Fluctuations in Coastal Aquifers using Hybrid Wavelet Packet-Support Vector Regression. *Cogent Eng.* **2015**, *2*, 999414. [CrossRef]
44. Shoaib, M.; Shamseldin, A.Y.; Melville, B.W.; Khan, M.M. A Comparison between Wavelet Based Static and Dynamic Neural Network Approaches for Runoff Prediction. *J. Hydrol.* **2016**, *535*, 211–225. [CrossRef]
45. Shoaib, M.; Shamseldin, A.Y.; Khan, S.; Sultan, M.; Ahmad, F.; Sultan, T.; Dahri, Z.H.; Ali, I. Input Selection of Wavelet-Coupled Neural Network Models for Rainfall-Runoff Modelling. *Water Resour. Manag.* **2019**, *33*, 955–973. [CrossRef]
46. Kasiviswanathan, K.S.; He, J.; Sudheer, K.P.; Tay, J. Potential Application of Wavelet Neural Network Ensemble to Forecast Streamflow for Flood Management. *J. Hydrol.* **2016**, *536*, 161–173. [CrossRef]
47. Peng, T.; Zhou, J.; Zhang, C.; Fu, W. Streamflow Forecasting using Empirical Wavelet Transform and Artificial Neural Networks. *Water* **2017**, *9*, 406. [CrossRef]
48. Seo, Y.; Kim, S.; Kisi, O.; Singh, V.P. Daily Water Level Forecasting using Wavelet Decomposition and Artificial Intelligence Techniques. *J. Hydrol.* **2015**, *520*, 224–243. [CrossRef]
49. Rajaee, T.; Nourani, V.; Zounemat-Kermani, M.; Kisi, O. River Suspended Sediment Load Prediction: Application of ANN and Wavelet Conjunction Model. *J. Hydrol. Eng.* **2011**, *16*, 613–627. [CrossRef]
50. Seo, Y.; Kim, S.; Kisi, O.; Singh, V.P.; Parasuraman, K. River Stage Forecasting using Wavelet Packet Decomposition and Machine Learning Models. *Water Resour. Manag.* **2016**, *30*, 4011–4035. [CrossRef]
51. Ramirez, M.C.V.; Velho, H.F.D.; Ferreira, N.J. Artificial Neural Network Technique for Rainfall Forecasting Applied to the Sao Paulo Region. *J. Hydrol.* **2005**, *301*, 146–162. [CrossRef]
52. Mandal, T.; Jothiprakash, V. Short-Term Rainfall Prediction using ANN and MT Techniques. *ISH J. Hydraul. Eng.* **2012**, *18*, 20–26. [CrossRef]
53. Chiang, Y.; Chang, F.; Jou, B.J.; Lin, P. Dynamic ANN for Precipitation Estimation and Forecasting from Radar Observations. *J. Hydrol.* **2007**, *334*, 250–261. [CrossRef]
54. Noori, N.; Kalin, L. Coupling SWAT and ANN Models for Enhanced Daily Streamflow Prediction. *J. Hydrol.* **2016**, *533*, 141–151. [CrossRef]
55. Prasad, R.; Deo, R.C.; Li, Y.; Maraseni, T. Input Selection and Performance Optimization of ANN-Based Streamflow Forecasts in the Drought-Prone Murray Darling Basin Region using IIS and MODWT Algorithm. *Atmos. Res.* **2017**, *197*, 42–63. [CrossRef]
56. Lallahem, S.; Mania, J.; Hani, A.; Najjar, Y. On the use of Neural Networks to Evaluate Groundwater Levels in Fractured Media. *J. Hydrol.* **2005**, *307*, 92–111. [CrossRef]
57. Daliakopoulos, I.; Coulibaly, P.; Tsanis, I. Groundwater Level Forecasting using Artificial Neural Networks. *J. Hydrol.* **2005**, *309*, 229–240. [CrossRef]
58. Piotrowski, A.P.; Napiorkowski, M.J.; Napiorkowski, J.J.; Osuch, M. Comparing various Artificial Neural Network Types for Water Temperature Prediction in Rivers. *J. Hydrol.* **2015**, *529*, 302–315. [CrossRef]
59. Zhu, S.; Hadzima-Nyarko, M.; Gao, A.; Wang, F.; Wu, J.; Wu, S. Two Hybrid Data-Driven Models for Modeling Water-Air Temperature Relationship in Rivers. *Environ. Sci. Pollut. Res.* **2019**, *26*, 12622–12630. [CrossRef]
60. Abramowitz, G.; Gupta, H.; Pitman, A.; Wang, Y.; Leuning, R.; Cleugh, H.; Hsu, K. Neural Error Regression Diagnosis (NERD): A Tool for Model Bias Identification and Prognostic Data Assimilation. *J. Hydrometeorol.* **2006**, *7*, 160–177. [CrossRef]
61. Hsu, K.; Gupta, H.; Gao, X.; Sorooshian, S.; Imam, B. Self-Organizing Linear Output Map (SOLO): An Artificial Neural Network Suitable for Hydrologic Modeling and Analysis. *Water Resour. Res.* **2002**, *38*, 1302. [CrossRef]
62. Tiwari, M.K.; Chatterjee, C. Development of an Accurate and Reliable Hourly Flood Forecasting Model using Wavelet-Bootstrap-ANN (WBANN) Hybrid Approach. *J. Hydrol.* **2010**, *394*, 458–470. [CrossRef]
63. Susnik, J.; Molina, J.; Vamvakeridou-Lyroudia, L.S.; Savic, D.A.; Kapelan, Z. Comparative Analysis of System Dynamics and Object-Oriented Bayesian Networks Modelling for Water Systems Management. *Water Resour. Manag.* **2013**, *27*, 819–841. [CrossRef]

64. Bakhshianlamouki, E.; Masia, S.; Karimi, P.; van der Zaag, P.; Susnik, J. A system dynamics model to quantify the impacts of restoration measures on the water-energy-food nexus in the Urmia lake Basin, Iran. *Sci. Total Environ.* **2020**, *708*, 134874. [CrossRef] [PubMed]
65. Gohari, A.; Mirchi, A.; Madani, K. System Dynamics Evaluation of Climate Change Adaptation Strategies for Water Resources Management in Central Iran. *Water Resour. Manag.* **2017**, *31*, 1413–1434. [CrossRef]
66. Ghashghaie, M.; Marofi, S.; Marofi, H. Using System Dynamics Method to Determine the Effect of Water Demand Priorities on Downstream Flow. *Water Resour. Manag.* **2014**, *28*, 5055–5072. [CrossRef]
67. Hassanzadeh, E.; Zarghami, M.; Hassanzadeh, Y. Determining the Main Factors in Declining the Urmia Lake Level by Using System Dynamics Modeling. *Water Resour. Manag.* **2012**, *26*, 129–145. [CrossRef]
68. Muetzelfeldt, R.; Massheder, J. The Simile Visual Modelling Environment. *Eur. J. Agron.* **2003**, *18*, 345–358. [CrossRef]
69. Van Vliet, M.T.H.; Ludwig, F.; Zwolsman, J.J.G.; Weedon, G.P.; Kabat, P. Global River Temperatures and Sensitivity to Atmospheric Warming and Changes in River Flow. *Water Resour. Res.* **2011**, *47*, W02544. [CrossRef]
70. Molina, J.L.; Zazo, S.; Rodriguez-Gonzalvez, P.; Gonzalez-Aguilera, D. Innovative Analysis of Runoff Temporal Behavior through Bayesian Networks. *Water* **2016**, *8*, 484. [CrossRef]
71. Zazo, S.; Macian-Sorribes, H.; Sena-Fael, C.M.; Martín-Casado, A.M.; Molina, J.L.; Pulido-Velazquez, M. Qualitative Approach for Assessing Runoff Temporal Dependence through Geometrical Symmetry. In Proceedings of the International Congress on Engineering. Engineering for Evolution (ICEUBI2019), Covilhã, Portugal, 28 November 2019.
72. Hao, Z.; Singh, V.P. Review of Dependence Modeling in Hydrology and Water Resources. *Prog. Phys. Geogr.* **2016**, *40*, 549–578. [CrossRef]
73. Chenoweth, T.; Hubata, R.; St Louis, R.D. Automatic ARMA Identification using Neural Networks and the Extended Sample Autocorrelation Function: A Reevaluation. *Decis. Support Syst.* **2000**, *29*, 21–30. [CrossRef]
74. Mohammadi, K.; Eslami, H.R.; Kahawita, R. Parameter Estimation of an ARMA Model for River Flow Forecasting using Goal Programming. *J. Hydrol.* **2006**, *331*, 293–299. [CrossRef]
75. Nelsen, R.B. *An Introduction to Copulas*; Springer Science & Business Media: Berlin/Heidelberg, Germany, 2007.
76. Friedman, J.H. Multivariate Adaptive Regression Splines. *Ann. Stat.* **1991**, *19*, 1–67. [CrossRef]
77. Put, R.; Xu, Q.S.; Massart, D.L.; Heyden, Y.V. Multivariate Adaptive Regression Splines (MARS) in Chromatographic Quantitative Structure-Retention Relationship Studies. *J. Chromatogr. A* **2004**, *1055*, 11–19. [CrossRef] [PubMed]
78. Craven, P.; Wahba, G. Smoothing Noisy Data with Spline Functions—Estimating the Correct Degree of Smoothing by the Method of Generalized Cross-Validation. *Numer. Math.* **1979**, *31*, 377–403. [CrossRef]
79. Adamowski, J.; Chan, H.F.; Prasher, S.O.; Sharda, V.N. Comparison of Multivariate Adaptive Regression Splines with Coupled Wavelet Transform Artificial Neural Networks for Runoff Forecasting in Himalayan Micro-Watersheds with Limited Data. *J. Hydroinf.* **2012**, *14*, 731–744. [CrossRef]
80. Sharda, V.N.; Prasher, S.O.; Patel, R.M.; Ojasvi, P.R.; Prakash, C. Performance of Multivariate Adaptive Regression Splines (MARS) in Predicting Runoff in Mid-Himalayan Micro-Watersheds with Limited Data. *Hydrol. Sci. J. J. Des. Sci. Hydrol.* **2008**, *53*, 1165–1175. [CrossRef]
81. Deo, R.C.; Kisi, O.; Singh, V.P. Drought Forecasting in Eastern Australia using Multivariate Adaptive Regression Spline, Least Square Support Vector Machine and M5Tree Model. *Atmos. Res.* **2017**, *184*, 149–175. [CrossRef]
82. Kisi, O.; Parmar, K.S. Application of Least Square Support Vector Machine and Multivariate Adaptive Regression Spline Models in Long Term Prediction of River Water Pollution. *J. Hydrol.* **2016**, *534*, 104–112. [CrossRef]
83. Kisi, O. Pan Evaporation Modeling using Least Square Support Vector Machine, Multivariate Adaptive Regression Splines and M5 Model Tree. *J. Hydrol.* **2015**, *528*, 312–320. [CrossRef]
84. Samadi, M.; Jabbari, E.; Azamathulla, H.M.; Mojallal, M. Estimation of Scour Depth below Free Overfall Spillways using Multivariate Adaptive Regression Splines and Artificial Neural Networks. *Eng. Appl. Comput. Fluid Mech.* **2015**, *9*, 291–300. [CrossRef]
85. Lee, T.; Salas, J.D.; Prairie, J. An Enhanced Nonparametric Streamflow Disaggregation Model with Genetic Algorithm. *Water Resour. Res.* **2010**, *46*, W08545. [CrossRef]

86. See, L.; Openshaw, S. A Hybrid Multi-Model Approach to River Level Forecasting. *Hydrol. Sci. J. J. Des. Sci. Hydrol.* **2000**, *45*, 523–536. [CrossRef]
87. Jain, A.; Kumar, A.M. Hybrid Neural Network Models for Hydrologic Time Series Forecasting. *Appl. Soft Comput.* **2007**, *7*, 585–592. [CrossRef]
88. Pearl, J. *Probabilistic Reasoning in Intelligent Systems: Networks of Plausible Inference*; Morgan Kaufmann: San Francisco, CA, USA, 1988.
89. Molina, J.L.; Pulido-Velazquez, D.; Luis Garcia-Arostegui, J.; Pulido-Velazquez, M. Dynamic Bayesian Networks as a Decision Support Tool for Assessing Climate Change Impacts on Highly Stressed Groundwater Systems. *J. Hydrol.* **2013**, *479*, 113–129. [CrossRef]
90. Molina, J.L.; Bromley, J.; Garcia-Arostegui, J.L.; Sullivan, C.; Benavente, J. Integrated Water Resources Management of Overexploited Hydrogeological Systems using Object-Oriented Bayesian Networks. *Environ. Model. Softw.* **2010**, *25*, 383–397. [CrossRef]
91. Castelletti, A.; Soncini-Sessa, R. Bayesian Networks and Participatory Modelling in Water Resource Management. *Environ. Model. Softw.* **2007**, *22*, 1075–1088. [CrossRef]
92. Said, A. The Implementation of a Bayesian Network for Watershed Management Decisions. *Water Resour. Manag.* **2006**, *20*, 591–605. [CrossRef]
93. Carmona, G.; Varela-Ortega, C.; Bromley, J. Supporting Decision Making Under Uncertainty: Development of a Participatory Integrated Model for Water Management in the Middle Guadiana River Basin. *Environ. Model. Softw* **2013**, *50*, 144–157. [CrossRef]
94. Jensen, F.V.; Nielsen, T.D. *Bayesian Networks and Decision Graphs*; Springer: New York, NY, USA, 2001.
95. Varis, O.; Fraboulet-Jussila, S. Water Resources Development in the Lower Senegal River Basin: Conflicting Interests, Environmental Concerns and Policy Options. *Int. J. Water Resour. Dev.* **2002**, *18*, 245–260. [CrossRef]
96. Nodelman, U.; Horvitz, E. *Continuous Time Bayesian Networks for Inferring Users' Presence and Activities with Extensions for Modeling and Evaluation*; Microsoft Research Technical Report MSR-TR-2003-97; Microsoft Research: Redmond, WA, USA, 2003.
97. Cain, J. *Planning Improvements in Natural Resources Management*; Centre for Ecology and Hydrology: Wallingford, UK, 2001; Volume 124, pp. 1–123.
98. Pena, J.M.; Lozano, J.A.; Larranaga, P. Unsupervised Learning of Bayesian Networks Via Estimation of Distribution Algorithms: An Application to Gene Expression Data Clustering. *Int. J. Uncertain. Fuzziness Knowl. Based Syst.* **2004**, *12*, 63–82. [CrossRef]
99. Genc, O.; Dag, A. A Machine Learning-Based Approach to Predict the Velocity Profiles in Small Streams. *Water Resour. Manag.* **2016**, *30*, 43–61. [CrossRef]
100. Patel, S.S.; Ramachandran, P. A Comparison of Machine Learning Techniques for Modeling River Flow Time Series: The Case of Upper Cauvery River Basin. *Water Resour. Manag.* **2015**, *29*, 589–602. [CrossRef]
101. Mousavi, S.J.; Ponnambalam, K.; Karray, F. Inferring Operating Rules for Reservoir Operations using Fuzzy Regression and ANFIS. *Fuzzy Sets Syst.* **2007**, *158*, 1064–1082. [CrossRef]
102. De Michele, C.; Salvadori, G. A Generalized Pareto Intensity-Duration Model of Storm Rainfall Exploiting 2-Copulas. *J. Geophys. Res. Atmos.* **2003**, *108*, 4067. [CrossRef]
103. Zhang, L.; Singh, V.P. Bivariate Flood Frequency Analysis using the Copula Method. *J. Hydrol. Eng.* **2006**, *11*, 150–164. [CrossRef]
104. Dutfoy, A.; Parey, S.; Roche, N. Multivariate Extreme Value Theory-A Tutorial with Applications to Hydrology and Meteorology. *Depend. Model.* **2014**, *2*, 30–48. [CrossRef]
105. Serinaldi, F.; Bardossy, A.; Kilsby, C.G. Upper Tail Dependence in Rainfall Extremes: Would we Know it if we Saw it? *Stoch. Environ. Res. Risk Assess.* **2015**, *29*, 1211–1233. [CrossRef]
106. Sarabia-Alzaga, J.; Gómez-Déniz, E. Construction of Multivariate Distributions: A Review of some Recent Results. *Stat. Oper. Res. Trans.* **2008**, *32*, 3–36.
107. Balakrishnan, N.; Lai, C. *Continuous Bivariate Distributions*; Springer Science & Business Media: Berlin/Heidelberg, Germany, 2009.
108. Singh, V.P. *Entropy Theory and Its Application in Environmental and Water Engineering*; John Wiley & Sons: New York, NY, USA, 2013.
109. Kalman, R.E. A New Approach to Linear Filtering and Prediction Problems. *Trans. ASME J. Basic Eng.* **1960**, *82*, 35–45. [CrossRef]

110. Reichle, R.H.; McLaughlin, D.B.; Entekhabi, D. Hydrologic Data Assimilation with the Ensemble Kalman Filter. *Mon. Weather Rev.* **2002**, *130*, 103–114. [CrossRef]
111. Evensen, G. Sequential Data Assimilation with a Nonlinear Quasi-Geostrophic Model using Monte-Carlo Methods to Forecast Error Statistics. *J. Geophys. Res. Oceans* **1994**, *99*, 10143–10162. [CrossRef]
112. Moradkhani, H.; Sorooshian, S.; Gupta, H.V.; Houser, P.R. Dual State-Parameter Estimation of Hydrological Models using Ensemble Kalman Filter. *Adv. Water Resour.* **2005**, *28*, 135–147. [CrossRef]
113. Weerts, A.H.; El Serafy, G.Y.H. Particle Filtering and Ensemble Kalman Filtering for State Updating with Hydrological Conceptual Rainfall-Runoff Models. *Water Resour. Res.* **2006**, *42*, W09403. [CrossRef]
114. Pathiraja, S.; Marshall, L.; Sharma, A.; Moradkhani, H. Detecting Non-Stationary Hydrologic Model Parameters in a Paired Catchment System using Data Assimilation. *Adv. Water Resour.* **2016**, *94*, 103–119. [CrossRef]
115. Shen, Z.; Tang, Y. A Modified Ensemble Kalman Particle Filter for Non-Gaussian Systems with Nonlinear Measurement Functions. *J. Adv. Model. Earth Syst.* **2015**, *7*, 50–66. [CrossRef]
116. Wang, S.; Huang, G.H.; Baetz, B.W.; Ancell, B.C. Towards Robust Quantification and Reduction of Uncertainty in Hydrologic Predictions: Integration of Particle Markov Chain Monte Carlo and Factorial Polynomial Chaos Expansion. *J. Hydrol.* **2017**, *548*, 484–497. [CrossRef]
117. Smith, P.J.; Beven, K.J.; Tawn, J.A. Detection of Structural Inadequacy in Process-Based Hydrological Models: A Particle-Filtering Approach. *Water Resour. Res.* **2008**, *44*, W01410. [CrossRef]
118. Dumedah, G.; Coulibaly, P. Evaluating Forecasting Performance for Data Assimilation Methods: The Ensemble Kalman Filter, the Particle Filter, and the Evolutionary-Based Assimilation. *Adv. Water Resour.* **2013**, *60*, 47–63. [CrossRef]
119. Galindo, M. An Alternative for Simultaneous Representations. *HJ-Biplot Qüestiió* **1986**, *10*, 13–23.
120. Gabriel, K.R. Biplot Graphic Display of Matrices with Application to Principal Component Analysis. *Biometrika* **1971**, *58*, 453–467. [CrossRef]
121. Kourgialas, N.N.; Dokou, Z.; Karatzas, G.P. Statistical Analysis and ANN Modeling for Predicting Hydrological Extremes Under Climate Change Scenarios: The Example of a Small Mediterranean Agro-Watershed. *J. Environ. Manag.* **2015**, *154*, 86–101. [CrossRef]
122. Praus, P. Water Quality Assessment using SVD-Based Principal Component Analysis of Hydrological Data. *Water SA* **2005**, *31*, 417–422. [CrossRef]
123. Syed, T.H.; Lakshmi, V.; Paleologos, E.; Lohmann, D.; Mitchell, K.; Famiglietti, J.S. Analysis of Process Controls in Land Surface Hydrological Cycle Over the Continental United States. *J. Geophys. Res. Atmos.* **2004**, *109*, D22105. [CrossRef]
124. Remesan, R.; Bray, M.; Mathew, J. Application of PCA and Clustering Methods in Input Selection of Hybrid Runoff Models. *J. Environ. Inform.* **2018**, *31*, 137–152. [CrossRef]
125. Sahan, T.; Ozturk, D. Investigation of Pb(II) Adsorption Onto Pumice Samples: Application of Optimization Method Based on Fractional Factorial Design and Response Surface Methodology. *Clean Technol. Environ. Policy* **2014**, *16*, 819–831. [CrossRef]
126. Thiele, J.E.; Haaf, J.M.; Rouder, J.N. Is there Variation Across Individuals in Processing? Bayesian Analysis for Systems Factorial Technology. *J. Math. Psychol.* **2017**, *81*, 40–54. [CrossRef]
127. Tang, X.; Dai, Y.; Sun, P.; Meng, S. Interaction-Based Feature Selection using Factorial Design. *Neurocomputing* **2018**, *281*, 47–54. [CrossRef]
128. Liu, Y.R.; Li, Y.P.; Huang, G.H.; Zhang, J.L.; Fan, Y.R. A Bayesian-Based Multilevel Factorial Analysis Method for Analyzing Parameter Uncertainty of Hydrological Model. *J. Hydrol.* **2017**, *553*, 750–762. [CrossRef]
129. Zhang, J.; Li, Y.; Huang, G.; Chen, X.; Bao, A. Assessment of Parameter Uncertainty in Hydrological Model using a Markov-Chain-Monte-Carlo-Based Multilevel-Factorial-Analysis Method. *J. Hydrol.* **2016**, *538*, 471–486. [CrossRef]
130. Zhou, Y.; Huang, G.; Zhu, H.; Li, Z.; Chen, J. A Factorial Dual-Objective Rural Environmental Management Model. *J. Clean. Prod.* **2016**, *124*, 204–216. [CrossRef]
131. Moreau, P.; Viaud, V.; Parnaudeau, V.; Salmon-Monviola, J.; Durand, P. An Approach for Global Sensitivity Analysis of a Complex Environmental Model to Spatial Inputs and Parameters: A Case Study of an Agro-Hydrological Model. *Environ. Model. Softw.* **2013**, *47*, 74–87. [CrossRef]

132. Frey, M.P.; Stamm, C.; Schneider, M.K.; Reichert, P. Using Discharge Data to Reduce Structural Deficits in a Hydrological Model with a Bayesian Inference Approach and the Implications for the Prediction of Critical Source Areas. *Water Resour. Res.* **2011**, *47*, W12529. [CrossRef]
133. Zhou, S.; Wang, Y.; Chang, J.; Guo, A.; Li, Z. Investigating the Dynamic Influence of Hydrological Model Parameters on Runoff Simulation using Sequential Uncertainty Fitting-2-Based Multilevel-Factorial-Analysis Method. *Water* **2018**, *10*, 1177. [CrossRef]

sustainability

MDPI

Review

Reviewing Arch-Dams' Building Risk Reduction Through a Sustainability–Safety Management Approach

Enrico Zacchei [1,2] and José Luis Molina [3,*]

1 Itecons—Institute for Research and Technological Development in Construction, Energy, Environment and Sustainability, Pedro Hispano Avenue, 3030-289 Coimbra, Portugal; enricozacchei@gmail.com
2 School of Civil Engineering of Bauru, São Paulo State University (UNESP), 14-01 Eng. Luís Edmundo Carrijo Coube Avenue, Bauru 17033-360, Brazil
3 IGA Research Group, Higher Polytechnic School of Ávila, University of Salamanca (USAL), 50 Avenue Hornos Caleros, 05003 Ávila, Spain
* Correspondence: jlmolina@usal.es

Received: 29 November 2019; Accepted: 31 December 2019; Published: 3 January 2020

Abstract: The importance of dams is rapidly increasing due to the impact of climate change on increasing hydrological process variability and on water planning and management need. This study tackles a review for the concrete arch-dams' design process, from a dual sustainability/safety management approach. Sustainability is evaluated through a design optimization for dams´ stability and deformation analysis; safety is directly related to the reduction and consequences of failure risk. For that, several scenarios about stability and deformation, identifying desirable and undesirable actions, were estimated. More than 100 specific parameters regarding dam-reservoir-foundation-sediments system and their interactions have been collected. Also, a summary of mathematical modelling was made, and more than 100 references were summarized. The following consecutive steps, required to design engineering (why act?), maintenance (when to act) and operations activities (how to act), were evaluated: individuation of hazards, definition of failure potential and estimation of consequences (harm to people, assets and environment). Results are shown in terms of calculated data and relations: the area to model the dam–foundation interaction is around 3.0 H_d^2, the system-damping ratio and vibration period is 8.5% and 0.39 s. Also, maximum elastic and elasto-plastic displacements are ~0.10–0.20 m. The failure probability for stability is 34%, whereas for deformation it is 29%.

Keywords: concrete arch-dams; stability scenarios; deformation scenarios; safety management; sustainability assessment

1. Introduction

There are many factors, largely controlled by the structures size, that hinder sustainability in the field of dam engineering. In this sense, the height of the blocks can reach more than 100 m and the crown length can reach more than 500 m [1,2]. Dams with these dimensions are called "super-high dams" [3,4]. Then, the presence of structural elements [5], and their interactions, with different functions that increase the difficulty of calculation and modelling, e.g., the cantilevers that support and distribute the vertical loads and the arches that distribute the horizontal loads. Finally, the interaction of dam, foundation, sediments and reservoir sub-systems, requires not only the knowledge of the structural and hydraulic engineering, but also, other engineering areas are involved.

Three aspects, namely geometry, behaviour, and materials, comprise the internal and intrinsic actions, which exclude the external actions and their uncertainties of probability and occurrence.

These uncertainties are called "random" and are related to the magnitude of variability and inherent randomness. Besides these types of uncertainties, there are the "epistemic" uncertainties that are related to the lack of knowledge of materials and models [6]. Random and epistemic uncertainties are studied in stochastic analyses, which are used to solve problems that cannot be deterministically solved because models are not known, or data are not available.

Due to the doubts of the input data, analyses, methodologies, and results, the concept of "risk" and quantitative risk assessment (QRA) is introduced through the following equation:

$$\text{Risk} = \int [P(L,E) \times P(R|L) \times C(L,R)] \quad (1)$$

where L = loads, E = events, and R = responses. P[R|L] is the conditional probability that R is true, given that L is true, and C stands for the consequences [7,8].

This integral is a measure of risk quantification based on the occurrence and probability of L, E and R, regarding the variability of extreme events, e.g., flooding, hurricanes, earthquakes, explosions. The interest of the concrete arch-dams is proven by the fact that several studies have been published since 1931 [9]. This interest has generated several codes/manuals/reports [10–14]. Furthermore, several academic works with the following goal have been published. First, there are researches about the definition of the shape (volume and area of concrete) optimization, aimed to minimize the cost and the impact of the dam body on the environment [15–23]. Then, publications addressing the analysis of the dam behaviour under seismic actions accounting the enormous importance of the structure [24–33]. Finally, there are studies that consider the fact that the dam body is linked with the foundation base, water reservoir, and soil sediments [34–39].

However, there are some aspects, described as follows, that are not well studied either synthetized or published in the literature. In this sense, the response estimation of arch-dams are not well studied or categorized, for example the effects of the non-uniform temperature variation due to the solar radiation and convective heat [30,40–44]. Furthermore, a good calibration between the theoretical and practical data is often difficult to obtain. In this sense, there is a lack of experimental tests made in the laboratory, which allow verifying the analytical and computational models. Also, there is a lack of practical experience of researchers and technical engineers do not easily accept the insights of researchers. In this sense, some cases about real concrete arch-dams are listed in Appendix A (see Table A1). Finally, but not least, there is a clear lack of academic papers that synthetize, integrate, and summarize most of the aspects involved in sustainability of concrete arch dam building. This review paper mainly aims to cover this deficiency, which comprises its main novelty too. This is performed herein by reviewing the existent knowledge on the development of sustainability and safety assessment through the study of structural stabilities/deformations and failure risk, respectively.

The rest of this paper is organized as follows: Section 2 shows a background about the data and mathematical modelling. Section 3 describes some main key findings about an operating system and the project variables in a managerial context [7,12,14]. Section 4 is dedicated to the materials and methodologies followed in this research, describing the structure gand content of the different stages. Regarding materials, Random Variables (RVs) are showed; on the other hand, methods such as Monte Carlo Simulation (MCS), sustainability assessment framework and seismic hazard assessment are described. Then, Section 5 comprises the description of results, largely addressing the sustainability assessment of structural stability and deformations. Finally, Section 6 is dedicated to show the main conclusions drawn from this research.

2. Data and Mathematical Modelling Background

The case study is the Rules Dam, which is situated on the Guadalfeo River in the Granada province, Southern Spain. It is a super-high concrete arch-gravity, formed by 32 blocks, with single curvature, 509 m of crown length and maximum height of the vertical cantilever H_d 132 m. The Down-Stream (DS) and Up-Stream (US) slope faces are 1:0.60 and 1:0.18, respectively. The capacity and area for the maximum operating level $H_{o,r}$ (i.e., water depth of 113 m) of the reservoir are 117.07 Hm^3 and 308 Ha, respectively. The area of the water basin is 1070 km^2 [1,2].

The whole system of concrete arch-gravity dams is composed by four sub-systems, i.e., dam, foundation, reservoir and sediments. Usually, only the dam-reservoir-foundation system is studied, and, in many analyses, sediments are not considered as a separated system, but they are included in the reservoir or foundation sub-system. The parameters of the sediments as well as the foundation are very complicated to estimate, unless specific analyses "in situ" are developed. Moreover, it is very complicated to model them because they are not visible without adequate means.

Considering the precedent studies of the authors about the dam [45–49], more of 100 technical data regarding the system dam-reservoir-foundation-sediments have been summarized and shown in the Appendix A (see Tables A2–A7). The subscripts represent the four parts of the system, i.e., d = dam, f = foundation, r = reservoir, and s = sediments.

2.1. Dam Sub-System

Concrete arch-gravity dams are designed to be stabilized by equilibrium forces (horizontals and verticals). Each section of the dam must be stable and independent of any other section.

The dam body is formed by several arch and cantilever units. Arch refers to a portion of the dam bounded by two horizontal planes. Arches have uniform or variable thickness, i.e., the arches may be designed so that their thickness increases gradually on both sides of the reference plane. Cantilever is a portion of the dam contained between two vertical radial planes [10].

The function of arches is to distribute the horizontal stresses along the dam body, whereas the function of cantilevers is to transmit the vertical stresses from the top to the bottom. Moreover, the arch has an important role respect to the stiffness which increases on the dam body.

Dam sub-systems can be modelled using several theories and models that are briefly mentioned as follows.

- Rigid body equilibrium and beam theory. The gravity method is based: (1) on rigid body equilibrium to determine the internal forces acting on the potential failure plane (joints and concrete-rock interface), and (2) on beam theory to compute stresses. The use of the gravity method requires several simplified assumptions regarding the structural dam behaviour and loads application [50].
- Membrane theory (tank structures). The behaviour of arch-dams can be imagined as being similar to the behaviour of storage tanks: an arch in plant is a part of the tank circumference. The function of the elements is analogue, i.e., arch-dams are formed by cantilever and arch units, whereas tanks are formed by meridional and circumferential units. The stresses in the tanks are: vertical compressive stress (meridional compression associated with hydrostatic and hydrodynamic pressures) and tensile hoop stress (circumferential stress) [35,46,51,52].
- Independent blocks model. Here, the dam's blocks are modelled as independent parts. Each block can be considered as a simple oscillator where the mass is the predominant parameter. This approximation is generally useful for estimating preliminary results [26,47].

Moreover, dam sub-systems can be modelled accounting the vertical joints, as follows.

- Monolithic model. The monolithic model ensures the continuity between adjacent blocks. The rigid connection between them is ensured by means of vertical joints, which are modelled by surface-based "tie" constraints that account the translational and rotational degrees of freedom [53–55]. Considering a series of monolithic, the model can be called "multi-monolithic model" [56].
- Surface-to-surface joints. The surface-to-surface joint model simulates the discontinuity between blocks along the contact surfaces. The contact model describes tangential and normal behaviour by adopting a coefficient of friction and contact pressures transmitted from surfaces [54,55].

- Solid elements joints. The solid element joint model simulates the joints, connected to the ashlars, as independent solid elements, separating the discontinuity surface and spacing the blocks. These joints are characterized by mechanical models (i.e., elastic or elasto-plastic model) [54,55].

2.2. Foundation Sub-System

Even if it is possible to analyse the four systems separately, it is too approximate to approach some aspects without considering the interactions. In this sense, the foundation sub-system is usually studied including the dam-foundation interactions.

The model that describes the dam base and top foundation contact is Mohr-Coulomb model. This model, used in the literature to evaluate base sliding [57], constitutes a simplified procedure to model a nonlinear single-degree-of-freedom system [58] and the failure mode under a reliability-based approach. This is performed as such due to the failure analysis of the dam–foundation interface being characterized by complexity, uncertainties on models and parameters, and a strong non-linear softening behaviour [59].

The foundation sub-systems can be modelled by a massed, massless, rigid, flexible model.

- The massed model ($m \neq 0$, $k = 0$) is composed by finite elements that form the foundation [24]. In 3D analysis, it consists of solid elements, of which, each one is an eight-node element. It is based upon an isoparametric formulation that includes nine bending modes [40]. For each element the density of the material is assigned. The massed model only accounts the weight of the elements in static analysis and the inertial force in dynamic analysis.
- The massless model ($m = 0$, $k \neq 0$) is composed by fixed joints (or nodal points). A joint is defined in three spatial coordinates x, y, z. It defines a joint individually, many joints on a line (or curve), surface or a three-dimensional region. The massless model accounts only material flexibility by elastic springs and forces. The foundation model should be extended to a large enough distance beyond which its effects on deflections and stresses of the dam become negligible [60]. It is possible to consider for the elastic modulus two cases: (i) the same modulus as the concrete and (ii) 1/5 the modulus of concrete [10].
- Rigid model ($k \rightarrow \infty$). Rigid foundation model neglects dam-foundation interaction and, in fact, neither stiffness nor mass of the foundation is accounted in overall coupled equation of motion. It can be modelled by elastic springs with very high stiffness (e.g., ~ 1.0×10^9 kN/m) or by fixed.
- The flexible model ($k \rightarrow 0$), conceptually, it is equal to the massless model because it is formed by a series of joints where are applied springs. An order of magnitude of the elastic spring can be ~ 1.0×10^6 kN/m.

2.3. Reservoir Sub-System

The main actions produced by water mass are the pressures, which can be static or dynamic pressures and act in horizontal or vertical directions. Reservoir sub-system can be approached by considering "rigid" or "flexible" dam, respectively. In this sense, it can be modelled as:

- Added mass, where the hydrodynamic pressures exerted on a dam, by an incompressible fluid, are considered [61]. The hydrodynamic pressures are the same as if a portion of the fluid body is forced to move back and forth with the dam and, that this "added mass" is confined in a volume bounded by a two-dimensional parabolic surface on the dam upstream side.
- Hydrodynamic interaction. Analytical equations for hydrodynamic response of dam-reservoirs considering compressibility effects during harmonic and arbitrary ground motions have been defined [62]. Effects of the deformability of the dam on hydrodynamic pressure have been introduced. The main limitation consists in considering the deformation by only the vibration fundamental mode of the structure [63].

A very popular modelling approach is the "acoustic elements". This model simulates the pressure distributions of the fluid considering the compressibility of the fluid through the "bulk modulus". To find a solution it is necessary to define appropriate boundary conditions, where the most important one takes place on the contact between fluid and structure [63–65]. Acoustic elements are used for modelling an acoustic medium undergoing small pressure changes. The solution in the acoustic medium is defined by a single pressure variable, which represents its degree of freedom [64,65].

2.4. Sediments Sub-System

Sediments can be modelled as a liquid (viscous model) or as a solid (elastic–plastic model). This is, because two cases should be considered: full and empty reservoir. In the first case, sediments are totally submerged, and therefore sediments can be considered in a more similar way to the liquid hydrodynamic behaviour. In contrast, in the second case, sediments can be dry (solid) or yet submerged (semi-solid) depending on the material of which sediments are made: if the predominant material is the sand soil, the liquid drains easily and thus sediments can be idealized as solid, whereas if it is made of clay soil, the liquid does not drain and so it can be idealized as a liquid.

Considering the two extreme cases, the liquid behaviour tends to the reservoir sub-system behaviour, whereas the solid tends to the foundation sub-system (liquid sediments → like reservoir sub system. Solid sediments → like foundation sub-system). The presence of sediments can affect the behaviour of the whole system. This is because, the reservoir bottom absorption affects the stiffness and damping ratio of the structure [34,66,67].

2.5. Interactions of Sub-Systems

By means of the aforementioned parameters of the four sub-systems, it is possible to define some parameters that account the interactions among sub-systems. By considering these values, it is possible to estimate some general relations that can be used to the design, for instance: (i) the area of rigid foundations under the dam can be estimated as ~ 3.0 H_d^2; (ii) the contribution of the damping ratio of each sub-system respect to the damping ratio of the system is $\xi_d = 0.05$ (26%), $\xi_f = 0.1$ (51%), $\xi_r = 0.005$ (3%), $\xi_s = 0.04$ (20%); (iii) the contribution of the vibration period of each sub-system respect to the system vibration period is $T_{1,d}$ (s) = 0.284 (40%), $T_{1,f} = 0.09$ (13%), $T_{1,r} = 0.314$ (45%), $T_{1,s} = 0.014$ (2%).

These percentages show the weight of each sub-system respect to the total response. However, it is important to note that these values refer to this specific case study or, more in general, to concrete arch-gravity dams under specific conditions.

Finally, a modelling process should be calibrated for accurately identifying the problem to be analysed. There is a closer correlation between models and types of analysis: The choice of a model (software) is based on the specific problem to be solved. Although nowadays, there are extremely complex models [68] that consider all the phenomena together, it is good to define and focus a specific problem aspect and then to converge and resolve it by using a unique model.

Each model is made to study a specific problem. It is important to consider all the parts of the whole system, but it is also necessary not to lose control of the parameters and their interactions.

3. Management Operating Systems

The managerial procedures that account for the risk analysis are studied in reliable papers [7,8,69] and guidelines [12,14]. Moreover, in the literature, it is possible to find several contributions regarding stability optimization for concrete arch-dams [17,18,22,36,45]. However, the search of a safety and no-safety domain by taking into account the stability and deformation of arch-dams in a managerial context, by considering some parameters (see Table 1 later) obtained from several data, has not been carried out. In this sense, this paper provides a novelty for the research.

Table 1. Probabilistic parameters (collected results from [47–49]).

Parameter	Unit	Details	Mean RV	SD	CV (%) [a]	Distribution
Maximum dead stress	kN/m^2	Stress at the heel dam (US face)	−2215.38	±221.54	10.0	N
Elastic displacement	mm	Top displacement of the central block	151.65	±30.33	20.0	N
Elasto-plastic displacement	mm	Top displacement of the central block	186.41	±37.28	20.0	N
Hydrostatic pressure	kN/m^2	For $H_{o,r}$ = 113 m	1107.40	±110.74	10.0	N
Hydrodynamic pressure	kN/m^2	Pressures for flexible dam accounting compressible water. First three modes are considered.	350.55	±35.06	10.0	N
Acceleration	cm/s^2	Horizontal PGA for a return period 1950 years	303.20	±60.64	20.0	N

Note: PGA = Peak Ground Acceleration. SD = Standard Deviation. N = Normal (Gaussian) distribution. [a] CV is the Coefficient of Variation defined by: CV = (SD/Mean RV) × 100.

The project management is formed by design phases, which are called "project baselines", "project procedures", and "project systems". Each phase contains several sub-phases listed in the Figure 1.

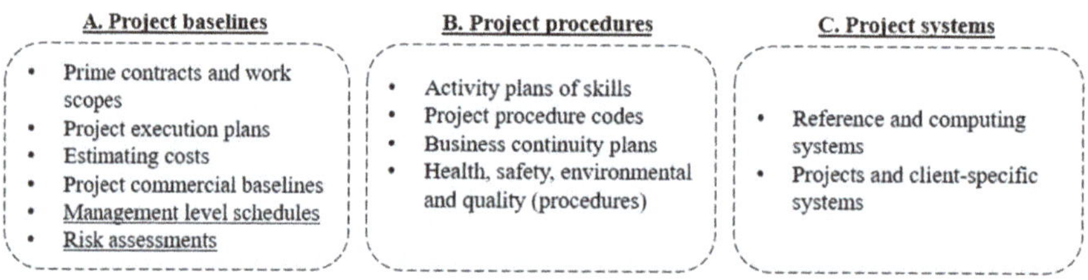

Figure 1. Operating system requirements.

In this paper, a particular attention about the "management level schedules" and "risk assessment" is considered; the former estimates the possible scenarios, whereas the latter defines the hazards.

In analyses there are different parameters/values that usually are adopted: deterministic parameters (DP), probabilistic parameters (PP), semi-probabilistic parameters (SPP), semi-deterministic parameters (SDP) and super-probabilistic parameters (SP2). SPP are the parameters obtained by combining DP and PP, whereas SDP are obtained by DP and SPP. SP2 are obtained by a probabilistic analysis, which are recalculated and re-estimated using one or more probabilistic approaches. Deterministic parameters are usually well known through the literature (papers, books, codes, guidelines), experience (real projects, academic works, research projects), and empirical experimentation (laboratory work, building sites). Probabilistic parameters are not known, and therefore are subject to aleatory (inherent randomness) and epistemic (lack of knowledge of materials/models) uncertainties, as have already been introduced.

4. Materials and Methods

4.1. Materials

This research comprises the analysis of probabilistic approaches which are the most reliable and precise ones for analysing the stability of dams. In this sense, these analyses are based on the definition of probability density functions (PDFs) through several random variables (RVs). The parameters used to develop the analysis in this paper are listed in Table 1. These parameters come from precedent studies [47–49], and here are considered as RVs to carry out a sustainability assessment, and are therefore plotted by a probabilistic distribution with a mean and standard deviation (SD).

4.2. Methods

4.2.1. Monte Carlo Simulation (MCS)

To estimate possible scenarios, MCS, which generates RVs, has been used in following way. Limit State (LS) function G(X) is defined. When the domain G(X) < 0, the LS is called "no safety", whereas when G(X) > 0, the LS is called "safety". The separation of both domains is given when G(X) = 0 (limit domain). Given a random variable vector $X = \{x_1, \ldots, x_j\} = \{x_i\}$ for a LS function G(X) and $f_X(x_i)$, which is the joint PDFs of x_i, the general probability x% that G(X) takes on a value less than 0 (called here probability of failure p_f) is [70,71]:

$$p_f = P[G(X) < 0] = \int_{\{X:\ G(X)<0\}} f_X(x_1, \ldots, x_j)dx_1 \ldots dx_j = \int_{\{X:\ G(X)<0\}} f_X(x_i)dx_i \tag{2}$$

Equation (2) represents the cumulative failure probability (CFP), which represents the area of the PDF within a defined interval.

By using MCS, Equation (2) can be rewritten as:

$$p_f = \int_{\{X:\ G(X)<0\}} I(x_i) f_X(x_i) dx_i \tag{3}$$

where $I(\cdot)$ is an indicator function, defined by:

$$I(x_i) = \begin{cases} 1, & for\ G(X) \le 0 \\ 0, & for\ G(X) > 0 \end{cases} \tag{4}$$

Finally, p_f can be considered as the mean value of $I(x_i)$, i.e., $\bar{I}(x_i) = E[I(x_i)]$, therefore Equation (3) becomes:

$$p_f = \frac{1}{N} \sum_{i=1}^{N} I(x_i) = \frac{N_f}{N} \tag{5}$$

where N is the number of simulations (or samples) and N_f is the number of simulations with $I(x_i) \le 0$. It is note that the result of p_f is more accurate when $N \to \infty$. In practice, samples required are 1×10^{N_k} where the choice of N_k is due to the computer power and available computational time.

4.2.2. Sustainability Assessment Framework

Sustainability has been assessed in this research from a double perspective. First, the perspective of temporal sustainability, closely related to the duration and useful life of arch dam infrastructures. This dimension is specifically articulated and assessed through the design parameters of "stability" and "deformation". Secondly, sustainability has been assessed from a safety perspective articulated through risk calculation. Consequently, in a broad scale, sustainability assessment is developed from a dual sustainability/safety management approach (Figure 5). On the other hand, in a detailed scale, the sustainability of concrete arch dams is evaluated from a design optimization perspective, specifically, for dams´ stability and deformation. Additionally, the safety perspective is directly related to the reduction and consequences of failure risk. For this, several scenarios about stability and deformation, identifying desirable and undesirable actions, were estimated. Quantitative results on both dimensions of sustainability are provided and explained in results section.

There are several types of actions that are generated either by human or by nature. These actions can be catalogued as either "environmental actions" or "human actions". All aspects regarding these actions are included in "impact matrices" where they are identified as "hazards". Several hazards can affect the durability of structures, e.g., environmental, social and economic impact; population and consumptions growing; climate change (temperature and humidity) [72]; flooding; hurricanes; explosions of blast waves in the terrorist attacks [73] or in demolitions [74]; seismic hazard [75]; corrosion [76,77].

Here, the last two hazards are introduced since are known by authors. However, in this analysis only seismic hazard assessment is considered.

Structures are subjected to internal and external stresses and deformations due to (1) excitation of masses by seismic vibrations or general dynamic loading by extreme events, and (2) the corrosion of the reinforced concrete (RC) elements.

Table 2 shows both hazards (as a succession: hazard → approach → scenarios), the relative approach and its scenario type.

Table 2. Identification of impacting hazards.

Hazard	Approach	Scenarios
Seismic hazard	Poisson [49]. Bayesian [78]	Probability of occurrence
Corrosion	Diffusion [79]. Reliability [70]	Probability of failure

Figure 2 shows the inter-combinations among the four sub-systems of the concrete arch-dams. It is possible to see all the possible combination among the dam-foundation, dam-sediments, dam-reservoir, foundation-sediments, foundation-reservoir and sediments-reservoir. By knowing the variables of the project, it is possible to treat the hazards in a practical way.

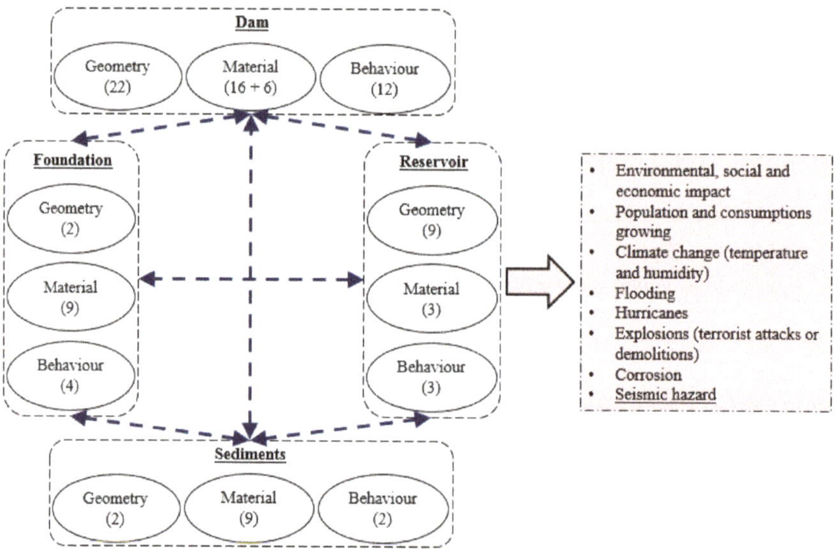

Figure 2. Inter-combinations of the sub-systems (the number in the brackets indicates the quantity of the used data provide in the Appendix A) and the impacting hazards.

4.2.3. Seismic hazard assessment

The seismic events are extreme events that may be accurately studied. The seismic hazard is usually estimated by using two approaches: probabilistic and deterministic. The former, probabilistic seismic hazard analysis (PSHA) is based on the Cornell method [80] and Poisson distribution [71]. To apply it, it is necessary to know seismogenic zones, i.e., zones where the earthquakes are equally likely and independent of each other at any location (e.g., in Spain [81]).

The probability that a ground motion parameter S exceeds the ground motion level S_0 in i-th source area is defined by Λ_i, which depends on: the PDF of the magnitude $f_m(m)$ and of the site-source distance $f_r(r)$, the standardization Normal distribution $f_\varepsilon(\varepsilon)$ [71] with the ground motion randomness ε and, the average annual rate of exceedance λ_c of an event with magnitude m described through the

Gutenberg–Richter trend line [82], which provides the ratio between the number of small and large events and the level of seismicity [83].

The probability is defined by:

$$\Lambda_i = \lambda_c \int_m \int_r \int_\varepsilon P[S > S_0 | m, r, \varepsilon] f_m(m) f_r(r) f_\varepsilon(\varepsilon) dm\, dr\, d\varepsilon \quad (6)$$

If the analysis involves more of one seismogenic zones (where N_s = number of seismogenic zones), the probability of exceedance is defined by:

$$\Lambda_{S_0} = P[S > S_0] = \sum_{i=1}^{N_s} \Lambda_i \quad (7)$$

Figure 3 shows some curves (as results example) in terms of accelerations vs. structural period (Figure 3a) and hazard contribution respect to the magnitude and fault-site distance (Figure 3b).

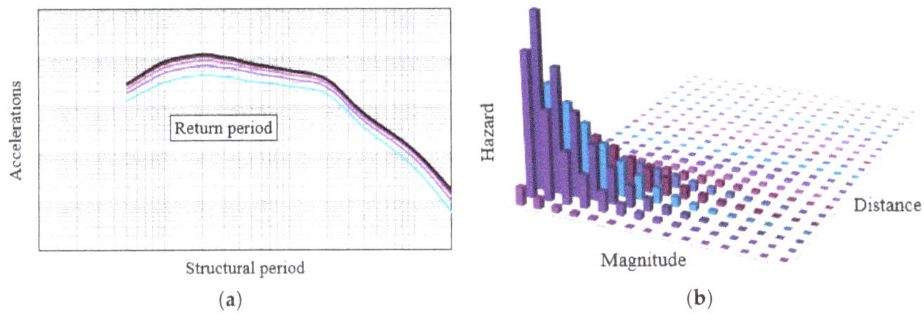

Figure 3. Example of hazard curves (**a**) and hazard contributions (**b**).

MCS is used to analyse the sustainability of the structure respect to the stability and deformations. LS function is written as the difference between the stable actions A_s, and unstable actions A_u: $G(X) = A_s(X) - A_u$. When $A_s < A_u$, $G(x) < 0$, the failure is achieved.

Figure 4a,b shows the generated MCS points, whereas Figure 4c–d shows an example how to identify the LS function (Figure 4c) and the PDF in 3D (Figure 4d). To the left of the intersection point (Figure 4c), between stable and unstable trend line, there is the "no safety" state ($G(X) < 0$), whereas to the right of this point there is the "safety" state ($G(X) > 0$). The PDF in the (x_i, x_{i+1}) point represents the value of the probability around (x_i, x_{i+1}) point in relation to the amplitude of this around (density).

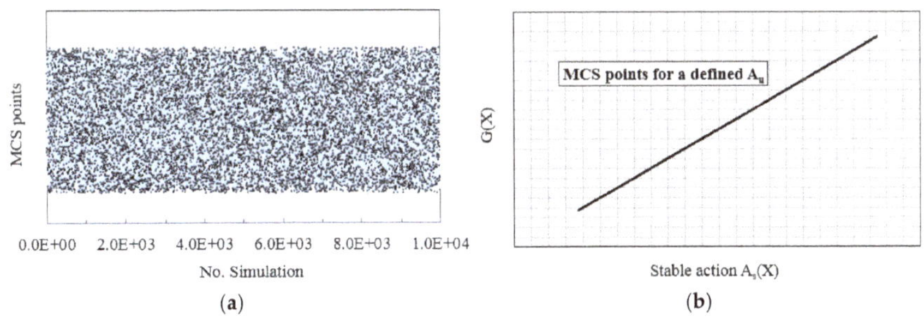

Figure 4. *Cont.*

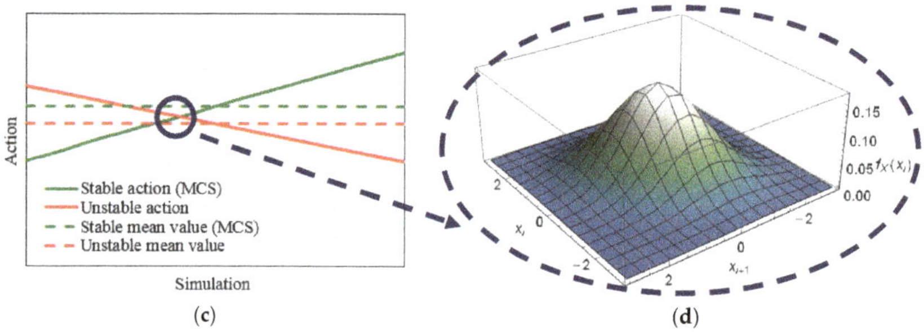

Figure 4. MCS points for 1×10^4 simulations in spread form (**a**) and linear form (**b**). Individuation of the LS (**c**) and PDF (**d**) respect to RVs for G(X) = 0.

Figure 5 shows the methodology by the flow chart used in the analysis. The flow chart is divided in two principal parts: general and specific part. In the first one, the process and operation phase are defined. Here, choices, decisions, individuation of the structure (issue), hazards, and the possible approaches are established. Then, the technical actions are analysed in terms of data and control of modelling and analyses. Here, a specific concrete arch-dam is individuated (case study), by defining sub-systems data, RVs, methods and approaches (if the modelling and analysis are not satisfactory and are not consistent to the individuated hazards, it is necessary to start over). Finally, scenarios are estimated in terms of stability and deformations of the dam by providing safety and no-safety domain (sustainability assessment) and probability of failure (safety assessment). The flow chart concluded by taking a final decision from managers and technical engineers.

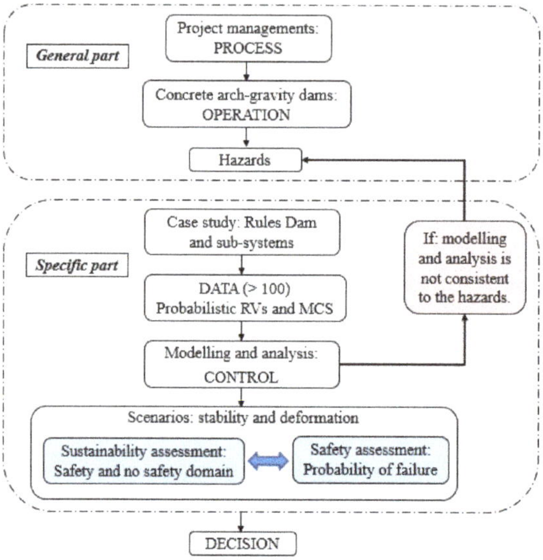

Figure 5. General methodology flow chart.

5. Results

5.1. Sustainability Assessment

Here, six scenarios to evaluate the sustainability assessment accounting the deformation and stability of concrete arch-dams are shown. Stable actions refer to the probabilistic parameters in Table 1. By knowing the mean RV and SD for each parameter it is possible to generate a several points by MCS.

To the left of the Figures 6 and 7 the trend lines of the stable and unstable actions are shown. The horizontal dashed line indicates the LS line (i.e., the mean line when the stable line intersects the unstable line). For the stable action, its logarithmic trend line is also plotted, which shows better the progress of an action that starts from zero and reaches its maximum value. The logarithmic trend intersects the unstable line before respect to the linear stable trend. This gap could represent a security factor that increase the "safety" LS. When the dashed horizontal line rises, the p_f increases and so the "no safety" state is more probable.

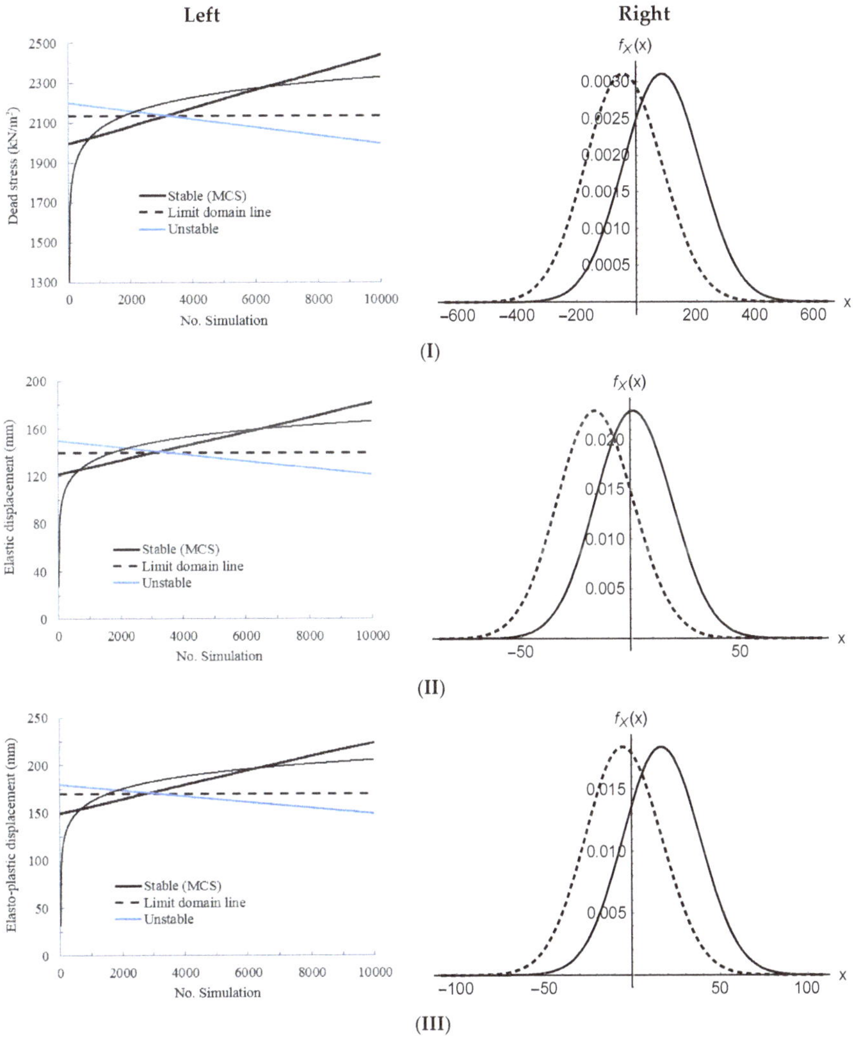

Figure 6. Three scenarios (**I–III**) regarding dams' deformation. Trend lines of stable and unstable actions vs. number of simulation (**left**); PDF when $A_s = A_u$ (**right**).

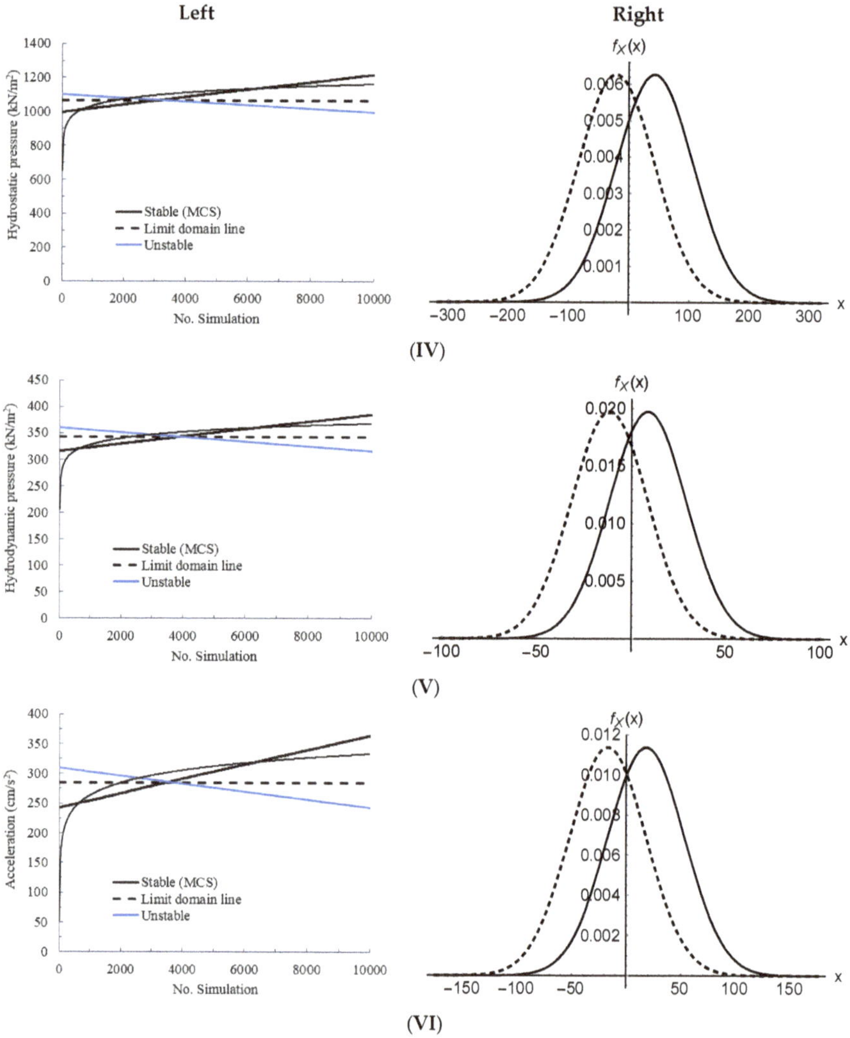

Figure 7. Three scenarios (**IV–VI**) regarding dams' stability. Trend lines of stable and unstable actions vs. number of simulation (**left**); PDF when $A_s = A_u$ (**right**).

To the right of the Figures 6 and 7, the PDFs when ($A_s = A_u$) are plotted. The solid curves represent the PDFs by mean RVs, whereas the dashed curves represent the PDFs by negative SDs.

5.2. Safety Assessment

Finally, the risk management model defined in literature [12,14] show the need of defining the undesirable event with the potential for harm or damage in these following steps: individuation of hazards → defining of potential for failure → estimating of consequences (harm to people, assets, environment). These steps are needed to design and justify engineering activities (why act?), to propose activities maintenance (when to act) and to tackle operations activities (how to act).

In this sense, the safety management assessment can be evaluated by quantifying the p_f. Table 3 and Figure 8 summarize the results in accordance to Figures 6 and 7.

Table 3. Identification of impacting hazards.

Scenario	Parameter	Unit	A_u	Mean of G(X)	SD of G(X)	p_f
I	Dead stress	kN/m^2	2130	85.4	128.33	0.3095
II	Elastic displacement	mm	140	11.32	17.40	0.3097
III	Elasto-plastic displacement	mm	170	16.71	21.67	0.2753
IV	Hydrostatic pressure	kN/m^2	1065	42.52	63.65	0.3053
V	Hydrodynamic pressure	kN/m^2	342	8.61	20.21	0.3791
VI	Acceleration	cm/s^2	285	18.14	35.10	0.3496

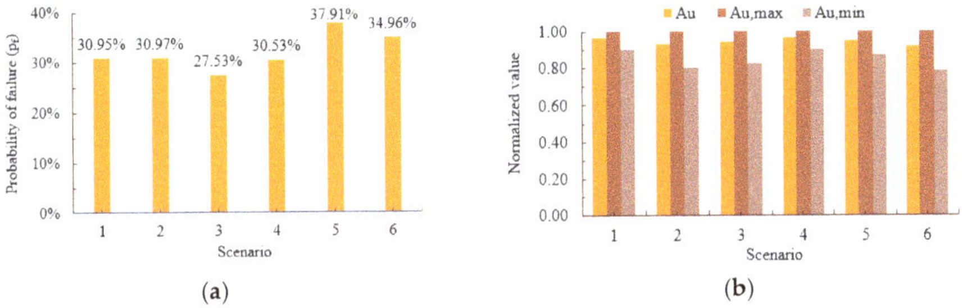

Figure 8. Estimates for risk management models in terms of p_f (**a**) and normalized values (**b**).

6. Summary

This paper mainly aimed to review the knowledge on the development of sustainability and safety assessment through the study of structural stabilities/deformations and failure risk consequences, respectively, for concrete gravity arch-dams.

In order to carry out the main analysis, several aspects have been defined: materials regarding the sub-systems (dam, foundation, reservoir, sediments) and their interactions; methods respecting to the operating systems of a project; deterministic and probabilistic variables; modelling and methodologies.

From precedent-specific studies of the authors investigating dam design, more than 10 theoretical modelling, 10 modelling types by software, more than 100 specific parameters, and more than 100 references are summarized.

This paper addresses and comprises critical aspects that are summarized as follows: (i) to show innovative approaches respecting to the enormous quantities of variables that are involved for concrete arch-dams; (ii) to provide numerical values of parameters to design concrete arch-dams; (iii) to show the project phases and methodologies; (iv) to estimate different scenarios respecting to the main actions on the dam system; (v) to contribute to the knowledge of the state-of-the-art about concrete arch dams.

The first results are shown in terms of new estimated data provided in the Appendix A. Other results concern the parameters of the interaction between dam–foundation–reservoir–sediments with respect to the area of rigid foundations under the dam (~ 3.0 H_d^2), the contribution of each sub-system damping ratio respect to the system damping ratio (8.5%), and the contribution of each sub-system vibration period respect to the system vibration period (0.393 s). These values are useful to estimate some general relations that can be used to aid design. Moreover, the maximum elastic and elasto-plastic displacements are of the order of ~ 0.10–0.20 m that, in relation to the maximum dam height, is $H_d/1000$, in accordance with the literature [6].

Furthermore, the sustainability assessment demonstrates that the mean probability of failure of the stability of dam body and its deformation is about 32%. In particular, that for stability is 34%, which is higher than for the deformation at 29%. These mean percentages are quite large because unstable actions have been taken. When the intersection point between the stable and unstable line rises, the p_f increases, and so the "no safety" state is more probable. However, this raises the level of attention during the design of a monitoring method for concrete arch-dams, and in this sense, risk management can be carried out satisfactory.

Author Contributions: Conceptualization, E.Z. and J.L.M.; methodology, E.Z. and J.-L.M.; software, E.Z.; validation, E.Z. and J.L.M.; formal analysis, E.Z.; investigation, E.Z.; resources, E.Z.; data curation, E.Z.; writing—original draft preparation, E.Z.; writing—review and editing, J.L.M.; visualization, E.Z. and J.L.M.; supervision, J.L.M.; funding acquisition, E.Z. All authors have read and agreed to the published version of the manuscript.

Funding: The study was funded by Coordination of the Improvement of Graduated Professionals (CAPES), Support Program for Foreign Students of Doctorate (PAEDEX) and Ibero-American University Postgraduate Association (AUIP) (reference number: 3.224.803.003.569).

Acknowledgments: The first author acknowledges the University of Salamanca to pay the rights (when applicable) to completely download all papers in the references.

Conflicts of Interest: The authors declare no conflict of interest.

Appendix A

Table A1. Some cases of real concrete-arch dams studied for scientific purposes.

Dam Name	Location	Researched Main Topics	Reference
Ertan Dam	Sichuan province, Southwestern China	Modal analysis. Seismic response	[84]
Tsankov Kamak Dam	Vacha River, Southwestern Bulgaria	Monitoring. Dam performance	[85]
Longyangxia Dam	Qinghai province, China	Dam-water-foundation interaction. Shock wave effects	[86]
Ridracoli Dam	Emilia Romagna, Italy	Modelling and reconstruction of dams. Unmanned aerial vehicle (UAV) photogrammetry	[54]
Lancang River Dam	Yunnan, China	Optimal sensor placement. Monitoring	[87]
Outardes 3 Dam	Quebec, Canada	Dam-reservoir-foundation interaction. Seismic analysis	[38]
Brezina Dam	Beyadh, Algeria west	Dam-water-foundation interaction. Sloshing effect	[35]
Shapai Dam	Sichuan province, China	Dam hazards. Seismic performance	[67]
Morrow Point Dam	Southwest Denver, Colorado	Shape optimal design. Fluid-structure interaction	[88]
Xiluodu Dam	Sichuan province, China	Excavation optimization design. Stability analysis	[89]
Rules Dam	Granada, Southern Spain	Probabilistic and deterministic seismic hazard. Dynamic analysis	[48]
Dagangshan Dam	Southwest China	Seismic damage. Joint opening. Artificial accelerograms	[26]
Jinping I Dam	Sichuan Province, China	Permeability of foundations. Behaviour of transient groundwater flow	[90]
Cabril Dam	Castelo Branco, Portugal	Seismic performance. Hydrodynamic pressures respect to the water level	[91]

Table A2. Collected data relative to dam sub-system.

Dam					
Geometry		**Material (Concrete)**		**Behaviour (Solid: Elasto-Plastic)**	
Blocks number [83]	32	Density ρ_d (kN/m^3) [92]	24	$T_{1,d}$ (s) [48]	0.284
US slope [83]	0.18	Volume (10^3 m^3) [83]	2051	$T_{2,d}$ (s) [48]	0.245
DS slope [83]	0.6	f_{cd} (MPa) [47]	47.5	$T_{3,d}$ (s) [48]	0.208
Base's max. length (m)	102 *	f_{cm} (MPa) [92]	58	MPMR for $T_{1,d}$ in x, y, x (%) [48]	45.1
Crown length (m)	10 *	σ_c (MPa)	38.46*	Mass (10^6 kg)	4830 *
Crown height (m)	7.5 *	E_{cm} (GPa) [47]	44.4	Stiffness (GN/m)	2406 *
Crown long. length (m) [83]	509	E_{ep} (GPa)	35.52*	Eq. inertia (m^4)	1,376,852 *
Max. height H_d (m) [83]	132	ε_{c1} (‰) [92]	2.45	Damping ratio ξ_d (%) [48]	5.0
Radius (m) [83]	500	ε_c (‰)	3.45 *	Blocks' eq. mean period (s)	0.262 *
Ange in plane (°)	71 *	Ductility (= $\varepsilon_c/\varepsilon_{c1}$)	1.408	Blocks' mean mass (10^6 kg)	130.56 *
Volume (10^3 m^3)	2291 *	Thermal expansion (10^{-6} 1/K) [92]	10	Blocks' mean eq. stiffness (MN/m)	75,089 *
Voids' volume (10^3 m^3)	239 *	ν_d [47]	0.19	Blocks' mean eq. inertia (m^4)	43,027 *
Long. area (10^3 m^2)	46 *	f_{ctd} (MPa) [47]	2.73	**Concrete crack model**	
Spillway's length (m) [83]	16.54	G_d (GPa)	9.92	ε_{lt} (‰) [47]	0.166
Min. block height (m)	7.0 *	c_d (kN/m^2) [63]	1000	a_c (m) [47]	0.484
Blocks' mean length (m) [83]	19.375	ϕ_d (°) [63]	55	w_c (μm) [47]	240.51
Min. block volume (m^3)	373 *			G_t (N/m) [47,93]	113.06
Max. block volume (10^3 m^3)	125 *			h_0 (m) [47,94]	1.35
Min. block long. area (m^2)	137 *			l_c (m) [47]	0.45
Max. block long. area (m^2)	2463 *				
Min. block trans. area (m^2)	19 *				
Max. block trans. area (m^2)	6624 *				

Note: * = Estimated value. US = Up-Stream. DS = Down-Stream. max. = Maximum. min. = Minimum. long. = Longitudinal. trans. = Transversal. f_{cd} = Design compressive strength. f_{cm} = Mean compressive strength at 28 days. σ_c = Compressive stress. E_{cm} = Secant modulus of elasticity. E_{ep} = Secant elasto-plastic modulus. ε_{c1} = Strain at peak stress. ε_c = Shortening strain. ν_d = Poisson's ratio of the concrete. f_{ctd} = Design tensile strength. G_d = Shear modulus. c_d = Cohesion of the concrete. ϕ_d = Angle of friction of the concrete. $T_{i,d}$ = Structural period for i-th mode. MPMR = Modal participating mass ratios. eq. = Equivalent. ε_{lt} = Limit dynamic tensile strain. a_c = Effective crack length. w_c = Characteristic micro-crack opening that propagate through the aggregates. G_t = Tension specific fracture energy. h_0 = Size of the element that model l_c for the linear analysis. l_c = Crack band width of the fracture.

Table A3. Collected data relative to foundation sub-system.

Foundation			
Material (Rock)		**Behaviour (Solid: Elastic)**	
Density ρ_f (kN/m^3) [48]	27.47	$T_{1,f}$ (s)	0.09 *
c_f (kN/m^2) [63]	45	Mass (10^3 kg)	205,175 *
ϕ_f (°) [63]	45	Stiffness (kN/m)	1.0 × 10^9 *
ν_f [47]	0.31	Damping ratio ξ_f (%)	10 *
G_f (GPa)	6.181 *	Geometry	
E_f (GPa) [47]	41.55	Radius of semicircle (m^2/m) [10]	27,355
$V_{s,f}$ (m/s)	1500 *	Area (m^2/m)	74,690 *
$E_{o,f}$ (GPa)	109.34 *		
$V_{p,f}$ (m/s)	6309 *		

Note: * = Estimated value. c_f = Cohesion of the foundation. ϕ_f = Angle of friction of the foundation. ν_f = Poisson's ratio of the foundation. G_f = Shear modulus. E_f = Elastic modulus of foundation. $V_{s,f}$ = Shear wave velocity in rock. $V_{p,f}$ = Compressive wave velocity. $E_{o,f}$ = Oedometric modulus. $T_{1,f}$ = Foundation's first period.

Table A4. Collected data relative to reservoir sub-system.

Reservoir			
Geometry		**Material (Water)**	
Operating level $H_{o,r}$ (m) [48]	113	Density ρ_r (kN/m^3) [49]	9.8
Operating level area (m^2/m) [11]	38,307	$V_{p,r}$ (m/s) [49]	1438
Flood level $H_{f,r}$ (m)	120 *	E_b (GPa)	2.026 *
Flood level area (m^2/m) [11]	43,200	**Behaviour (Liquid: Viscous)**	
DS Operating level (m)	5.0 *	$T_{1,r}$ for $H_{o,r}$ (m) [95]	0.314
Capacity for $H_{o,r}$ (Hm^3) [83]	117.07	$T_{1,r}$ for $H_{f,r}$ (m) [95]	0.334
Area for $H_{o,r}$ (Ha) [83]	308	Damping ratio ξ_r (%) [48]	0.5
Water basin area (km^2) [83]	1070		
Spillway capacity (m^3/s) [83]	2987		

Note: * = Estimated value. DS = Down-Stream. $V_{p,r}$ (or C_r) = Compressive wave velocity. E_b = Bulk modulus of reservoir. $T_{1,r}$ = Reservoir's first period.

Table A5. Collected data relative to sediments sub-system.

Sediments			
Material		**Behaviour (Semi-Solid: Visco-Elastic)**	
Density ρ_s (kN/m^3)	13 *	$T_{1,s}$ (s) [95]	0.014
c_s (kN/m^2)	20 *	Damping ratio ξ_s (%)	4.0 *
ϕ_s (°)	20 *	**Geometry**	
ν_s	0.45 *	Area (m^2/m) [11]	75.0
$G_{d,s}$ (GPa)	0.81 *	Height H_s (m)	5.0 *
$E_{d,s}$ (GPa)	0.27 *		
$V_{s,s}$ (m/s)	25 *		
$V_{p,s}$ (m/s)	1450 *		
$E_{o,s}$ (GPa)	2.73 *		

Note: * = Estimated value. c_s = Cohesion of the sediments. ϕ_s = Angle of friction of the sediments. ν_s = Poisson's ratio of the sediments. $G_{d,s}$ = Shear modulus. $E_{d,s}$ = Elastic modulus. $V_{s,s}$ = Shear wave velocity in sediments. $V_{p,s}$ = Compressive wave velocity in sediments. $E_{o,s}$ = Oedometric modulus. $T_{1,r}$ = Sediments' first period.

Table A6. Parameters accounting the interactions.

Sub-Systems' Combination	Parameter	Value
Dam + foundation + reservoir + sediments	Damping ratio (%) [48]	8.5
	Vibration period (s) [48]	0.393
Dam + foundation (rigid)	Impedance ratio	0.853 *
	Area (m^2/m)	~ 3.0 H_d^2 *
Dam + reservoir	Vibration period (s)	0.37*
Foundation + reservoir	q	5.655×10^{-5} *
	α [47]	0.85
Reservoir + sediments	q	5.199×10^{-4} *
	α	0.144 *

Note: * = Estimated value. q = Admittance coefficient. α = Wave reflection.

Table A7. Modelling types.

Model	Input	Output	Dimension	Description	Software
FEM [24,96]	Elements. Joints. Material properties. Loads	Stresses. Deformations. Modal parameters (e.g. frequency, modal participating mass ratio)	2D/3D	Discretization of an area or volume in mesh. A function is performed on each mesh and so the calculus is extended over the whole structure	[97]
Gravity method [29,98]	Loads. Geometry. Material properties	Stresses. Pressures. Stabilities	2D	It based on the rigid body equilibrium and beam theory. It performs stability analyses for hydrostatic loads and seismic loads	[50]
Numerical [47,99]	Differential equations. Boundary and initial conditions	Displacements. Velocities. Accelerations	2D/3D	By using interpolation function, it is possible to solve partial differential equations under specific conditions	[100]
Variational [90,101]	Functionals. Boundary and initial conditions	Optimum shape. Modal parameters (eigenvalues and eigenvectors)	2D	Through functionals, it is possible to find the maximum and minimum solutions	[100]
Analytical [95,102]	Analytical equations	Stresses. Pressures	1D	Substituting specific numerical values in the equations it is possible to find the solutions	[103]
BEM [19,53] [a]	Differential equations	Displacements. Velocities. Accelerations	2D/3D	It is a numerical computational method that solves partial differential equations under specific conditions	[64,65]
UAV photogrammetry [54,104]	Drones. Sensors	Geometry. Photogrammetry	3D	Geodetic survey of a study site by creating a detailed point cloud. It provides measurements from photographs	[105]
Geometric [9,106]	Measures. Quotes	Geometrical and architectural design	2D/3D	Plotting of drawings through heights, lengths and thicknesses	[107]
Experimental	Measures. Quotes. Tools. Laboratory	Simulations. Calibrations	3D	Reproduction of a structure with scaled dimensions respect to the real project	N/A
Rendering [b]	Measures. Quotes. Imagens	Photos. Animations	3D	Generation of 3D reconstructions by algorithms that define the colour and size of each point of the input image	[108]

Note: FEM = Finite Element Method. BEM = Boundary Element Method. UAV = Unmanned Aerial Vehicle. N/A = Not applicable. [a] Coupled BEM-FEM is used to study the fluid-structure interactions [19]. Also, accurate computation of fluid-structure nonlinear interaction is analysed by the immersed boundary method (IBM) proposed in [41]. [b] The reader can refer to specific bibliographies in the area of the design and/or architecture.

References

1. Inventory of Dams and Reservoirs (SNCZI). Available online: http://sig.mapama.es/snczi/visor.html (accessed on 1 February 2019).
2. Spanish Association of Dams and Reservoirs (SEPREM). Available online: http://www.seprem.es/index.php (or: http://www.seprem.es/st_pe_f/JDFerrovial/PRESA_DE_RULES.pdf) (accessed on 1 February 2019).
3. Huang, H.; Chen, B.; Liu, C. Safety monitoring of a super-high dam using optimal kernel partial least squares. *Math. Probl. Eng.* **2015**, *2015*, 571594. [CrossRef]
4. Shi, Z.; Gu, C.; Qin, D. Variable-intercept panel model for deformation zoning of a super-high arch dam. *SpringerPlus* **2016**, *5*, 898–917. [CrossRef] [PubMed]
5. Zhang, Q.L.; Wang, F.; Gan, X.Q.; Li, B. A field investigation into penetration cracks close to dam-to-pier interfaces and numerical analysis. *Eng. Fail. Anal.* **2015**, *57*, 188–201. [CrossRef]
6. Hariri-Ardebili, M.A.; Saouma, V.E. Seismic fragility analysis of concrete dams: A state-of-the-art review. *Eng. Struct.* **2016**, *128*, 374–399. [CrossRef]

7. Hariri-Ardebili, M.A. Risk, reliability, resilience (R^3) and beyond in dam engineering: A state-of-the-art review. *Int. J. Disaster Risk Reduct.* **2018**, *31*, 806–831. [CrossRef]
8. Altarejos-García, L.; Escuder-Bueno, I.; Serrano-Lombillo, A.; De Membrillera-Ortuño, M.G. Methodology for estimating the probability of failure by sliding in concrete gravity dams in the context of risk analysis. *Struct. Saf.* **2012**, *36–37*, 1–13.
9. Savage, J.L.; Houk, I.E. Checking arch dam design with models. *Civ. Eng.* **1931**, *1*, 695–699.
10. U.S. Army Corps of Engineers (USACE). *Arch Dam Design, Manual No. 1110-2-2201*; USACE: Washington, DC, USA, 1994.
11. U.S. Army Corps of Engineers (USACE). *Theoretical Manual for Analysis of Arch Dams*; Technical Report ITL-93-1; USACE: Washington, DC, USA, 1993.
12. International Commission on Large Dams (ICOLD). *Selecting Seismic Parameters for Large Dams*; Guidelines, Bulletin 148; ICOLD: Paris, France, 2016.
13. Federal Guidelines for Dam Safety (FEMA). *Earthquake Analyses and Design of Dams*; FEMA: Washington, DC, USA, 2005.
14. International Commission on Large Dams (ICOLD). *Dam Safety Management: Operational Phases of the Dam Life Cycle*; Bulletin 154; ICOLD: Paris, France, 2017.
15. Pouraminian, M.; Ghaemian, M. Multi-criteria optimization of concrete arch dams. *Sci. Iran.* **2017**, *24*, 1810–1820. [CrossRef]
16. Kaveh, A.; Ghaffarian, R. Shape optimization of arch dams with frequency constraints by enhanced charged system search algorithm and neural network. *Int. J. Civ. Eng.* **2015**, *13*, 102–111.
17. Seyedpoor, S.M.; Gholizadeh, S. Optimum shape design of arch dams by a combination of simultaneous perturbation stochastic approximation and genetic algorithm methods. *Adv. Struct. Eng.* **2008**, *11*, 500–510. [CrossRef]
18. Seyedpoor, S.M.; Salajegheh, J.; Salajegheh, E.; Gholizadeh, S. Optimal design of arch dams subjected to earthquake loading by a combination of simultaneous perturbation stochastic approximation and particle swarm algorithms. *Appl. Soft Comput.* **2011**, *11*, 39–48. [CrossRef]
19. Mahani, A.S.; Shojaee, S.; Salajegheh, E.; Khatibinia, M. Hybridizing two-stage meta-heuristic optimization model with weighted least squares support vector machine for optimal shape of double arch dams. *Appl. Soft Comput.* **2015**, *27*, 205–218. [CrossRef]
20. Zhang, X.F.; Li, S.Y.; Chen, Y.L. Optimization of geometric shape if Xiamen arch dam. *Adv. Eng. Softw.* **2009**, *40*, 105–109. [CrossRef]
21. Akbari, J.; Ahmadi, M.T.; Moharrami, H. Advances in concrete arch dams shape optimization. *Appl. Math. Model.* **2011**, *35*, 3316–3333. [CrossRef]
22. Shouyi, L.; Lujun, D.; Lijuan, Z.; Wei, Z. Optimization design of arch dam shape with modified complex method. *Adv. Eng. Softw.* **2009**, *40*, 804–808.
23. Gholizadeh, S.; Seyedpoor, S.M. Shape optimization of arch dams by metaheuristics and neural networks for frequency constraints. *Sci. Iran.* **2011**, *18*, 1020–1027. [CrossRef]
24. Mirzabozorg, H.; Varmazyari, M.; Ghaemian, M. Dam-reservoir-massed foundation system and travelling wave along reservoir bottom. *Soil Dyn. Earthq. Eng.* **2010**, *30*, 746–756. [CrossRef]
25. Lamea, M.; Mirzabozorg, H. Simulating structural responses of a generic AAR-affected arch dam considering seismic loading. *Sci. Iran.* **2018**, *25*, 2926–2937. [CrossRef]
26. Wang, J.T.; Jin, A.Y.; Du, X.L.; Wu, M.X. Scatter of dynamic response and damage of an arch dam subjected to artificial earthquake accelerograms. *Soil Dyn. Earthq. Eng.* **2016**, *87*, 93–100. [CrossRef]
27. Hariri-Ardebili, M.A.; Furgani, L.; Meghella, M.; Saouma, V.E. A new class of seismic damage and performance indices for arch dams via ETA method. *Eng. Struct.* **2016**, *110*, 145–160. [CrossRef]
28. García, F.; Aznárez, J.J.; Padrón, L.A.; Maeso, O. Relevance of the incidence angle of the seismic waves on the dynamic response of arch dams. *Soil Dyn. Earthq. Eng.* **2016**, *90*, 442–453. [CrossRef]
29. Furgani, L.; Imperatore, S.; Nuti, C. Analisi sismica delle dighe a gravità: Dal semplice al complesso, se necessaio. In Proceedings of the XIV Convegno ANIDIS, L'Ingeneria Sismica, Bari, Italy, 18–22 September 2011.
30. Hariri-Ardebili, M.A.; Kianoush, M.R. Integrative seismic safety evaluation of a high concrete arch dam. *Soil Dyn. Earthq. Eng.* **2014**, *67*, 85–101. [CrossRef]
31. Hariri-Ardebili, M.A.; Mirzabozorg, H.; Kianoush, R. Comparative study of endurance time and time history methods in seismic analysis of high arch dams. *Int. J. Civ. Eng.* **2014**, *12*, 219–236.

32. Chen, B.F.; Yuan, Y.S. Nonlinear hydrodynamic pressures on rigid arch dams during earthquakes. In Proceedings of the 12 WCEE 2000, 12th World Conference on Earthquake Engineering, Auckland, New Zeland, 30 January–4 February 2000.
33. Schultz, M.; Huynh, P.; Cvijanovic, V. Implementing nonlinear analysis of concrete dams and soil-structure interaction under extreme seismic loading. In Proceedings of the 34th Annual USSD Conference, San Francisco, CA, USA, 7–11 April 2014.
34. Khosravi, S.; Heydari, M.M. Modelling of concrete gravity dam including dam-water-foundation rock interaction. *World Appl. Sci. J.* **2013**, *22*, 538–546.
35. Alegre, A.; Oliveira, S.; Ramos, R.; Espada, M. Resposta sísmica de barragens abobada. Estudo numérico sobre a influência da cota de água na albufeira. In Proceedings of the Encontro Nacional Betão Estrutural—BE2018, Lisbon, Portugal, 7–9 November 2018.
36. Seyedpoor, S.M.; Salajegheh, J.; Salajegheh, E. Shape optimal design of arch dams including dam-water-foundation rock interaction using a grading strategy and approximation concepts. *Appl. Math. Model.* **2010**, *34*, 1149–1163. [CrossRef]
37. Mirzabozorg, H.; Kordzadeh, A.; Hariri-Ardebili, M.A. Seismic response of concrete arch dams including dam-reservoir-foundation interaction using infinite elements. *Electron. J. Struct. Eng.* **2012**, *12*, 63–73.
38. Proulx, J.; Paultre, P. Experimental and numerical investigation of dam-reservoir-foundation interaction for a large gravity dam. *Can. J. Civ. Eng.* **1997**, *25*, 90–105.
39. Akköse, M.; Adanur, S.; Bayraktar, A.; Dumanoğlu, A.A. Elasto-plastic earthquake response of arch dams including fluid-structure interaction by the Lagrangian approach. *Appl. Math. Model.* **2008**, *32*, 2396–2412. [CrossRef]
40. Hariri-Ardebili, M.A.; Seyed-Kolbadi, S.M. Seismic cracking and instability of concrete dams: Smeared crack approach. *Eng. Fail. Anal.* **2015**, *52*, 45–60. [CrossRef]
41. Demirel, E. Numerical simulation of earthquake excited dam-reservoirs with irregular geometries using an immersed boundary method. *Soil Dyn. Earthq. Eng.* **2015**, *73*, 80–90. [CrossRef]
42. Li, Q.; Guan, J.; Wu, Z.; Dong, W.; Zhou, S. Equivalent maturity for ambient temperature effect on fracture parameters of site-casting dam concrete. *Constr. Build. Mater.* **2016**, *120*, 293–308. [CrossRef]
43. Shi, N.; Chen, Y.; Li, Z. Crack risk evaluation of early age concrete based on the distributed optical fiber temperature sensing. *Adv. Mater. Sci. Eng.* **2016**, *2016*, 4082926. [CrossRef]
44. Jin, F.; Chen, Z.; Wang, J.; Yang, J. Practical procedure for predicting non-uniform temperature on the exposed face of arch dams. *Appl. Therm. Eng.* **2010**, *30*, 2146–2156. [CrossRef]
45. Zacchei, E.; Molina, J.L. Shape optimization of double-arch dams by using parameters obtained through Bayesian estimators. *Iran. J. Sci. Technol. Trans. Civ. Eng.* **2018**, *43*, 649–662. [CrossRef]
46. Zacchei, E.; Molina, J.L. Artificial accelerograms to estimate damage of dams by using failure criteria. *Sci. Iran.* **2018**, in press. [CrossRef]
47. Zacchei, E.; Molina, J.L.; Brasil, M.R. Nonlinear degradation analysis of arch-dam blocks by using deterministic and probabilistic seismic input. *J. Vib. Eng. Technol.* **2019**, *7*, 301–309. [CrossRef]
48. Zacchei, E.; Molina, J.L.; Brasil, M.R. Seismic hazard and structural analysis of the concrete arch dam (Rules dam on Guadalfeo River). *Procedia Eng.* **2017**, *199*, 1332–1337. [CrossRef]
49. Zacchei, E.; Molina, J.L.; Brasil, M.R. Seismic hazard assessment of arch dams via dynamic modelling: An application to the Rules Dam in Granada, SE Spain. *Int. J. Civ. Eng.* **2019**, *17*, 323–332. [CrossRef]
50. Leclerc, M.; Léger, P.; Tinawi, R. *CADAM, Version 1.4.14*; CRSNG/Hydro-Québec/Alcan: Montréal, QC, Canada, 2004.
51. Zacchei, E.; Brasil, M.R. Seismic action on oil storage tanks: Induced pressures, total response and state of buckling. *Int. J. Modeling Simul. Pet. Ind.* **2017**, *10*, 45–53.
52. European Committee for Standardization (CEN). *Design of steel structures—Part 1–6: Strength and stability of shell structures*; EN 1993-1-6:2007; European Committee for Standardization (CEN): Brussels, Belgium, 2007.
53. Furgani, L.; Imperatore, S.; Nuti, C. Seismic assessment methods for concrete gravity dams. In Proceedings of the 15 WCEE, 15th World Conference on Earthquake Engineering, Lisbon, Portugal, 24–28 September 2012.
54. Buffi, G.; Manciola, P.; Grassi, S.; Barberini, M.; Gambi, A. Survey of the Ridracoli dam: UAV-based photogrammetry and traditional topographic techniques in the inspection of vertical structures. *Geomat. Nat. Hazards Risk* **2017**, *8*, 1562–1579. [CrossRef]

55. Buffi, G.; Manciola, P.; De Lorenzis, L.; Cavalagli, N.; Comodini, F.; Gambi, A.; Gusella, V.; Mezzi, M.; Niemeier, W.; Tamagnini, C. Calibration of finite element models of concrete arch-gravity dams using dynamical measures: The case of Ridracoli. *Procedia Eng.* **2017**, *199*, 110–115. [CrossRef]
56. Binici, B.; Arici, Y.; Aldemir, A.; Akman, A. Comparisons of two and three dimensional nonlinear dynamic analyses results of a roller compacted concrete dam. *Res. Dev. Pract. Struct. Eng. Constr.* **2012**, *12*, 13–20.
57. Fenves, G.L.; Chávez, J.W. Evaluation of earthquake induced sliding in gravity dams. In Proceedings of the 11 WCEE Eleventh World Conferences on Earthquake Engineering, Acapulco, Mexico, 23–28 June 1996.
58. Basili, M.; Nuti, C. A simplified procedure for base sliding evaluation of concrete gravity dams under seismic actions. *Int. Sch. Res. Netw.* **2011**, *2011*, 413057. [CrossRef]
59. Altarejos-García, L.; Escuder-Bueno, I.; Morales-Torres, A. Advances on the failure analysis of the dam-foundation interface of concrete dams. *Materials* **2015**, *8*, 8255–8278. [CrossRef] [PubMed]
60. Hariri-Ardebili, M.A.; Seyed-Kolbadi, S.N.; Kianoush, M.R. FEM-based parametric analysis of a typical gravity dam considering input excitation mechanism. *Solid Dyn. Earthq. Eng.* **2016**, *84*, 22–43. [CrossRef]
61. Chwang, A.T.; Housner, G.W. Hydrodynamic pressures in sloping dams during earthquakes. Part 1. Momentum method. *J. Fluid Mech.* **1978**, *87*, 335–341. [CrossRef]
62. Chakrabarti, P.; Chopra, A.K. Earthquake analysis of gravity dams including hydrodynamic interaction. *Earthq. Eng. Struct. Dyn.* **1973**, *2*, 143–160. [CrossRef]
63. Furgani, L. Verifiche Sismiche di Dighe in Calcestruzzo. Ph.D. Thesis, University of Roma Tre, Rome, Italy, 2014.
64. ABAQUS 6.11. *Abaqus/CAE User's Manual*; Dassault Systèmes, Simulia: Johnston, RI, USA, 2014.
65. ABAQUS 6.14. *Analysis User's Guide, Volume IV: Elements*; Dassault Systèmes, Simulia: Johnston, RI, USA, 2014.
66. Fenves, G.; Chopra, A.K. Effects of reservoir bottom absorption on earthquake response of concrete gravity dams. *Earthq. Eng. Struct. Dyn.* **1983**, *11*, 809–829. [CrossRef]
67. Lin, G.; Wang, Y.; Hu, Z. Hydrodynamic pressure on arch dam and gravity dam including absorption effect of reservoir sediments. *Mater. Sci. Eng.* **2010**, *10*, 012234. [CrossRef]
68. De Biagi, V.; Chiaia, B. Complexity and robustness of frame structures. *Int. J. Solids Struct.* **2013**, *50*, 3723–3741. [CrossRef]
69. Escuder-Bueno, I.; Mazzà, G.; Morales-Torres, A.; Castillo-Rodríguez, J.T. Computational aspects of dam risk analysis: Finding and challenges. *Engineering* **2016**, *36*, 319–324. [CrossRef]
70. Nogueira, C.G.; Leonel, E.D.; Coda, H.B. Probabilistic failure modelling of reinforced concrete structures subjected to chloride penetration. *Int. J. Adv. Struct. Eng.* **2012**, *4*, 10–23. [CrossRef]
71. Ross, S.M. *Probability and Statistics for Engineers and Scientists*; Pearson: Apogeo, Italy, 2008.
72. Bastidas-Arteaga, E. Reliability of reinforced concrete structures subjected to corrosion-fatigue ad climate change. *Int. J. Concr. Struct. Mater.* **2018**, *12*, 10. [CrossRef]
73. Alañón, A.; Cerro-Prada, E.; Vázquez-Gallo, M.J.; Santos, A.P. Mesh size effect on finite-element modelling of blast-loaded reinforced concrete slab. *Eng. Comput.* **2018**, *34*, 649–658. [CrossRef]
74. Masoero, E.; Darò, P.; Chiaia, B.M. Progressive collapse of 2D framed structures: An analytical model. *Eng. Struct.* **2013**, *54*, 94–102. [CrossRef]
75. Lin, P.; Huang, B.; Li, Q.; Wang, R. Hazard and seismic reinforcement analysis for typical large dams following the Wenchuan earthquake. *Eng. Geol.* **2015**, *194*, 86–97. [CrossRef]
76. Bastidas-Arteaga, E.; Chateauneuf, A.; Sánchez-Silva, M.; Bressolette, P.; Schoefs, F. A comprehensive probabilistic model of chloride ingress in unsaturated concrete. *Eng. Struct.* **2011**, *33*, 720–730. [CrossRef]
77. Carrara, P.; De Lorenzis, L.; Bentz, D.P. Chloride diffusivity in hardened cement paste from microscale analyses and accounting for binding effects. *Model. Simul. Mater. Sci. Eng.* **2016**, *24*, 1–44. [CrossRef]
78. Yazdani, A.; Kowsari, M. Bayesian estimation of seismic hazards in Iran. *Sci. Iran. A* **2013**, *20*, 422–430.
79. Zacchei, E.; Nogueira, C.G. Chloride diffusion assessment in RC structures considering the stress-strain state effects and crack width influences. *Constr. Build. Mater.* **2019**, *201*, 100–109. [CrossRef]
80. Cornell, C.A. Engineering Seismic Risk Analysis. *Bull. Seismol. Soc. Am.* **1968**, *58*, 1583–1606.
81. Instituto Geológico y Minero de España (IGME). ZESIS: Base de Datos de Zonas Sismogénicas de la Península Ibérica y Territorios de Influencia para el Cálculo de la Peligrosidad Sísmica en España. 2015. Available online: http://info.igme.es/zesis (accessed on 1 February 2019).

82. Zhan, Z. Gutenberg-Richter law for deep earthquakes revisited: A dual-mechanism hypothesis. *Earth Planet Sci. Lett.* **2017**, *461*, 1–7. [CrossRef]
83. Kijko, A.; Smit, A. Extension of the Aki-Utsu b-value estimator for incomplete catalogs (short note). *Bull. Seismol. Soc. Am.* **2012**, *102*, 1283–1287. [CrossRef]
84. Yang, J.; Jin, F.; Wang, J.T.; Kou, L.H. System identification and modal analysis of an arch dam based on earthquake response records. *Soil Dyn. Earthq. Eng.* **2017**, *92*, 109–121. [CrossRef]
85. Prakash, G.; Sadhu, A.; Narasimhan, S.; Brehe, J.M. Initial service life data towards structural health monitoring of a concrete arch dam. *Struct. Control Health Monit.* **2018**, *25*, e2036. [CrossRef]
86. Ghanaat, Y.; Hall, R.L.; Redpath, B.B. Measuement and computation of dynamic response of arch dams including interaction effects. *J. Seismol. Earthq. Eng.* **2000**, *2*, 1–19.
87. Zhu, K.; Gu, C.; Qiu, J.; Liu, W.; Fang, C.; Li, B. Determining the optimal placement of sensors on a concrete arch dam using a quantum genetic algorithm. *J. Sens.* **2016**, *2016*, 2567305. [CrossRef]
88. Hamidian, D.; Seyedpoor, S.M. Shape optimal design of arch dams using an adaptive neuro-fuzzy inference system and improved particle swarm optimization. *Appl. Math. Model.* **2010**, *34*, 1574–1585. [CrossRef]
89. Fan, Q.; Zhou, S.; Yang, N. Optimization design of foundation excavation for Xiluodu super-high arch dam in China. *J. Rock Mech. Geotech. Eng.* **2015**, *7*, 120–135. [CrossRef]
90. Chen, Y.; Hong, J.; Tang, S.; Zhou, C. Characterization of transient groundwater flow through a high arch dam foundation during reservoir impounding. *J. Rock Mech. Geotech. Eng.* **2016**, *8*, 462–471. [CrossRef]
91. Amina, T.B.; Mohamed, B.; André, L.; Abdelmalek, B. Fluid-Structure Interaction of Brezina Arch Dam: 3D Modal Analysis. *Eng. Struct.* **2015**, *84*, 19–28. [CrossRef]
92. European Committee for Standardization (CEN). *Design of Concrete Structures—Part 1-1: General Rules and Rules for Buildings*; EN 1992-1-1:2004; European Committee for Standardization (CEN): Brussels, Belgium, 2004.
93. Guan, J.; Li, Q.; Wu, Z.; Zhao, S.; Dong, W.; Zhou, S. Minimum specimen size for fracture parameters of site-casting dam concrete. *Constr. Build. Mater.* **2015**, *93*, 973–982. [CrossRef]
94. Omidi, O.; Lofti, V. Seismic plastic-damage analysis of mass concrete blocks in arch dams including contraction and peripheral joints. *Soil Dyn. Earthq. Eng.* **2017**, *95*, 118–137. [CrossRef]
95. Millán, M.A.; Young, Y.L.; Prévost, J.H. The effects of reservoir geometry on the seismic response of gravity dams. Part 1: Analytical model. *Earthq. Eng. Struct. Dyn.* **2002**, *36*, 1441–1459.
96. Su, H.; Li, J.; Wen, Z.; Fu, Z. Dynamic non-probabilistic reliability evaluation and service life prediction for arch dams considering time-varying effects. *Appl. Math. Model.* **2016**, *40*, 6908–6923. [CrossRef]
97. *Sap2000, Version 16.0.0 Plus*; Computers and Structures, Inc.: Walnut Creek, CA, USA; New York, NY, USA, 2013.
98. Durieux, J.H.; van Rensburg, B.W.J. Development of a practical methodology for the analysis of gravity dams using the non-linear finite element method. *J. S. Afr. Inst. Civ. Eng.* **2016**, *58*, 2–13. [CrossRef]
99. Clough, R.W.; Penzien, J. *Dynamics of Structures*, 3rd ed.; McGraw-Hill: New York, NY, USA, 2003.
100. *Wolfram Mathematica, Version 11 Student Edition*; Wolfram Research, Inc.: Champagne, IL, USA, 2017.
101. Koh, H.M.; Kim, J.K.; Park, J.H. Fluid-structure interaction analysis of 3-D rectangular tanks by a variationally coupled BEM-FEM and comparison with test results. *Earthq. Eng. Struct. Dyn.* **1998**, *27*, 109–124. [CrossRef]
102. Datei, C. Costruzioni Idrauliche (Dighe). Available online: http://www.manualihoepli.it/media/doc/H_3-10_Dighe_Datei_85aEd.pdf (accessed on 1 February 2019).
103. *Microsoft Office 365. Excel*; Microsoft: Redmond, WA, USA, 2016.
104. Ridolfi, E.; Buffi, G.; Venture, S.; Manciola, P. Accuracy analysis of a dam model from drone surveys. *Sensors* **2017**, *17*, 1777. [CrossRef]
105. CloudCompare. *Version 2.10.2*. 2019. Available online: https://www.danielgm.net/cc/ (accessed on 1 February 2019).
106. Fanelli, M.; Lombardi, G. Practice and theory of arch dams. In Proceedings of the International Symposium on Arch Dams, Nanjing, China, 17–20 October 1992.
107. *AutoCAD, Version 2010*; Autodesk, Inc.: San Rafael, CA, USA, 2010.
108. Mash3D Factory S.r.l. Rome, Italy. Available online: http://www.mash3dfactory.com/ (accessed on 1 February 2019).

MDPI
St. Alban-Anlage 66
4052 Basel
Switzerland
Tel. +41 61 683 77 34
Fax +41 61 302 89 18
www.mdpi.com

Sustainability Editorial Office
E-mail: sustainability@mdpi.com
www.mdpi.com/journal/sustainability

www.ingramcontent.com/pod-product-compliance
Lightning Source LLC
Chambersburg PA
CBHW061434210926
51850CB00008B/1512
* 9 7 8 3 0 3 9 4 3 2 0 2 8 *